TECHNOLOGIES IN
FOOD PROCESSING

TECHNOLOGIES IN FOOD PROCESSING

Edited by
Harish Kumar Sharma, PhD
Parmjit S. Panesar, PhD

Apple Academic Press Inc.	Apple Academic Press Inc.
3333 Mistwell Crescent	9 Spinnaker Way
Oakville, ON L6L 0A2 Canada	Waretown, NJ 08758 USA

© 2018 by Apple Academic Press, Inc.

First issued in paperback 2021

Exclusive worldwide distribution by CRC Press, a member of Taylor & Francis Group

No claim to original U.S. Government works

ISBN-13: 978-1-77463-141-6 (pbk)
ISBN-13: 978-1-77188-651-2 (hbk)

Library and Archives Canada Cataloguing in Publication

Technologies in food processing / edited by Harish Kumar Sharma, PhD, Parmjit S. Panesar, PhD.

Includes bibliographical references and index.
Issued in print and electronic formats.
ISBN 978-1-77188-651-2 (hardcover).--ISBN 978-1-315-14719-2 (PDF)

1. Food industry and trade--Technological innovations.
I. Panesar, P. S. (Parmjit Singh), editor II. Sharma, Harish Kumar, editor

| TP370.T43 2018 | 664 | C2018-902903-X | C2018-902904-8 |

Library of Congress Cataloging-in-Publication Data

Names: Sharma, Harish K., editor. | Panesar, P. S. (Parmjit Singh), editor.

Title: Technologies in food processing / editors, Harish Kumar Sharma, PhD, Parmjit S. Panesar, PhD.

Description: Toronto ; New Jersey : Apple Academic Press, 2018. | Includes bibliographical references and index.

Identifiers: LCCN 2018023159 (print) | LCCN 2018025881 (ebook) | ISBN 9781315147192 (ebook) | ISBN 9781771886512 (hardcover : alk. paper)

Subjects: LCSH: Food processing.

Classification: LCC TP370.6 (ebook) | LCC TP370.6 .T43 2018 (print) | DDC 338.4/7664--dc23

LC record available at https://lccn.loc.gov/2018023159

Apple Academic Press also publishes its books in a variety of electronic formats. Some content that appears in print may not be available in electronic format. For information about Apple Academic Press products, visit our website at **www.appleacademicpress.com** and the CRC Press website at **www.crcpress.com**

CONTENTS

ABOUT THE EDITORS

Harish Kumar Sharma, PhD
Professor and Former Dean (R&C), Food Engineering and Technology Department, Sant Longowal Institute of Engineering and Technology (SLIET), Longowal, India

Harish Kumar Sharma, PhD, is a Professor and Former Dean (R&C) of the Food Engineering and Technology Department at the Sant Longowal Institute of Engineering and Technology (SLIET), Longowal, India. He mainly works in the area of food engineering and technology, and his research deals primarily with the process development, value addition, and exploration of newer quality control methods. He has published more than 170 papers in national/international journals, has authored four books, and is the editor of four books. He has contributed over 20 chapters to books by national and international publishers and has represented India in several different foreign countries, including Israel, Canada, China, Switzerland, Germany, France, Singapore, etc., in different programs. He has successfully handled a number of projects from government agencies and industry in the capacity of principal investor or co-principal investor. He also successfully transferred technology to the industry and has made significant contributions toward newer process development and has imparted professional guidance to industries. Dr. Sharma is a member/life member of various associations and state and national committees and has acted as a referee and editorial board member for different journals.

Parmjit S. Panesar, PhD
Professor and Head, Department of Food Engineering and Technology, Sant Longowal Institute of Engineering and Technology, Longowal, Punja, India

Parmjit S. Panesar, PhD, is a Professor and Head of the Department of Food Engineering and Technology at Sant

Longowal Institute of Engineering and Technology, Longowal, Punjab, India. His research is focused in the area of food biotechnology, especially prebiotics, probiotics, microbial enzymes, application of immobilized cells/enzymes in different bioprocesses, and value addition of agro-industrial wastes. He has successfully completed seven research projects and has published more than 100 international/national scientific papers, 50 book reviews in peer-reviewed journals, and 25 chapters and has authored/edited six books. He is a member of the editorial advisory boards of several national/international journals. He has received advanced training from several different institutes/labs, including the International Centre for Genetic Engineering & Biotechnology, Italy, and Chembiotech Labs, University of Birmingham Research Park, Birmingham, United Kingdom. Dr. Panesar was also awarded a Young Scientist Fellowship by the Punjab State Council for Science & Technology, India. In 2005, he was awarded a BOYSCAST (Better Opportunities for Young Scientists in Chosen Areas of Science & Technology) fellowship by the Department of Science and Technology (DST), Government of India. He has visited several countries, including USA, UK, Canada, Switzerland, Italy, China, Singapore, Iran, Thailand, etc.

LIST OF CONTRIBUTORS

U. S. Annapure
Food Engineering and Technology Department, Institute of Chemical Technology, Matunga, Mumbai–400019, India

Kiran Bains
Department of Food and Nutrition, Punjab Agricultural University, Ludhiana, 141004, India

C. Khodifad Bhargavbhai
Department of Agricultural Process Engineering, College of Agricultural Engineering and Technology, Anand Agricultural University, Godhra (Gujarat), India

Suheela Bhat
Department of Food Engineering and Technology (FET), Sant Longowal Institute of Engineering and Technology, Longowal, Sangrur, Punjab, India, 148106, E-mail: Suheelabhat1@gmail.com; Charanjiv_cjs@yahoo.co.in; profh.sharma27@gmail.com

R. Chandrakala
Department of Food Science and Technology, National Institute of Food Technology Entrepreneurship and Management (NIFTEM), Kundli, Haryana, India

K. Chauhan
Carbohydrate and Protein Biotechnology Laboratory, Department of Biotechnology, Punjabi University, Patiala–147002, Punjab, India

Rohant Kumar Dhaka
Department of Food Science and Technology, National Institute of Food Technology Entrepreneurship and Management (NIFTEM), Kundli, Haryana, India

Pushpa Dhami
Department of Food and Nutrition, Punjab Agricultural University, Ludhiana, 141004, India

Maninder Kaur
Department of Food Science and Technology, Guru Nanak Dev University, Amritsar, India

Rupinder Kaur
Food Biotechnology Research Laboratory, Department of Food Engineering & Technology, Sant Longowal Institute of Engineering & Technology, Longowal 148106, Punjab, India, E-mail: 29.rupinderkaur@gmail.com

Pragati Kaushal
Department of Food Science and Technology, Punjab Agricultural University, Ludhiana, Punjab, India, E-mail: Pragati_gndu@yahoo.co.in; profh.sharma27@gmail.com

Robinka Khajuria
School of Bioengineering and Biosciences, Lovely Professional University, Phagwara, Punjab, India, 144411, E-mail: robinkakhajuria@gmail.com

Navneet Kumar
Department of Agricultural Process Engineering, College of Agricultural Engineering and Technology, Anand Agricultural University, Godhra (Gujarat), India

P. Kumar
Department of Food Engineering and Technology Sant Longowal Institute of Engineering and Technology, Longowal, Sangrur, Punjab, 148106, India

Vivek Kumar
Harcourt Butler Technical University, Kanpur–208002, Uttar Pradesh, India

Mudasir Ahmad Malik
Department of Food Engineering and Technology, Sant Longowal Institute of Engineering and Technology Longowal, Sangrur, Punjab, 148106, India, E-mail: malikfet@gmail.com

Parmjit S. Panesar
Food Biotechnology Research Laboratory, Department of Food Engineering & Technology,Sant Longowal Institute of Engineering & Technology, Longowal 148106, Punjab, India

M. A. Parray
Department of Food Engineering and Technology Sant Longowal Institute of Engineering and Technology, Longowal, 148106, India

Manjeet Prem
Department of Farm Machinery and Power Engineering, College of Agricultural Engineering and Technology, Anand Agricultural University, Godhra (Gujarat), India

Sneh Punia
Department of Food Science and Technology, Chaudhary Devi Lal University, Sirsa, India

Charanjiv Singh Saini
Department of Food Engineering and Technology (FET), Sant Longowal Institute of Engineering and Technology, Longowal, Sangrur, Punjab, India, 148106; E-mail: Charanjiv_cjs@yahoo.co.in; profh.sharma27@gmail.com

Kawaljit Singh Sandhu
Department of Food Science and Technology, Chaudhary Devi Lal University, Sirsa, India, Tel: +91-1666-247124, Fax: +91-1666-248123, E-mail: kawsandhu@rediffmail.com

Rajender S. Sangwan
Centre for Innovative and Applied Bioprocessing, C-127, Phase-VIII, Industrial Area, S.A.S. Nagar, Mohali-160071, Punjab, India

Harish Kumar Sharma
Department of Food Engineering and Technology (FET), Sant Longowal Institute of Engineering and Technology, Longowal, Sangrur, Punjab, 148106, India

Loveleen Sharma
Department of Food Engineering and Technology (FET), Sant Longowal Institute of Engineering and Technology, Longowal, Sangrur, Punjab, 148106, India

R. P. Singh
Carbohydrate and Protein Biotechnology Laboratory, Department of Biotechnology, Punjabi University, Patiala-147002, Punjab, India

R. S. Singh
Carbohydrate and Protein Biotechnology Laboratory, Department of Biotechnology, Punjabi University, Patiala–147 002, Punjab, India. E-mail: rssingh11@lycos.com; rssbt@pbi.ac.in

Gisha Singla
Centre for Innovative and Applied Bioprocessing, C-127, Phase-VIII, Industrial Area, S.A.S. Nagar, Mohali-160071, Punjab, India

Anil Kumar Siroha
Department of Food Science and Technology, Chaudhary Devi Lal University, Sirsa, India

Anjali Srivastava
Harcourt Butler Technical University, Kanpur-208002, Uttar Pradesh, India

Ashutosh Upadhyay
Department of Food Science and Technology, National Institute of Food Technology Entrepreneurship and Management (NIFTEM), Kundli, Haryana, India

S. A. Wani
Department of Food Engineering and Technology Sant Longowal Institute of Engineering and Technology, Longowal, Sangrur, Punjab, 148106, India

LIST OF ABBREVIATIONS

AC	amylose content
AFLP	amplified fragment length polymorphism
AOA	antioxidant activities
BCAA	branched chain amino acids
BD	bulk density
BGRFs	β-glucan-rich fractions
BIS	Bureau of Indian Standard
BOD	bio-chemical oxygen demand
BR	broken rice
BSA	bovine serum albumin
BSE	bovine spongiform encephalopathy
BSG	basil seed gum
CA	citric acid application
CAFW	custard apple fruit wine
CAPP	custard apple pulp powder
CAPS	cleaved amplified polymorphic sequence
CD	circular dichroism
COD	chemical oxygen demand
CSPI	Centre for Science in the Public Interest
CTE	critical tracking events
CVD	cardio-vascular disease
DBD	dielectric barrier discharges
DDGSs	distiller's dried grains with soluble
DE	degree of esterification
DPn	degree of polymerisation
DSE	desugared sugar cane extract
EAAs	essential amino acids
EAN-UCC	European Article Numbering- Uniform Code Council
EDTA	ethylene diamine tetra acetic acid
EHE	enzyme-heat-enzyme
EU	European Union
FA	fly ash

FDA	Food and Drug Administration
FFA	free fatty acid
FINS	forensically informative nucleotide sequence
FOS	fructo-oligosaccharides
FSMA	Food Safety Modernization Act
FTIR	Fourier transform infrared spectroscopy
FTS	food traceability system
FV	final viscosity
GFL	general food law
GHGs	greenhouse gases
GM	genetically modified
GMOs	genetically modified organisms
GPC	gel permeation chromatography
GRAS	generally recognised as safe
HAQN	hydroxyanthraquinoid
HDPE	high density poly ethylene
HDPEV	high density polyethylene under vacuum
HFCS	high fructose corn syrup
HFS	high fructose syrup
HHP	high hydrostatic pressure
HMT	heat moisture treatment
HPC	hydroxy propyl cellulose
HPMC	hydroxyl propyl methylcellulose
HTST	high temperature-short time
IAAs	indispensable amino acids
ICP	inductively coupled plasma
ICP-MS	inductively coupled plasma Mass spectrometry
IMF	intermediate moisture foods
IMO	iso-malto-oligosaccharides
KDE	key data elements
LAB	lactic acid bacteria
LAF	laminated aluminum foil
LDPE	low density poly ethylene
LMW	low-molecular-weight
MAP	modified atmospheric packaging
MC	methylcellulose
MCA	metal chelating values
MIP	microwave induced plasma

MP	mushroom powder
MRI	magnetic resonance imaging
MRPs	Maillard reaction products
NGS	next generation sequencing
NIS	near infrared spectroscopy
NMR	nuclear magnetic resonance
NSI	nitrogen solubility index
OAC	oil absorption capacity
OPP	oriented polypropylene
OSA	octenyl succinic anhydride
PCA	Peanut Corporation of America
PCA	processed cheese analogue
PCR	polymerase chain reaction
PCR-RFLP	polymerase chain reaction-linked restriction fragment length polymorphism
PEF	pulsed electric fields
PP	polypropylene
PPO	polyphenol oxidase
PPW	potato processing waste
PT	pasting temperature
PV	peak viscosity
RBO	rice bran oil
RF	radio frequency
RHA	rice husk ash
RMSE	root mean square error
RSM	response surface methodology
RTE	ready-to-eat
SBP	steamed before pearling
SBP	sugar beet pulp
SEM	scanning electron microscopy
SM	skim milk
SME	specific mechanical energy
SNF	solid not fat
SNIF	site specific natural isotope fractionation
SNP	single nucleotide polymorphism
SP	swelling power
SPF	sweet potato flour
SSC	soluble solids content

SSPs	shelf stable products
SSRR	species-specific repeat region
SV	setback viscosity
TPC	total phenolic content
TRFLP	terminal restriction fragment length polymorphism
TV	trough viscosity
UPC	universal product codes
UV–NIR	ultraviolet near infrared
WAC	water absorption capacity
WAI	water absorption index
WPC	whey protein concentrate
WPI	whey protein isolate
WSI	water solubility index
XRD	x-ray diffraction

PREFACE

The world has witnessed an unprecedented population growth. The global population increased from about 1.6 billion people in 1900 to 7.3 billion today. The demand for processed food is therefore expected to increase substantially. Therefore, the need of different processing techniques becomes extremely important. With increase in awareness about the quality of the foods, processed products with improved quality and better taste along with safety are now the important aspects, to be dealt with.

Technology to process the foods should have distinguished features in which the nutrients could be preserved, anti-nutrients and toxins could be eliminated, vitamins and minerals could be added, waste and product loss could be reduced and bioactive constituents could be retained to provide health benefits. Modern techniques can be used to process the heat sensitive and other foods, which were earlier considered difficult to process and improve the quality of life for individual by offering modified food as per the requirement. It is therefore important to understand different techniques and their applications in foods. The book, "Technologies in Food Processing" consists of 14 chapters, which are contributed by experts in their area of research.

Chapter 1 deals with the application of ohmic heating, an alternate thermal processing method in which heat generation occurs within the process material itself that prevents its dependency on heating surfaces. Such phenomenon provides a great potential to food processing especially in processing of heat sensitive foods. Cold plasma can be used for surface modification of substrate to change the physical and functional properties of the material and renders minimum thermal effects on nutritional and sensory quality parameters of food with no chemical residues. Chapter 2 discusses cold plasma in different foods and its effect on the nutritional qualities and toxicology.

The various applications of biotechnology in food processing have been discussed Chapter 3. The Chapter deals with the role of biotechnology in production of fermented foods, alcoholic beverages, enzymes, food additives and functional foods. The application of newer biotechnological tools such the development of biosensors and genetic engineering has also been covered.

Chapter 4 deals with the commercial production of HFS by conventional multienzymatic hydrolysis of starch and subsequent isomerization of dextrose into fructose by glucose isomerase. However, product yield by this method is not more than 42%. In the present chapter, to attain the best yield, enzymatic approaches are elaborated for the preparation of HFS from starch/ inulin. Chapter 5 deals with modification of food proteins using gamma irradiation and the effect of irradiation on the structural and functional properties. The chapter also deals with the changes in films due to the Gamma irradiation, which affects the protein either directly or indirectly through radiolysis of water molecules.

The purpose of edible coatings is to restrain migration of moisture, oxygen, carbon dioxide, or any other solute materials and serve as a carrier for food additives like antimicrobial or antioxidants and decrease the decay without affecting quality of the food. Chapter 6 deals with the functions and components of edible coating and different types of coatings based on the proteins and polysaccharides along with the applications. Chapter 7 discusses the rapid increase in the demand for natural colorants due to the toxic effects of synthetic coloring and the issues associated with synthetic coloring agents and the various sources of natural pigments that can serve as an alternative to synthetic dyes.

Hurdle technology can effectively preserve foods without compromising the microbial stability and safety. Chapter 8 deals with mechanism, role, different types of hurdles, and the various application areas of hurdle technology in the food industry. Chapter 9 discusses different varieties and production of mushroom, nutritional composition, medicinal properties, processing, packaging and storage along with technologies of different mushroom based products such as mushroom soup powder, mushroom nuggets, mushroom ketch-up, mushroom candy, mushroom based pasta/noodles, mushroom biscuit/cookie/cake/bread, mushroom *mathri* and *papad*, mushroom chips, etc.

Chapter 10 deals with the extrusion cooking and the changes occurred in the form of physical, functional and nutritional properties during extrusion cooking.

Custard apple is one of the delicious fruits with pleasant flavor, mild aroma, sweet taste with low glycemic index and significant amounts of vitamin C, thiamine, magnesium, potassium and dietary fiber. Chapter 11 elaborates compositional and functional parameters of custard pulp and the methodology of number of value added products viz. ice cream, toffee, milk shake, ready to serve beverage, nectar, etc., from the custard apple. The chapter also deals with the case study dealing with the kinetic changes in color during ripening.

Pearl millet is prone to rancidity therefore the conversion of pearl millet into different value added products are discussed in Chapter 12. The chapter also describes the functional and antioxidant properties of pearl millet flour and its value addition in starch and flour. The starch properties and various methods of modifications of starch have also been discussed in detail. Chapter 13 explores the unrecognized potential of agro-industrial waste and highlights the different uses of agro-industrial waste targeted towards improved human health and environmental sustainability.

Chapter 14 focuses on the relevance of traceability systems for food supply chain, particularly as a sustainable solution. Reader will find basic understanding of meaning and scope of traceability along with certain case studies. New traceability enabling techniques regarding food authenticity like RFID, DNA based methods, spectroscopy tools, etc., have also been discussed.

The book can be useful for the students, academicians, researchers and other interested professionals working in the field/allied field. There are books available in the field but this book is designed in such a way that it deals with the main important aspects related to the technique or produce processing. It is therefore expected that this book will serve as a resource book and find a unique place. The text in the book is standard work and therefore can be used as a source of reference. The editors would appreciate to receive any comments/information for the future course of actions.

CHAPTER 1

OHMIC HEATING IN FOOD PROCESSING

SUHEELA BHAT,[1] CHARANJIV SINGH SAINI,[1] and
HARISH KUMAR SHARMA[1]

*Department of Food Engineering and Technology (FET),
Sant Longowal Institute of Engineering and Technology, Longowal,
Sangrur, Punjab, 148106, India, E-mail: Suheelabhat1@gmail.com;
Charanjiv_cjs@yahoo.co.in; profh.sharma27@gmail.com*

CONTENTS

1.1 INTRODUCTION

Ohmic heating is also referred as resistance heating, electro-heating, and joule heating. It is considered as an alternate heat treatment, especially for liquids and pumpable particles. Unlike other heating methods, it does not heat the heating surface. In the past few years, the use of electrical energy in the treatment of foods has gained scope in food industries. Advances in the utilization of electrical energy provide an opportunity to produce high-quality, shelf-stable products. This alternate heat treatment technology has an advantage of fast and uniform heating due to the fact that heat generation occurs within the process material itself. Unlike conventional heating where heating of the product is dependent on the heating of heating surfaces or the heating medium, the generation of heat inside the material has an advantage of reducing fouling problem and thermal damage to the product.

The rate of heat generation in ohmic heating is a critical factor, and the success of heating is directly related to it. Moreover, other factors that affect heat generation in ohmic heating are as follows: electric conductivity of the food material, strength of the electric field created, resistance time, and the technique by which food is directed to the heating system. Although ohmic heating has wide application in food industries, the technology is most suitable for processing of liquids and particulates, because it requires moving particles and atomic ions to form the body of the conductor to pass current. Passage of heat through such medium allows to simultaneously heat liquids and solids.

1.2 FUNDAMENTAL PRINCIPLE OF OHMIC HEATING

Like other electromagnetic-based methods, ohmic heating shares similar working principle of heat transfer. The basic principle of ohmic heating lies in the conversion of electrical energy into heat energy, and the method of processing is mainly thermal with additional electrical effect. This conversion of electrical energy into heat increases the temperature of the food material that allows its successful processing like blanching, thawing, sterilization, or pasteurization. Such direct conversion of electrical energy into heat has an advantage of uniform heating due to direct dissipation of heat into the product. Passage of the electric current through a conductor releases heat. A generator is required to supply AC current to the electrodes at both ends. Increase in the electric field strength, E, and conductivity linearly increases the heating rate that can be adjusted by varying the gap between the electrodes. The correlation among heat, current, resistance, and temperature can be represented as follows:

$$H \alpha I^2 .R.T$$

where H: heat, I: current, R: resistance, and T: temperature.

The major difference between ohmic heating and other thermal processing methods is that the food processed during ohmic heating is in direct contact with heating surfaces that allows uniform heating and prevent stack burning and build up. Moreover, parameters like frequency and voltage can be varied according to the requirements of the food material and the processing method.

1.3 OHMIC HEATER DESIGN

An ohmic heater can be designed in number of ways, and the design may vary greatly from one manufacturer to another. There are, however, several basic needs for the design of equipment that will remain the same for each design. A generator is required to supply the necessary electricity and power required for the successful operation of electricity. Two electrodes are linked with power supply, and the food material to be processed is kept in contact with these electrodes; finally, the electrode gap is adjusted according to the size of the ohmic heater system. The basic outline of an ohmic heater is shown in Figure 1.1.

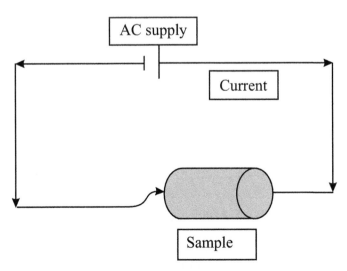

FIGURE 1.1   Basic outline of an ohmic heater.

1.3.1 POWER SOURCE

An electrical power source with a variable output of voltage and frequency is necessary for the generation of AC current in the ohmic heater. The electric circuit is purely resistive (only resistors are connected to a battery), and the energy from the source is continually dissipated totally in the form of heat. The prerequisite for the successful operation and proper maintenance of an ohmic heating system is an inert material for electrodes that prevents interaction with the processed material. Moreover, a thermostat is necessary for keeping the requisite temperature within the desired range and under control.

1.3.2 ELECTRODE

The design of the electrode is crucial as the material used for its manufacturing needs to be examined carefully. Normally, materials that are inert and of high grade are recommended for designing electrodes due to problems faced earlier in the food industry that resulted in contamination of processed food material due to dissolution of the electrode material caused by the electrolytic effect. However, presently, compatible food grade material is used for manufacturing electrodes that prevents metallic dissolution. Graphite and aluminum were used earlier but are now limited in food industries; stainless steel may be sometimes recommended for the manufacture of electrodes. Figure 1.2 shows the basic design of an electrode for heat processing of liquids.

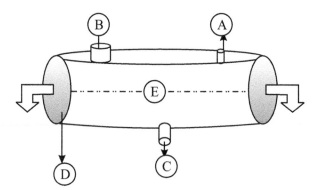

FIGURE 1.2   Electrode for ohmic heating of liquids (A = used to insert a multiprobe; B = the input port; C = the output port; D = distance between the electrode and the exit port; and E = rubber closure).

1.3.3   ELECTRODE GAP

The electrode gap is not any definite and fixed space and may vary according to the size of the ohmic heating system. The type of ohmic heater may also influence the gap between electrodes, but most often, transverse and collinear field modes are operated. In the transverse mode operation, the material to be processed is always placed at right angles to the electric field. However, in the collinear mode operation, the product flow is parallel to the electric field and flux (Stiring, 1987).

1.3.3.1   Transverse Configuration

This type of configuration includes electrodes that are placed either co-axial or plane, and the product flow is directed always at right angles to the electric field and parallel to electrodes. This type of configuration has few disadvantages as well. Products processed in this type of configuration act as earth through which a large amount of current leakage occurs due to the presence of the electrode in the close vicinity of inlet and outlet sources of the ohmic heater. Overheating and stack burning are observed at the outer surfaces of electrodes due to nonuniform phase to neutral density (Varghese et al., 2014). This disadvantage restricted its use for fluids containing no particles, and hence, another type of configuration that is more suitable to fluids with lesser or no particles is collinear configuration.

1.3.3.2   Collinear Configuration

In contrast to the transverse mode, the material flow is directed in parallel fashion to the electric field as it moves from one electrode to another. The collinear design is a better option for high conductivity and offers wider electrode spacing. The electrodes can be positioned in the fluid stream or as collars around a pipe that provides a fully unrestricted flow channel. However, this design may require high voltage supply.

1.4   EFFECT OF FREQUENCY AND AMPLITUDE

Typically, ohmic heaters operate at low frequencies, greater than 20 Hz and up to 50–60 Hz. Lima et al. (1999) studied the effect of frequency and amplitude effect on turnip. While the electric field strength during ohmic heating can be precisely controlled, the behavior of plant tissue for electrical conductivity is very complex and depends on many other parameters like electro-thermal condition of the material, required temperature, and the frequency of the tissue cell membrane (Kulshrestha and Sastry, 2006; Lebovka et al., 2007). Electrochemical and metallic dissolution is a common problem noticed at lower frequencies; however, many researchers suggested that by increasing the frequency, this problem can be significantly reduced (Samaranayake et al., 2005). Usually, to minimize undesirable electrochemical reactions between electrodes and the food product, an increase in the alternating electric field frequency is required. Increased frequency may result in increased opposition by the circuit to the current, which is known as electrical impedance. Higher the electrical impedance, higher will be the extraction efficiency and heating rate. As studied in the case of tulips by Lima et al. (1999), increased frequency caused high rate of microbial reduction. In contrast, lower frequencies may allow the current to take longer time to penetrate the particulates, thereby preventing cell membranes from pore formation that in turn decreases microbial destruction (Imai et al., 1995).

1.4.1 AMPLITUDE

Amplitude refers to the direct measurement of degree of change in wave from its equilibrium state (compression and rarefaction). Amplitude measures the variation in air molecules, with high pressure resulting in high amplitude and vice-versa. The amplitude of ohmic heating is always related to time; higher amplitude reduces the time as it is actually a force that is forced over an area to cover. All amplitude variations revolve around the change in magnitude from lower to higher limit and is measured in Newton per square meter

1.4.2 ELECTRICAL CONDUCTIVITY

Electrical conductivity is one of the most important key parameters that affect the ohmic heating process. This property of ohmic heating itself is dependent on several parameters like food material composition, present ionic concentration, and amount of unbound water (Lima and Sastry, 1999).

The concentration of ions directly influences the electric conduction, but the presence of nonionic and nonpolar substances decreases electric conductivity (Sastry and Palaniappan, 1992; Wang and Sastry, 1993). Many factors that play a role in electrical conductivity include the electric field strength, breakdown of ionic forces, temperature, and structural properties of the food material at the microstructural level (Parrott, 1992). The material to be processed in ohmic heating should be of conductive nature and should contain large number of ions (Zoltai and Swearingen, 1996). It was observed that voltage gradient has a significant role in increasing electrical conductivity of materials, but many studies conducted by Wang and Sastry (1993) concluded that electrical conductivity increased with increase in temperature at constant voltage gradient. Food materials with porous and fibrous structures affect the electrical conduction and tend to decrease it. Many researchers found that preheating of the food material before ohmic heating has a positive effect on the electrical conductivity of foods possibly due to proper dissipation of ionic concentration. Ionic concentration may sometimes be increased by the addition of NaCl. In the case of liquid foods processed ohmically, the current is mainly carried through ionic charges, and higher concentration of ions results in higher electrical conductivity. Hence, pH is one of the important factors to be considered while processing liquid foods ohmically.

1.5   TEMPERATURE DISTRIBUTION DURING OHMIC HEATING

Ohmic heating as discussed is an alternate thermal processing technique. Like other thermal treatments, the processing occurs by heat generation; however, heat production does not occur like that in conventional thermal methods. Heat generation in case of ohmic heating occurs within the process material itself unlike conventional thermal processing where the heating of the process material is dependent on heating surfaces. Basic source of heat generation within the process material during ohmic heating is the electric current. As observed from Joules laws, heat production is directly proportional to square of the current induced. Heat generation in a system is a function of many parameters, but the most important is conductivity of the processed material. The rate of heat generation depends on strength of the electric field produced by the current flow. One of the prerequisites for foods processed by ohmic heating is that the food should be conductive in nature and should contain a large number of ions, so as to direct the current in the processing medium. This could be the main reason for suitability of ohmic heating to liquid foods with particulates. Liquid medium acts as a passing medium that organizes microparticulates like ions and electrons to allow the passage of the current inside the food material. The induced current induces the electric field, and particles present in the medium are accelerated by such induction. The accelerated charged particles (ions in motion) result in random collisions, and the resulting collisions of such accelerated particles lead to generation of thermal energy in a system. Conversion of kinetic energy into thermal energy is the basic mechanism of heat generation in the ohmic system. Such random phenomenon takes place inside the processing medium itself and hence results in temperature rise inside the process material and not in heating surfaces. Prevention or minimization of heating the heating surfaces reduces the problem of deposit formation. Deposit formation has always been an issue in food processing industries due to poor hygiene and microbial contamination concern, especially in diary industry. Tham et al. (2009) observed that fouling of surface deposition in dairy processing is mainly due to thermal denaturation of β-lactoglobulins. Such problem can be greatly minimized in ohmic heating, which not only prevents the processing material from contamination but also surfaces from pitting and corrosion.

1.6 OHMIC HEATING PROCESS SENSITIVITY

Ohmic heating is a well-known technology to shorten the heating time of processed material; however, validation of this technology in terms of uniform heating of the liquid and particulate system is an effective tool to determine process sensitivity. Success of a process is directly dependent on the sensitivity of that process and how output of that process is divided, quantitatively and qualitatively (Chen, 2010). Not only input parameters of the heating system, like voltage and frequency are important for a successful and effective process, but shape, size, conductivity, and orientation of particulates in the process material are critical factors as well. De Alwis et al. (1989) reported that the heating capacity of an elongated particle placed in a fluid perpendicular to the electrode was much higher than that of the resulting fluid, although the electrical conductivity of elongated particles was much lesser than that of fluids. However, the same particle when placed parallel to the electrode did not heat for a longer time. These results confirmed that besides other parameters, the shape and direction of the heat flow and the reception of heat are also critical parameters; however, limited scientific literature is available on the heat flow and heating rate of particulates in the ohmic heating system. Chen et al. (2010) selected few critical control points in the ohmic heating process and compared the sensitivity after every critical point. Process temperature, target heating value, and conductivity of both liquids and particulate foods were taken as critical control points. The results showed that electrical conductivity is the most sensitive critical control point that affected both product temperature and target lethality. Increased electrical conductivity of the fluid will result in decreased process temperature for both fluids as well as particulates present. However, if particles have higher conductivity than the fluid, it may result in simultaneous collision without movement of the current that generates exothermic heat and increases temperature of both fluid and particulates. Hence, the results conclude that during ohmic heating, it can be possible to achieve much increased temperature if both phases (liquids and particles) have the same electrical conductivity or particles have higher electrical conductivity than fluids. Moreover, the sensitivity of the process requires proper monitoring of particle size, particle and fluid temperature, concentration of particles, and the expected lethality value.

1.7 OHMIC HEATING TECHNOLOGY IN FOOD PROCESSING

Heat processing has been used in the food industry since ages and is considered as one of the most successful modes of processing in food industries for preservation, processing and many more aspects of biomaterials. Conventional heating has also played a successful role in the preservation and processing of certain foods due to inactivation of microbial flora at high temperature. These treatments are mainly dependent on thermal conduction and convection methods. Despite of effectiveness of conventional heating from microbial standpoint, non uniform heat distribution result in uneven texture. Such quality defects reduce market value of processed foods. Processing of complicated materials has improved by introduction of new technologies like ohmic heating, microwave, and radio frequency. These heating methods generate heat inside the food and depend less on thermal conduction and convection and so cause fewer temperature gradients.

Besides processing, ohmic heating can be used in many applications like thawing, blanching, and alternate peeling. This technique is best suited for liquid foods. Ohmic heating is also a very suitable alternate method for the treatment of proteinaceous foods because these foods tend to denature during thermal treatment. Ohmic heating prevents coagulation of these foods by reducing processing time and is also referred to as flash heating. Such heating methods allow successful treatment of sensitive foods without coagulation.

Dry beans were blanched for different length of times (35 and 50 s) in a water bath at 90°C after soaking (12 h) in tap water. Heating curves of solid-liquid mixtures (53 g of beans/100 g of mixture) of pretreated beans and salt solution (1 g/100 mL) were determined in a static ohmic heating device by applying a constant voltage of 100 V. The results showed that the ohmic heating rate of beans increased with increasing the blanching time as a direct result of increase in electrical conductivity. Optimal pretreatment conditions, which allowed the ohmic heating of both phases of the solid-liquid mixture at the comparable rate, were found when beans were blanched for 50 s at 90°C. Beans pretreated under optimal blanching conditions and sodium chloride solution (1 g/100 mL) (53 g of beans/100 g of mixture) were then subjected to both ohmic and in-container sterilization processes. The solid-liquid mixture was sterilized at 121°C for 10, 15, and 20 min. The experimental results showed that, regardless the thermal method and treatment conditions, initial microbial spoilage was reduced up to below the detection limit of the method (<10 cfu/mL). When compared to the beans sterilized by conventional treatment, the samples treated by the ohmic method appeared to have attractive appearance, with a lower split degree. The study of the chemical composition revealed a higher protein concentration in the sample processed by ohmic heating than in those treated by the conventional method. Microbial destruction during ohmic heating occurs due to thermal effect. Additional nonheating effect may also be observed due to build-up of charge on molecules that subsequently results in electroporation and microbial destruction. In another study, different electric field strengths were applied to orange and tomato juices (25–40 V/cm). Treatment time for each trial was also varied. A linear increase was observed in temperature of ohmic heating as field strength and processing time were increased. Orange juice showed lesser temperature rise than tomato juice at all gradients. Microbial reduction was also observed to be higher at increased field strength. Treatment time also has a significant role in microbial reduction and resulted in 5 log reduction in microbial load with increased treatment beyond 60°C. Another report stated that ohmic heating is currently used for processing fruit juices with added sugar syrup and buffalo milk in many developed countries (Icier et al., 2006). Buffalo milk was also processed ohmically and analyzed microbiologically for total aerobic count (TPC), mold and yeast, coliform count, and *Escherichia coli* (Manish et al., 2014).

Streptococcus thermophilus is the largely studied microorganism in milk and milk products as it is highly heat resistant and considered sometimes as a reference microbe to check the inactivation efficiency of treatment process. *S. thermophilus* was studied by Shivmurti et al. (2014) in milk products to investigate the rate of inactivation through both conventional and ohmic heating. Besides microbial load reduction, quality assessment of milk was also taken into consideration. The results showed that ohmic heating not only decreased the particular microbial count but also decreased D value resulting in faster destruction.

Certain studies showed the ohmic processing effect on selected parameters was comparable to that of conventional treated samples, but the only difference observed was in reduction of processing time (Table 1.1). Shivmurti et al. (2014) studied chemical properties such as acidity (as lactic acid), fat, solid not fat (SNF), protein, and total solids of buffalo milk before and after ohmic and conventional heating. Milk treated by the ohmic heating technology showed acidity, fat, SNF, protein and total solids to be very much comparable to those found in milk treated by the conventional heating technology process. The only difference observed in both processing methods was time reduction in ohmic heating was about 18% as compared to conventional heating. Besides liquid foods, solid and semi-solid foods were processed ohmically to study the voltage gradient effect on peach and apricot purees.

Wang and Sastry (1993) reported that the major difficulty in processing of foods containing particulates is variation in temperature between the solid and liquid phase of processed material due to difference in electrical conductivities of both phases. Hence, the successful processing system requires minimal variation in electrical conductivities of both phases. Sarang et al. (2008) reported processing of low acid foods having different particulates. In such foods, some conductive fluids were added to increase ion strength in order to increase electrical conductivity. In chicken chow mein, sauce was added and was found to be more effective for carrying electric current than rest of the components like spices and condiments. These studies concluded that blanching of particulates in some conductive liquids like sauce has a positive effect on increasing the conduction of the processed material. Moreover, pretreatments like blanching also showed a positive effect in processing foods during ohmic heating.

TABLE 1.1 Effect of Ohmic Processing on Different Foods

Food sample	Ohmic heating conditions	Effect of ohmic heating	References
Salsa	Frequency: 20 KHz	Enhanced bacterial and virucidal effect was observed	Kim and Kang, (2017)
Mashed potatoes	Frequency: 50 Hz, Temperature: 20–85°C, Time: 10–30 min	Rapid and uniform pasteurization was observed than conventional blanching	Guo et al., (2017)
Pomegranate juice	Voltage: 7.5–12.5 V/cm, Temperature: 10–65°C Time: 7–25 min	Concentration time was reduced by 56 %	Icier et al., (2017)
Bread dough	Temperature: 35°C Time: 20 min Voltage: 5–50 V	Expansion ratio was achieved in shorter time due to reduction in lag phase	Gally et al., (2017)
Shrimps	Temperature : 72°C, Time: 30–40 s	Reduced cooking time by 50 % with less color change	Diana et al., (2016)
Baby foods	Frequency: 25 KHz Temperature: 129°C Time: 11 min	Essential and non-essential amino acids remain constant after heating	Mesias et al., (2016)
Black rice bran	Voltage: 50–200 V/cm Temperature: 105°C Time: 1 min	High yield and concentration of bioactive components was achieved	Loypimai et al., (2015)
Blueberry pulp	Frequency: 60 Hz Voltage: 0–240 V Temperature: 90°C Time: 2 min	Percentage degradation was low when processed at lower voltage conditions	Sarkis et al., (2013)
Apples	Voltage: 13 V/cm Temperature: 30°C Time: 90 min	Polyphenol oxidase was completely inactivated with better color retention and microbial stability	Moreno et al., (2013)
Fruit deserts	Temperature: 20–105°C Time: 12.36 min	Ohmic heated deserts had very negligible effect on formation of 5-Hydroxyl methyl furfurals and furfurals as compared to conventional heating	Louarme and Billaud, (2012)
Apple and cloudberry juice	Frequency: 50 Hz Temperature: 70°C Time: 150 min	Enzyme inactivation was achieved at much faster rates	Jakob et al., (2010)

TABLE 1.1   (Continued)

Food sample	Ohmic heating conditions	Effect of ohmic heating	References
Meat balls	Temperature: 25–140°C Frequency: 60 Hz Voltage: 15 V/cm	Meat balls showed increased conductivity that led to effective processing	Sarang et al., (2008)
Tofu	Temperature: 70°C Time: 10 min	Reduced synersis rate by 21.8 %	Wang et al., (2007)

1.7.1   OHMIC BLANCHING

Vegetables and fruits are considered to be one among the most perishable foods with short shelf-life. Fruits and vegetables are largely consumed as fresh; however, if these products are supposed to be processed, they need to be stored at low temperature. The only drawback that low temperature storage offers to fruits and vegetables is loss of color, flavor, and nutrients. To eliminate such drawbacks, blanching becomes prerequisite. Blanching is usually carried out before processing to reduce initial microbial load, inactivate enzymes for better color retention, and eliminate air from cellular spaces for prevention of oxidative degradation of nutrients (Bhat et al., 2016). Since long, conventional hot water blanching is being successfully used in food industries, but due to high temperature, fruits and vegetables may suffer leaching of solutes, leading to lesser yield. Moreover, high temperature use results in degradation of nutrient profile at high levels. Alternatively, ohmic blanching has been used by many researchers to reduce the extent of solute loss and nutrient loss and better color retention with lesser blanching time and temperature. Jaworska et al. (2010) found that conventional blanching was unsuitable for mushrooms that were meant to be frozen due to loss of texture, nutrients, and organoleptic properties like flavor and aroma. Icier et al. (2006) found that ohmic blanching has the shortest critical inactivation time of 54 s for peroxidase enzyme in case of pea puree. Besides enzyme inactivation, better color retention was found in ohmic blanched samples than in conventional blanched samples. Ohmic blanching is also considered as the most accurate blanching method before freezing of fruits and vegetables. Yaari and Nussinovitch (2014) found that browning was prevented in samples during freezing and also during rehydration due to ohmic blanching of potato slices. It was concluded that ohmic blanching of fresh tissue without much rise in temperature reduced the problem of browning after rehydration. Ohmic blanching was assisted with other nonthermal methods for effective processing. Khuenpet et al. (2017) studied extraction of inulin from tuber powder of artichoke by using a combination of ohmic and ultrasonication treatments. The results concluded that inulin yield was higher in ohmic-assisted blanching followed by ultrasonication than in conventional blanching.

1.7.2   NUTRITIONAL IMPACT OF OHMIC HEATING

Heating is an effective method of processing and preservation since times, either by conventional or ohmic heating. The major concern in heating a food material is preservation of its nutritional status besides achieving safety. Allowing minimum damage or prevention of loss of vitamins and other nutrients remains to be the paramount concern in all heat-processing stages. Although total prevention of loss of most sensitive nutrients in food is not a guaranteed process, especially during thermal processing, a reduction in loss was achieved by shifting to alternate thermal methods like ohmic heating. Sarkis et al. (2013) reported that an anthocyanin in blueberry degraded to lesser extent during ohmic heating than during conventional heating, provided that voltage and solid content are kept low. Mercali et al. (2012) stated that vitamin C of acerola pulp was much retained during ohmic heating than during conventional heating at low voltage. Maintenance of voltage during the ohmic heating process also plays an important role for retention of nutritional components like vitamin C. Heat degradation rate of ascorbic acid in orange juice was much lower when processed with ohmic heating at the electric field strength of 40 V/cm (Vikram et al., 2005).

The basic difference between conventional and ohmic heating technology that allows attaining high standards with lesser nutrient damage involves difference in degradation parameters for microbial spores than that of biochemical reactions. Microbial spores always tend to be destroyed faster once target temperature is reached than the reactions involved in biochemical processes of food. This benefit allows reaching the microbial safety target quickly and uniformly without allowing occurrence of biochemical reactions. Moreover, volumetric heating and rapid cooling in the ohmic heating system do not allow such chemical reactions to proceed, and even if some reactions may take place, they result in lesser losses than any other heating methods. Occasionally, ohmic heating may result in better nutrient profile of foods than that of untreated ones due to volumetric heating and shorter time that sometimes proved to be an advantage to phenolic compounds in the food system. This is because phenolics are hidden in lower layers and require mild heat for extraction. Prolonged heat and high temperature result in subsequent degradation of high heat-sensitive phenolics. Ohmic heating allows the benefit of rapid heat and flash cooling without raising temperature to the level of degradation due to heat generation within the process material itself. As soon as AC is removed, the products do not remain hot for a long time due to the fact that ohmic heating does not heat the heating surfaces, thus preventing degradation of extracted compounds in the medium.

1.7.3   OHMIC HEATING AND MICROBIAL INACTIVATION

Ohmic heating is found to be an effective method to achieve high microbial safety (USA-FDA, 2000). Till date, no special microorganism has been found to have particular resistance to ohmic heating, and the effect on microbial destruction is somewhat like the conventional heating method with additional damage to the microbial cell wall due to electrical effect. This additional effect was confirmed by many scientists on the basis of observed lower D and Z values as compared to those in conventional methods (Pereira et al., 2007). Inactivation kinetics of many microorganism like *Bacillus licheniformis, E. coli*, and *Zygosaccharomyces bailii* were studied during ohmic heating treatment, and it was observed that these microorganisms showed decline in D value when treated at temperatures higher than 500C (Palaniappan and Shastry, 1992; Cho et al., 1999). The decrease in processing time with satisfactory microbial inactivation may be possibly due to the electrolytic effect of ohmic heating (Table 1.2).

Studies suggested that a mild electroporation-type mechanism is induced by ohmic heating that involves changes in the micropore structure of tissue cells and increases permeability of membranes. The phenomenon has been confirmed by various studies (Sarang et al., 2008; Shastry, 2009).

Ohmic heating in twin steps becomes more effective in causing microbial cell death, provided that the second stage is operated with subsequent holding. Studies have also confirmed that leakage of intracellular constituents was found to be enhanced during ohmic heating than during any other method of heating, which leads to increased death rate of microbes. Ohmic heating can be discussed mainly by two possible reasons: the first reason is

TABLE 1.2 Effect of Frequency on Inactivation of Different Microorganisms in Different Foods

Microorganism	Frequency	Food product	Mechanism	References
Coliform	200 Hz	Buffalo milk	Increased cell permeability	Manish et al., (2014)
Escherichia coli 0157:H7	20–45 KHz	Orange juice	Cell membrane bursting	Park and Kang, (2013)
Listeria monocytogens	20 KHz	Tomato juice	Increased cell permeability	Lee et al., (2012)
Salmonella enteritidis	60 KHz	Skim milk	Disruption of cellular membrane	Juliane et al., (2005)
Streptococcus thermopiles	20 KHz	Yogurt	Mild electroporation seized growth due to generation of internal heat	Yoon et al., (2002)
Saccharomyces cerevisiae	20 KHz	Korean rice beer	Leakage of cellular material	Yoon et al., (2002)
Bacillus Subtilis	50 KHz	Water	Formation of pores in cell wall due to build up charge	Sastry, (1998)

the combination of huge charge and development of pores on the cell wall of microorganisms that causes lesions and cell death, and the second reason is normal heating with uniform heat distribution that allows heat to sanitize the material uniformly and does not allow any build-up around the leftover places that may later turn the substance unsafe (USA-FDA, 2000).

Due to these reasons, ohmic heating is found to be very effective in the reduction of microbial load as compared to conventional heating at even lesser temperature-time combinations as that of conventional heating. Apple juice treated with both conventional and ohmic heating at varying temperature–time combinations showed the maximum reduction of microbial load, especially of *E. coli, Salmonella enterica*, and *Listeria monocytogenes*. The results for maximum reduction in ohmic heating were justified due to induced electrical effect at a field strength of 30 V/cm and 60 V/cm (Perk and Kanj, 2013). Apple juice was further investigated for inactivation rate at different temperature-time combinations by transmission electron microscopy and Propidium iodide uptake (PI) to check the permeability of cell membrane and damage rate by pore formation during ohmic heating. Ohmic heating above

600C and 10 s resulted in electroporation of microbial cell membranes, leading to proportional death with increased time and temperature. The results concluded that although pasteurization by conventional means is an effective process for the reduction of microbial load, ohmic pasteurization can result in both pasteurization and better quality maintenance of foods due to better microbial reduction in lesser temperature-time combinations.

Chronological sequence of salient features and application of the ohmic heating technique in food processing is shown in Table 1.3.

1.8   ADVANTAGES AND DISADVANTAGES

Major advantages of the ohmic heating process include the following:

- Lesser environmental pollution than other methods of heating.
- Lesser energy consumption than conventional thermal methods.
- Heats liquids and particulates simultaneously and uniformly (Fryer, 1993).
- Lowers the problem of stack burning of foods to be processed.
- Requires very less initial cost for setup.
- In combination with other heating or nonheating methods, ohmic heating results in the production of safer and better product processing (Brunton, 2005).
- Decreases cooking time of food materials.
- Meets microbial safety standards at lesser temperature-time combinations than any other heating method (Tucker and Withers, 1994; Pereira et al., 2007).
- Maintenance cost of ohmic heater set up is very minimal.
- Results in lesser damage to the product quality due to rapid heating and no dependency for heating on heating surfaces.

1.8.1   DISADVANTAGES

Food material containing fat globules are very difficult to process by ohmic heating due to the property of fat to act as inert material that does not allow to conduct heat (Rahman, 1999). The concentration of fat globules present in the food material also has greater influence on processing. If these globules are larger in quantity and scattered throughout the processed material, then

TABLE 1.3 Chronological Sequence of Ohmic Heating Application in Food Processing

Product	Application	Salient features	References
Rice	Cooking behavior	Lesser energy consumption and higher quality product in terms of swelling capacity.	Kanjanapongkul, (2017)
Grape and orange juice	Pasteurization	Xanthophylls and carotenoid profile was retained after ohmic heating at high temperatures	Achir et al., (2016)
Sugarcane juice	Shelf life assessment	No yeast and mold growth was observed after ohmic heating treatment	Saxena et al., (2016)
Beef	Cooking	Lesser cooking losses and better color retention	Tian et al, (2016)
Bottle gourd Juice	Reduction in microbial load and enzymatic activity	Faster reduction in microbial load and enzymatic activity at lesser temperature–time combinations	Bhat et al, (2016)
Black rice bran	Extraction of anthocyanin for production of natural colorant	Higher yield of anthocyanin contents	Loypimia et al., (2015)
Tomatoes	Peeling	Absence of residues, decreased processed time and reusable media	Ngasri and Sastry, (2015)
Meat	Thawing	Ohmic processed meat balls provide less weight loss and shorter thawing time	Duygu and Umit, (2015)
Blue mussel	Effect on texture, microbial and mineral content	Reduced cadmium content in products processed with ohmic heating	Bastias et al., (2015)
Meat balls	Processing	Firm and uniform microstructure	Engchuan et al., (2014)
Acerola pulp	Processing and effect of electric field frequency	Higher color changes and ascorbic acid degradation at low frequency	Mercalli et al., (2014)

TABLE 1.3 (Continued)

Product	Application	Salient features	References
Tomato juice	Inactivation kinetics	Significant lower D value of microbes with better color retention	Somavat et al., (2013)
Blueberry pulp	Anthocyanin degradation rate	Lower voltage levels decrease degradation rate	Sarkis et al., (2013)
Orange and pineapple juice	Temperature estimation for processing	Vitamin C content loss was insignificant during ohmic heating	Tumpanuvatr and Jittanit, (2012)
Seawater	Evaporation	Evaporation rate was higher at high field strength	Assiry, (2011)
Apple fruit	Production of highly impregnated fruit particles in short dehydration time	Homogenous and equilibrated fruit and syrup temperature.	Allali et al., (2009)
Juice	Inactivation of spores and enzymes	Reduced inactivation time, improved stability and safety	Loypimai et al., (2009)
Stew type	Production of space food and military ration	Lesser energy consumption for reheating food in retortable pouches	Jun et al., (2007)
Rice bran	Stabilization and oil extraction	Decrease in free fatty acid value and increase in oil extraction yield	Lakkakula et al., (2004)
Strawberry	Electrical conductivity measurement	Lesser ascorbic acid degradation	Teixeira et al., (2004)
Potato slices	Blanching and extraction	Enhanced moisture loss and increase in yield	Wang and Shastry, (2000)
Cauliflower	Influence on precooking of cauliflower	Increased firmness with proper sterilization	Eliot et al., (1999)
Shrimp blocks	Thawing	Ohmic heating lead to the thawing of shrimp without increase in moisture content.	Roberts et al., (1996)

the food material cannot be successfully processed by ohmic heating; however, if these globules are lesser in concentration and present in high electrical conduction areas, then these globules can be bypassed and food can be processed, but the success of this operation cannot be guaranteed in terms of microbial safety due to lesser heat interaction with fat globules (Sastry, 1992).

1.9 CONCLUDING REMARKS

Ohmic heating has proved to be a novel alternate thermal technique that has manifold benefits than conventional heating due to generation of heat within the system and not by an indirect method. Ohmic heating is especially advantageous in processing semi-solid, particulate foods but can also be used to process solid foods. Production of high-quality product, lesser energy consumption, processing at lesser temperature-time combination, environmental friendly, and better energy efficiency are some of the benefits of ohmic heating over conventional heating. The area has also got tremendous scope to work upon the control on rate of heat, modification of the electrode material to reduce corrosion, etc.

KEYWORDS

- antimicrobial effect
- food quality
- heater design
- nutritional quality
- ohmic heating
- process utility

REFERENCES

Allali, H., Marchal, L., & Vorobiev, E., (2009). Blanching of strawberries by ohmic heating: effects on the kinetics of mass transfer during osmotic dehydration. *Food Bioprocess Technol.*, *3*, 406–414.

Achir, N., Mayer, C. D., Hadjal, T., Madani, K., Pain, J. P., & Dornial, M., (2016). Pasturization of citrus juices with ohmic heating to preserve the carotenoid profile. *Innovative Food Sci. Emerging Technol.*, *33*, 397–404.

Assiry, A. M., (2011). Application of ohmic heating technique to approach near ZLD-during the evaporation process of seaweed. *Adv. Water Desalin.*, *280*(1), 217–223.

Bastais, J. M., Moreno, J., Pia, C., Reyes, J., Quevuda, R., & Munoz, O., (2015). Effect of ohmic heating on texture, microbial load, cadmium and lead content of Chilean blue mussel. *Innovative Food Sci. Emerging Technol.*, *30*, 98–102.

Bhat, S., Saini, C. S., Manish. K., & Sharma, H. K., (2016). Effect of thermal and alternate thermal processing on bottle gourd juice. *J. food process. preserv.*, DOI: 10.1111/jfpp.12911.

Brunton, N. P., Lyng, J. G., Li, W., Cronin, D. A., Morgan, D. B., & McKenna., (2005). Effect of Radio frequency on texture, color and sensory properties of a comminuted pork meat product. *Food Res. Int.*, *38*(3), 337–344.

Catro, I., Teixeira, J. A., Salengke, S., Sastry, S. K., & Vicente, A. A., (2004). Ohmic heating of strawberry products electrical conductivity measurements and ascorbic acid degradation kinetics, *5*(1), 27–36.

Chen, C., Andelrakim, K., & Beckerick, (2010). Sensitivity analyzis of continous ohmic heating process for multipurpose foods. *Food Eng.*, *98*(2), 257–265.

Cho, H. Y., Yousef, A. E., & Sastry, S. K., (1999). Kinetics of inactivation of *Bacillus subtilis* spores by continuous or intermittent ohmic and conventional heating. *Biotechnol. Bioeng.*, *62*(3), 368–372.

De Alwis, A. A. P., Halden, K., & Fryer, P. G., (1989). Shape and conductivity effect in ohmic heating foods. *Chem. Eng. Res. Des.*, *67*, 159–168.

Diana, L., Erica, T., James, G. L., & Cristina, A., (2016). The potential of ohmic heating as an alternative to steam for heat processing shrimps. *Innovative Food Sci. Emerging Technol.*, *37*, 329–335.

Duygu, B., & Umit, G., (2015). Application of ohmic heating system in meat thawing. *Procedia Soc. Behav. Sci.*, *193*, 2822–2828.

Eliot, S. C., Goullieux, A., & Pain, J., (1999). Processing of cauliflower by ohmic heating, influence of precooking on firmness. *J. Sci. Food Agric.*, *79*, 1406–1412.

Engchuan, W., Jittanit, W., & Garanjanagoonchorn, W., (2014). The ohmic heating of meat balls : modeling and quality determination. *Innovative Food Sci. Emerging Technol.*, *23*, 121–130.

Fryer, P. J., deAlwis, A. A. P., Koury, E., Stapley, A. G. F., & Zhang, L., (1993). Ohmic processing of liquid-solid matrix mixture, heat generation and energy convection effects. *Food Eng.*, *18*, 101–125.

Gally, T., Rouaud, O., Jury, V., Havet, M., Oge, A., & Le-Bail, A., (2017). Proofing of bread assisted by ohmic heating. *Innovative Food Sci. Emerging Technol.*, *39*, 55–62.

Guo, W., Llave, Y., Jin, Y., Fukuoka, M., & Sakai, N., (2017). Mathematical modeling of ohmic heating of two component foods with non uniform electric properties at high frequencies. *Innovative Food Sci. Emerging Technol.*, *39*, 63–78.

Icier, F., Yildiz, H., & Baysal, T., (2006). Peroxidase inactivation and color changes during ohmic blanching of pea puree. *Food Eng.*, *74*(3), 424–429.

Icier, F., Yildiz, H., Sabanci, S., Cevik, M., & Cokgezme, O. M., (2017). Ohmic heating assisted vacuum evaporation of pomegranate juice: electrical conductivity changes. *Innovative Food Sci. Emerging Technol.*, *39*, 241–246.

Imai, T., Uemura, K., Ishida, N., Yoshizaki, S., & Noguchi, A., (1995). Ohmic heating of Japanese white raddish *Rhaphanus sativus L. Int J. Food Sci. Technol.*, *30*(4), 461–72.

Jakob, A., Bryjak, J., Wojtowicz, H., Annus, J., & Polakovic, M., (2010). Inactivation kinetics of food enzymes during ohmic heating. *Food chem.*, *123*(2), 369–376.

Jaworska, G., Pagon, K., Bernas, E., & Maciejaskez, I., (2010). Camparison of texture of fresh and preserved Agricus bisporus and Bolitus edilus mashrooms. *Food Chem., 123,* 269–376.

Juliane, F., Noel, G., Elodie, L., & Romain, J., (2006). Continous processing of skim milk by a combination of pulsed electric fields and conventional heat treatments: does a synergistic effect on microbial inactivation exist?. *Lait., 86*(3), 203–211.

Jun, S., Sastry, S., & Samaranayake, C., (2007). Migration of electrode components during ohmic heating of foods in retort pouches. *Innovative Food Sci. Emerging Technol., 8,* 237–243.

Kanjanapongkul, K., (2017). Rice cooking using ohmic heating: determination of electrical conductivity, water diffusion and cooking energy. *Food Eng., 192,* 1–10.

Khuenpet, K., Fukuoka, M., Jattinat, W., & Sirisansaneeyakul, (2017). Spray drying of inulin component extracted from Jerusalem artichoke tuber powder using conventional and ohmic ultrasound heating for extraction procedure. *Food Eng., 194,* 67–78.

Kim, S. S., & Kang, D. H., (2017). Synergist effect of carvacrol and ohmic heating for inactivation of *E. Coli O157:H7, S. typhimurium, L. monocytogens and MS 2* bacteriophage in salsa. *Food control., 73*(B), 300–305.

Kulshrestha, S., & Sastry, S., (2006). Low-frequency dielectric changes in cellular food material from ohmic heating: effect of end point temperature. *Innovative Food Sci. Emerging Technol., 7*(4), 257–62.

Lakkakula, N. R., Lima, M., & Walker, T., (2004). Rice bran stabilization and rice bran oil extraction using ohmic heating. *Bioresour. technol., 92*(2), 157–161.

Lebovka, N. I., Shynkaryk, M., & Vorobiev, E., (2007). Modern electric field treatment of sugarbeet tissue *Biosyst. Eng., 96*(1), 47–56.

Lee, S. Y., Sagong, H. J., Ryu, S., & Kang, D. H., (2012). Effect of continuous ohmic heating to inactivate *E. coli, Salmonella typhimurium* and *Listeria monocytogens* in orange juice and tomato juice. *Appl. Microbial., 112*(4), 723–31.

Lima, M., Heskett, B. F., & Sastry, S. K., (1999). The effect of frequency and waveform of electrical conductivity and temperature profiles of turnip tissues. *J. Food Process Eng., 22,* 41–54.

Lima, M., & Sastry, S. K., (1999). Effect of ohmic heating on hot air drying and juice yield.. *Food Eng., 41,* 115–119.

Louarme, L., & Billaud, C., (2012). Evaluation of ascorbic acid and sugar degradation products during fruit dessert processing under conventional or ohmic heating treatment. *LWT—Food Sci. Technol., 49*(2), 184–187.

Loypimai, A., Moonggarm, A., Chottano, P., & Moontree, T., (2015). Ohmic heating assisted extraction of anthocyanin from black rice bran to prepare natural colorant. *Innovative Food Sci. Emerging Technol., 27,* 102–110.

Loypimai, P., Moonggarm, A., & Chottano, P., (2009). Effects of ohmic heating on lipase activity, bioactive compounds and antioxidant activity of rice bran. *Aust. J. Basic Appl. Sci., 3*(4), 3642–3652.

Manish, K., Jyoti., & Abid, H., (2014). Effect of ohmic heating of buffalo milk on microbial quality and texture of paneer. *Asian J. Dairy Food Res., 33*(1), 9–13.

Merceli, G. D., Sarkis, J. R., Jaeschke, D. P., & Tessaro, I. C., (2012). Physical properties of acerola and blueberry pulp. *Food Engg., 106,* 283–289.

Merceli, G. D., Schwartz, S., Marczak, L. D. F., Tessaro, I. C., & Sastry, S., (2014). Ascorbic acid degradation and color changes in acerola pulp during ohmic heating: effect of electric field frequency. *Food Engg., 123,* 1–7.

Mesias, M., Wagner, M., George, S., & Francisco, J., (2016). Impact of conventional sterilization and ohmic heating on the amino acid profile in vegetable baby foods. *Innovative Food Sci. Emerging Technol.*, 34, 24–28.

Moreno, J., Simpson, R., Pizarro, N., Pavez, C., Dorvil, F., Petzold, G., & Bugueno, G., (2013) Influence of ohmic heating/osmatic dehydration treatments on polyphenoloxidase inactivation, physical properties and microbial stability of apples (cv. Granny smith). *Innovative Food Sci. Emerging Technol.*, *20*, 198–207.

Ngasri, P. W., & Sastry, S. K., (2015). Effect of ohmic heating on tomato peeling. *LWT – Food Sci. Technol.*, *61*(2), 269–274.

Park, I. I. K., & Kanj, D. H., (2013). Effect of electro permeabilization by ohmic heating for inactivation of E-Coli 0157: H7, *Salmonella enterica Serover Typimurium,* and *Listeria monocytogens* in buffered peptone water and apple juice. *J. Appl. Environ. microbiol.*, *79*(23), 7122–7129.

Palaniappan, S., & Sastry, S. K., (1992). Effect of electroconductive heat treatments and electrical pretreatment on thermal death kinetics of selected microorganisms. *Biotechnol. Bioeng.*, *39*, 225–232.

Pereira, R., Martins, J., Mateis, C., Teixeira, J. A., & Vicente, A. A., (2007). Death kinetics of Escherichia coli in goat milk and *Bacillus licheniformis* in cloudberry jam treated by ohmic heating. *Chem. Pap.*, *61*(2), 121–126.

Rahman, M. S., (1999). In Rahman, M. S., (Ed.), *Handbook of Food Preservation*; Dekker, New York, pp. 521–532.

Roberts, J., Balaben, M., & Luziriaga, D., (1996). Automated ohmic thawing of shrimp blocks. *Sea food Sci. technol Soc. of Amer.*, 72–81.

Samaranayake, C. P., Satry, S. K., & Zhang, H., (2005). Electrode and pH effect on electrochemical reaction during ohmic heating . *J. Electroanal. Chem.*, *577*, 125–135.

Sarang, S., Sastry, S., & Knipe, L., (2008). Electrical conductivity of fruits and meats during ohmic heating. *Food Eng.*, *87*(3), 351–6.

Sarkis, J. R., Jaeschke, D. P., Tessaro, I. C., & Merczak, L. D. F., (2013). Effect of ohmic and conventional heating on anthocyanin degradation during the process of blueberry pulp. *LWT – Food sci. technol.*, *51*(1), 79–85.

Sastry, S. K., (1992). A model for heating of liquid-particle mixtures in a continuous flow ohmic heater. *J. Food Process Eng.*, *15*, 263–278.

Sastry, S. K., & Palaniappan, S., (1992). Ohmic heating of liquid-particle mixture. *Food technol.*, *46*(12), 68–72.

Saxena, J., Makroo, H. M., & Srivastava, B., (2016). Optimization of time electric field combination for PPO inactivation in sugarcane juice by ohmic heating and its shelf-life assessment. *LWT – Food Sci. Technol.*, *71*, 329–338.

Shivmurti, S., Rinkita, P., Harshit, P., & Smit, P., (2014). Ohmic Heating is an Alternative Preservation Technique: A Review. *Global J. Sci. Frontier Resr: E Interdiciplinary.*, *14*(4), ISSN: 2249–4626.

Somavat, R., Hussein, H. M. H., & Sastry, S. K., (2013). Inactivattion kinetics of *bacillus coagulam* spores under ohmic and conventional heating. *LWT – Food Sci. Technol.*, *54*(1), 194–198.

Stiring, R., (1987). Ohmic heating-A new process for food industry. *Food Engg.*, *15*, 21–48.

Tham, H. J., Chen, X. D., & Young, B., (2009). 3D simulation of ohmic heating of milk solutions in laminar annular flow. *Chem. Prod. Proc. Modelling.*, *4*(3), 1934–1937.

Teixeira, J. A., Salangke, S., Sastry, S. K., & Vicente, A. A., (2004). Ohmic heating of strawberry products : electric conductivity measurement and ascorbic acid degradation kinetics. *Innovative Food Sci. Emerging. Technol.*, *5*, 27–36.

Tian, X., Wu, W., Yu, Q., Hou, M., Jai, F., Li, X., & Dia, R., (2016). Quality and proteome changes of beef *M. Longissimus dorsi* cooked using a water bath and ohmic heating process. *Innovative Food Sci. Emerging Technol.*, *34*, 259–266.

Tucker, G. S., & Withers, P. M., (1994). Determination of residence time distribution of non settling food particles in viscous food carrier fluids using Hall effect sensors. *J. Food Process Eng.*, *17*, 401–422.

Tumpanuvatr, T., & Jittanit, W., (2012). The temperature prediction of some botanical beverages, concentrated juice and puree of orange and pineapple during ohmic heating. *Food Engg.*, *113*(2), 226–233.

USA-FDA., (2000). United States of America, Food and Drug Administration, Center for Food Safety and Applied Nutrition. Kinetics of microbial inactivation for alternative food processing technologies: ohmic and inductive heating.

Varghese, K. S., Pandey, M. C., Radhakrishna, K., & Bawa, A. S., (2014). Technology, applications and modelling of ohmic heating: A review. *J of Food Sci. Technol.*, *51*(10), 2304–2317.

Vikram, V. B., Ramesh, M. N., & Prapulla, S. G., (2005). Thermal degradation of kinetics of nutrients in orange juice heated by electromagnetic and conventional methods. *Food Engg.*, *69*, 31–40.

Wang, L. J., Li, D., Tatsumi, E., Liu, Z. S., Chen, X. D., & Li, L.T., (2007). Application of two stage ohmic heating to tofu processing. *Chemi Eng. Process. Proc. Intensftn.*, *46*(5), 486–490.

Wang, W., & Sastry, S., (1993). Salt diffusion into vegetable tissue as a pretreatment for ohmic heating: electrical conductivity profiles and vacuum infusion studies. *Food Eng.*, *20*, 299–309.

Wang, W., & Sastry, S., (2000). Effects of thermal and electrotermal pretreatments on hot air drying rate of vegetable tissue. *J Food Process Eng.*, *23*(4), 299–319.

Wen, G., Yvan, L., Yinzhe, J., Mike, F., & Noboru, S., (2017). Mathematical modeling of ohmic heating of two component foods with non-uniform electric properties at high frequencies. *Innovative Food Sci. Emerging Technol.*, *39*, 63–78.

Yaari, Y. Z., & Nussinovitch, A., (2014). Browning prevention in rehydrated freeze dried non-blanched potato slices by electric treatment. *LWT – Food Sci. Technol.*, *56*(1), 194–199.

Yoon, S. W., Lee, C. Y. J., Kim, K. M., & Lee, C. H., (2002). Leakage of cellular material from *Saccharomyces cerevisiae* by ohmic heating. *J. Microbiol. Biotechnol.*, *12*, 183–188.

Zoltai, P., & Swearingen, P., (1996). Product development considerations for ohmic processing. *Food Technol.*, *50*, 263–266.

CHAPTER 2

APPLICATION OF COLD PLASMA IN FOOD PROCESSING

U. S. ANNAPURE

Food Engineering and Technology Department, Institute of Chemical Technology, Matunga, Mumbai–400019, India

CONTENTS

2.1 INTRODUCTION

Plasma is known as the fourth state of matter, and it contains many active species like electrons, ions, free radicals, excited and ground state atoms, and a large number of nonionized neutral molecules. In food processing, cold plasma is widely used for microbial inactivation, enhancement of germination rate of seeds, starch modification, enzymatic inactivation, and reduction in cooking time of rice (Misra et al., 2011, 2016; Thirumdas et al., 2015a, b). The term plasma was first time used by Irving Langmuir in 1928 to define the plasma as the fourth state of matter. Bogaerts et al. (2002) stated that the plasma can be grouped into high temperature (or fusion plasma) and low temperature (or gas discharges). The authors have reported that in high temperature plasma, all the plasma species are at the same temperature (local thermal equilibrium), and in the cold plasma, the temperature of all the plasma species are not the same (nonlocal thermal equilibrium). The cold plasma contains low temperature particles like neutral molecules, atomic species, and relatively high temperature electrons; hence, it does not affect the sensitive materials that come in contact with it. Cold plasma has the least shadow effect, which means that the plasma flows around the entire product (Banu et al., 2012).

2.2   PLASMA SOURCES AND CHEMISTRY

Plasma state can be achieved by the supply of energy in different forms like thermal and electric fields, radio and microwave frequencies, and magnetic fields (Thirumdas et al., 2017b). Among these, electric fields are the most commonly used energy source in generating plasma for technological applications (Wan et al., 2009). DC low-pressure plasma requires costly vacuum equipment for ionization and breakdown of gaseous molecules at normal voltages. Recently, with advances in technology, atmospheric cold plasma has been developed that can be directly applied to both solid and liquid food materials. The various plasma apparatus used for the generation of cold plasma includes dielectric barrier discharges (DBD), corona glow discharges, radio frequency (RF), atmospheric glow discharge inductively coupled plasma (ICP), microwave-induced plasma (MIP), and gliding arc discharge (Bogaerts et al., 2002; Thirumdas et al., 2015a). Among the different plasma sources, plasma jet and DBD are widely used in food research and are commercially available (Misra et al., 2016).

In plasma chemistry, ionization of gas used for plasma generation is the most important element along with other factors like free mean path, types of collisions, and electron energy distribution (Fridman, 2008). During plasma generation, the common gas phase reactions are excitation, de-excitation, ionization, dissociations, fragmentations, etc. Misra et al. (2016) reported that plasma chemistry depends on several factors such as feed gas composition, humidity, power, the voltage applied, and surrounding phase. Oxygen and nitrogen are introduced in the biopolymer when exposed to O_2, air, N_2, or NH_3 gas cold plasmas (Morent et al., 2011). The chemical composition of cold plasma of nitrogen, carbon dioxide, and nitrogen gas mixtures are dominated by O^-, O_2^-, O^+, N^+, N_2^-, NO^+, and CO_2^+ ions (Wan et al., 2009). Atomic oxygen and hydroxyl radicals are found to be most reactive due to their versatility in covalent bonding with many different compounds. In the presence of nitrogen and oxygen gas during plasma generation, energetic electron collision takes place and results in chain of reactions that form nitrogen oxides (Misra et al., 2016).

2.3 APPLICATIONS

2.3.1 MICROBIAL INACTIVATION OF FUNGAL SPECIES

Aflatoxins are the most toxic, carcinogenic, and teratogenic compounds among the known mycotoxins and are mainly produced by *Aspergillus flavus* and *A. parasiticus* (Ellis et al., 1991). Aflatoxins belong to the polyketide class of secondary metabolites produced by toxigenic strains of *A. flavus* and *A. parasiticus,* and are synthesized by enzymes encoded within a large gene cluster (Yabe and Nakajima, 2004). There are four major aflatoxin types: B1, B2, G1, and G2 that are designated based on their blue and yellow-green fluorescence. *A. flavus* produces aflatoxin B1 and B2, while *A. parasiticus* produces B1, B2, G1, and G2 (Eaton and Groopman, 1994). The total plate count in plasma-treated groundnuts reduced drastically when compared to the untreated samples (Table 2.1). Initially, the total plate count was observed to be 2.1×10^3 colony forming unit (cfu)/g, and yeast and mold counts were found to be 4×10^2 cfu/g. The total microbial load reduced by one log within 5 min of treatment, and the maximum log reduction was found for 60 W-15 min samples. A significant difference was observed in the reduction of microorganisms after the treatment. An increase in efficiency of inactivation was observed with the increase in power applied and treatment time. There was 99.94% and 99.52% reduction in total plate count and yeast

TABLE 2.1   Microbial Content of Cold Plasma-Treated Groundnuts

Power (W)	Treatment time (min)	Total Plate Count (cfu/g)	Yeast and mold Count (cfu/g)
Control		$2.1 \times 10^3 \pm 0.7 \times 10^3$	$4 \times 10^2 \pm 0.5 \times 10^2$
40	5	$4.5 \times 10^2 \pm 0.2 \times 10^2$	$3.4 \times 10^2 \pm 0.1 \times 10^2$
	10	$2.2 \times 10^2 \pm 0.1 \times 10^2$	$8 \times 10^1 \pm 0.2 \times 10^1$
	15	$6 \times 10^1 \pm 0.5 \times 10^1$	$4 \times 10^1 \pm 0.1 \times 10^1$
60	5	$1.5 \times 10^2 \pm 0.1 \times 10^2$	$8 \times 10^1 \pm 0.8 \times 10^1$
	10	$5 \times 10^1 \pm 0.1 \times 10^1$	$2 \times 10^1 \pm 0.2 \times 10^1$
	15	$1.1 \times 10^0 \pm 0.05 \times 10^0$	1.9×10^0

All the data are expressed as mean ± standard deviations.
(adapted from Devi et al., 2017).

and mold growth, respectively, for the 60 W-15 min-treated samples at the end of 20 min. Many mechanisms have been considered to be responsible for microbial inactivation by different researchers. Montie et al. (2000) considered membrane lipid alteration caused by fatty acid peroxide formation and protein oxidation as a possible path to affect the cell. During plasma generation, the oxygen present in the air plasma results in the formation of lethal species like O^*, O_2, and O_3, which leads to inactivation of microorganisms due to the formation of peroxides (Bermudez-Aguirre et al., 2013). Because of the accumulation of charged particles on the outer cell membrane, electrostatic forces can then lead to rupture of the cell membrane and subsequently cause cell death (Laroussi et al., 2003).

The groundnuts were artificially inoculated with *A. parasiticus* spores at loads of 1.45×10^3 cfu/g and exposed to cold plasma treatment for various times in multiples of 3 min until complete inactivation of pores was achieved. The reduction in cfu/g of *A. parasiticus* at 40 W and 60 W is given in Figure 2.1.

The inactivation of *A. parasiticus* is effective at higher power level (60 W) than at 40 W; the optimum time for inhibition at 60 W was 6 min earlier than for 40 W treated samples. The rate of reduction of spores was found to be high up to 15 min of treatment, and in the last 5 min of treatment, only one log reduction was observed. From the SEM micrograph images (Figure 2.2), it is clearly observable that the spore coat was disrupted and ruptured when treated.

The increase in power supplied and treatment time resulted in formation of much larger pores on the surface of the fungal spores. The effect of

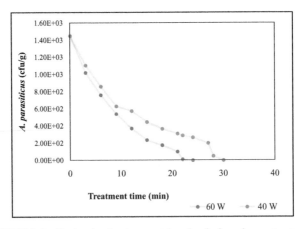

FIGURE 2.1   Reduction in *A. parasiticus* load after plasma treatment.

a) 2500X   b) 5000X

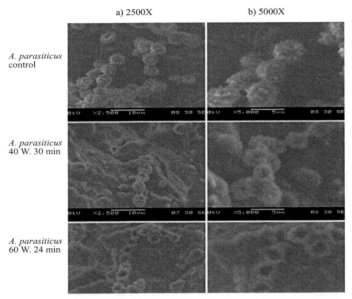

FIGURE 2.2 SEM images of A. *parasiticus* showing disintegration of the spore wall (adapted from Devi et al., 2017).

inactivation was also dependent on the surface structure of spores. When the plasma interacts with the surface of the spores, it causes formation of small pores on the surface due to electric fields and subsequently promotes death of the spores (Bermudez-Aguirre et al., 2013).

2.3.2 COOKING AND TEXTURAL PROPERTIES OF BROWN RICE

Rice (*Oryza sativa* L.) is the staple food for nearly half of the world's population and the third leading cereal in production. The consumption of brown rice is decreased across the world due to its dark color, hard texture, chewiness, and prolonged cooking time. A significant difference ($p<0.05$) was observed in contact angles on the application of plasma. The decrease in contact angles was observed with increase in plasma power and treatment time. For untreated rice samples, the initial contact angle was found to be 100.5°, which was decreased to 87.4° after plasma treatment for 90 s. There was a decrease of 45% in the contact angle at 50 W-10 min (Figure 2.3).

The decrease in the contact angle is due to increase in the hydrophilic nature of the substrate after the plasma treatment (Thirumdas et al., 2015b).

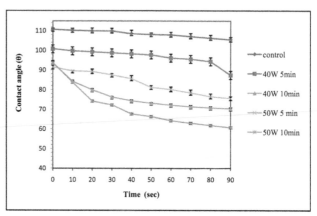

FIGURE 2.3 Contact angle of cold plasma-treated brown rice (adapted from Thirumdas et al., 2016a).

A similar decrease in the contact angle after the plasma treatment was observed by Sarangapani et al. (2016, 2017). Another reason for decrease in the contact angle may be due to the increase in the effective surface area (as seen from SEM, Figure 2.4) for contact after air plasma treatment.

Air plasma treatment increased the surface area for contact, thereby causing decrease in the contact angle. The extent of decrease in the contact angle was observed more in 50W-10 min-treated samples, which may be due to more etching by the reactive species with higher plasma power and treatment time when compared with other samples. It is well known that plasma treatment causes an increase in the hydrophilic nature of the substrate by incorporation of polar functional groups. The results of cooking properties of plasma-treated samples are shown in Figure 2.5. Cooking time taken for different samples varied from 21.1 to 29.1 min. The decrease in cooking time was observed with

FIGURE 2.4 SEM micrographs of brown rice surface showing etching (adapted from Thirumdas et al., 2016a).

increase in plasma power and time, which may also be due to surface etching of plasma-treated samples as compared to the untreated samples. The reactive species generated in the plasma etched the outermost fibrous bran layer opening, which facilitates easy penetration of water into the rice grain, thus resulting in shorter cooking time. The decrease in cooking time may also be explained by the incorporation of polar groups between the starch molecules.

Plasma treatment can significantly affect the uptake of water. A significant difference was observed for water uptake after the plasma treatment. Samples with the least cooking time were found to have higher water uptake. Water uptake of the untreated and plasma-treated brown rice samples varied from 2.22 to 2.36 g/g, as shown in Figure 2.6. Water uptake is directly related to the reduction in cooking time and surface modification of the

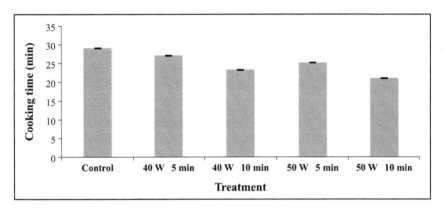

FIGURE 2.5   Cooking time of cold plasma-treated brown rice.

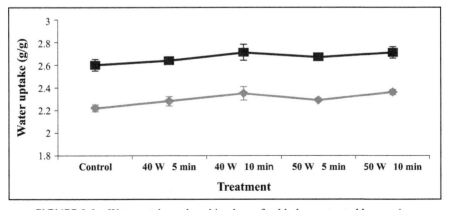

FIGURE 2.6   Water uptake and cooking loss of cold plasma-treated brown rice.

grains (Mohapatra and Bal, 2006). Cooking loss increased in plasma-treated samples as compared to that in the control sample, which may be due to degradation of starch molecules by the plasma species. During the cooking operation, the fragmented starch particles were easily leached into the surrounding water. Similar results were reported when brown rice was treated with gamma irradiations (Sabularse et al., 1991).

From the change in water uptake and leaching of the particles after the plasma treatment, textural changes in the brown rice are expected. The hardness and adhesiveness are the important parameters that are considered for texture evaluation and consumer acceptability (Chen et al., 2012). Significant differences were observed between the treated and untreated samples for hardness (Table 2.2). A similar decrease in hardness was also reported by Thirumdas et al. (2015b), suggesting that leaching components can be responsible for a decrease in hardness and an increase in adhesiveness of cooked rice samples. The decrease in hardness was observed with an increase in time and power of plasma treatment. Chewiness tends to decrease with an increase in plasma power and treatment time. The decrease in hardness is directly correlated to chewiness, which shows that lesser work is required to chew the rice. Control sample had a cohesiveness of 0.11, which was decreased to 0.09 in plasma-treated brown rice. The decrease in hardness showed a reduced cohesiveness for different rice varieties.

TABLE 2.2 Textural Properties of Cold Plasma-Treated Brown Rice

	Control	40 W 5 min	40 W 10 min	50 W 5 min	50 W 10 min
Hardness (N)	40.47±0.3	38.82±0.1	33.10±0.5	33.97±0.2	30.09±0.2
Cohesiveness	0.11±0.01	0.09±0.01	0.09±0.01	0.09±0.01	0.09±0.01
Adhesiveness (g mm)	2.52±0.14	4.31±0.20	4.4±0.26	4.1±0.10	3.81±0.18
Springiness (mm)	0.09±0.01	0.12±0.01	0.18±0.01	0.09±0.01	0.11±0.01
Gumminess	4.91±0.5	3.4±0.1	2.92±0.3	2.97±0.4	2.68±0.4
Chewiness	0.409±0.03	0.412±0.01	0.348±0.08	0.296±0.02	0.30±0.04

All the data are expressed as mean ± standard deviations.
Adapted from Thirumdas et al. (2016a).

2.3.3   STARCH MODIFICATION USING COLD PLASMA

Chemical, physical, and enzymatic modification methods are generally used for starch modification (Luo et al., 2008). Amylose content of cold plasma-treated starches and untreated starch differed significantly (Table 2.3). The amylose content of starches varied from 22.83% to 29.93% after the treatment. The decrease in the amylose content was found after the plasma treatment as compared to the native starch. A decreasing trend was observed with increase in plasma power and treatment time, which may be due to depolymerization of the amylose chains caused by the bombardment of the reactive species of the plasma (Thirumdas et al., 2017b). It is well known that application of cold plasma to starches leads to depolymerization and decreases the average molecular weight of starches. The nature of starch and type of gas used for glow discharge plasma could lead to a different extent of starch depolymerization (Lii et al., 2002). A similar decrease in the amylose content was reported in starches treated with ultrasound and irradiation (Sujka and Jamroz, 2013). Physical methods used for starch modification predominantly affect the amorphous regions of the starch granules rather than the crystalline regions (Luo et al., 2008).

Iodine-blue complex values of all starches sample are given in Table 2.3. It was observed that the absorbance was increased from 0.10 to 0.13 after the plasma treatment. The increase in absorbance might be due to the changes in the surface morphology of the starch granules caused by the plasma species, which easily facilitates the chemicals to penetrate the starch molecules due to plasma etching. A similar result was reported by Zhu (2012) in the case of irradiated starch molecules. Juliano et al. (1968) reported that the lower values of iodine-blue complex are inherent for those starches with higher amylose content.

TABLE 2.3   Amylose, Iodine-Blue, and Syneresis Values of Cold Plasma-Treated Starches

Treatment	Amylose (%)	Iodine-Blue value (640 nm)	Syneresis (%)
Untreated	29.93 ± 0.31	0.10 ± 0.00	19.02 ± 0.75
40 W 5 min	25.42 ± 0.43	0.12 ± 0.00	19.87± 0.40
40 W 10 min	23.15 ± 0.35	0.12 ± 0.00	19.57 ± 0.34
60 W 5 min	22.83 ± 0.60	0.13 ± 0.00	19.70 ± 0.15
60 W 10 min	24.39 ± 0.51	0.12 ± 0.00	19.19 ± 0.56

All the data are expressed as mean ± standard deviations.
(adapted from Thirumdas et al., 2017a).

A slight change in pH (7.42 to 6.96) of an aqueous rice starch solution was observed as a consequence of cold plasma treatment (Figure 2.7). Decrease in the pH of starch solution was increased after the plasma treatment, which is in agreement with the data reported by Lii et al. (2002), thus indicating the formation of chemical groups with acidic characteristics, like carboxyl, carbonyl, and peroxide groups. This shows that the oxidation of starch granules occurred when exposed to the plasma treatment. Starch hydrolysis decreased to 87.02% for the samples treated at 60 W-10 min from 91.60% for the untreated sample (Figure 2.8). The decrease in the degree of starch hydrolysis might be due to the decrease in pH of the starch after the treatment.

Results of water and fat absorption of the starch samples showed an increasing trend after the plasma treatment (Figure 2.9). The water absorption was increased from 2.38 g/g for the untreated sample to 2.58 g/g for the

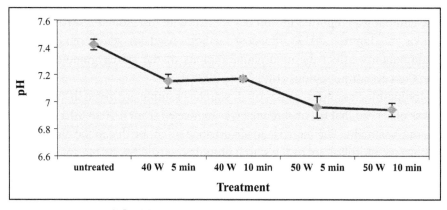

FIGURE 2.7   pH of cold plasma-treated starch.

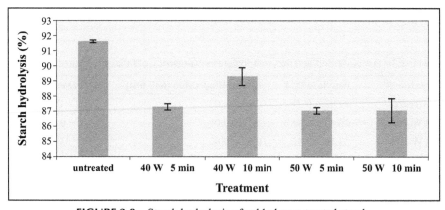

FIGURE 2.8   Starch hydrolysis of cold plasma-treated starch.

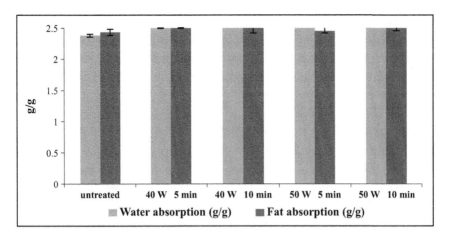

FIGURE 2.9   Water absorption and fat absorption of cold plasma-treated starches.

samples treated at 40 W-10 min. It is a known fact that plasma treatment has been used for quite some time to increase the hydrophilic nature of the treated materials, depending upon the nature of gas used.

Pasting viscosity of the treated and untreated starch analyzed using rapid visco analyzer (RVA) is shown in Figure 2.10. Pasting temperature (PT) denotes the temperature at which the viscosity suddenly rises during the continuous heating process. The PT of the untreated sample (78.02°C) was decreased to 76.68°C (40 W-5 min), 77.16°C (60 W-5 min), and 77.11°C for the samples treated with 40 W-10 min and 60 W-10 min. Peak viscosity

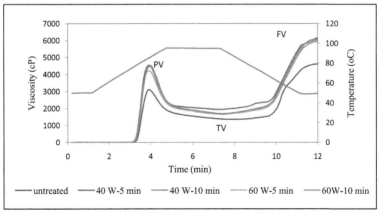

FIGURE 2.10   RVA of cold plasma-treated rice starches (adapted from Thirumdas et al., 2017a).

(PV) of starches increased after the plasma treatment. According to Pal et al. (2016), cross-linking of starch molecules induced by plasma oxidation resulted in higher PV than that for the untreated sample. The modification or substitution of starch takes place in the amorphous region of amylose and amylopectin and only outer lamellar layers of crystalline regions (Chen et al., 2004). Like PV, the trough viscosity (TV) was also found to be increased after the treatment. TV of the untreated sample was 1360 cP, which increased to 1943 cP (40 W-5 min), 1659 cP (40 W-10 min), 1674 cP (60 W-5 min), and 1697 cP (60 W-10 min) for the treated samples. The breakdown and final viscosity (FV) of plasma-treated samples were found to be much higher than those for the untreated sample. The breakdown viscosity was (1753 cP, 2599 cP, 2813 cP, 2554 cP, 2863 cP) and final viscosity was (4610 cP, 6150 cP, 5932 cP, 5987 cP, 5995 cP) for untreated and treated samples at 40 W-5 min, 40 W-10 min, 60 W-5 min, and 60 W-10 min, respectively. The increase in FV during the cooling process is due to the beginning of retrogradation of amylase, which causes re-formation of bonds between the molecules of the chains.

2.3.4   GERMINATION CHARACTERISTICS OF MUNG BEANS

Plasma enhanced the seed germination and seedling growth rates, which might be associated with the increased water uptake of seeds (Ling et al., 2014). The speed of germination for all the cold plasma-treated mung bean seeds increased by a maximum of 20%. Researchers have suggested that cold plasma treatment promotes germination and seedling growth. Atmospheric cold plasma treatment for 1 min strongly improved chickpea seed germination (89.2%) and speed of germination (Mitra et al., 2012). Few reports confirmed that the cold plasma has the ability to enhance seed germination and seedling growth (Sera et al., 2008; Bormashenko et al., 2012). The plasma treatment can increase the roughness of the seed surface, and finally lead to increase in the hydrophilicity of the seed by changing the chemical structure, which may account for the increase in water uptake by the seed (Bormashenko et al., 2012). Sprout length of all the treated samples was considerably higher than that of the control samples. Mitra et al. (2012) reported that 3 min of atmospheric cold plasma treatment on chickpea resulted in a maximum increase of 51.1% in the sprout length during germination. Jiang et al. (2014) also reported 9% increase in root length of plasma-treated wheat grains compared to that of untreated ones. Reactive oxygen and nitrogen species produced by the cold plasma may have a significant effect on abscisic acid catabolism and gibberellin synthesis (Mitra et al., 2012) (Table 2.4).

The increase or decrease in the activity of enzymes depends upon the type of plasma apparatus used, nature of feed gas, and the power applied (Misra et al., 2016). The amylase activity of cold plasma-treated mung beans during germination at 0, 12, and 24 h is shown in Figure 2.11a. At 12 h and 24 h of germination, plasma-treated samples showed higher amylase activity than the control sample. The increase in the enzyme activity is due to faster imbibition of water in the plasma-treated samples. The enzymatic activity of alpha-amylase was increased by 126% in the germinating brown rice after the cold plasma treatment (Chen et al., 2016). Similar kinds of results were obtained by Yin et al. (2005) in cold plasma-treated tomatoes. The author reported an increase in alpha amylase-activity due to the accumulation of soluble sugars, which was related to their metabolism. Ling et al. (2014) observed an increase in soluble sugar content in cold plasma-treated soy beans. Similar to amylase activity, cold plasma-treated mung beans showed an increase in protease activities during germination (Figure 2.11c). At 12 h and 24 h, all the plasma-treated samples showed significantly higher protease activity than the control sample. Yin et al. (2005) reported an increase in the protease enzyme activity after the plasma treatment. The increase in protease activity also accounts for the increased

TABLE 2.4 Germination Characteristics of Cold Plasma-Treated Mung Beans

Power	Time (min)	Germination rate (%)		Germination speed (Seeds/day)	Sprout length (cm)	
		24 h	48 h		24 h	48 h
Control	-	73.40 ± 2.77^a	86.70 ± 2.89^a	16.00 ± 0.00^a	1.45 ± 0.07^a	3.13 ± 0.04^a
40 W	10	100.00 ± 0.00^b	100.00 ± 0.00^b	20.00 ± 0.00^b	1.76 ± 0.04^b	3.85 ± 0.06^b
	15	100.00 ± 0.00^b	100.00 ± 0.00^b	20.00 ± 0.00^b	1.79 ± 0.06^b	3.92 ± 0.1^b
	20	100.00 ± 0.00^b	100.00 ± 0.00^b	20.00 ± 0.00^b	1.81 ± 0.01^b	4.23 ± 0.07^c

All the data are expressed as mean ± standard deviations.
(adapted from sadhu et al., 2016).

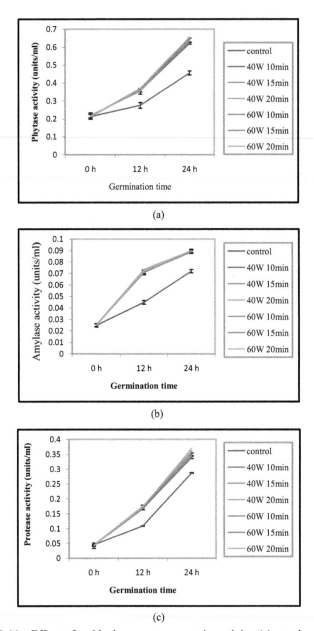

FIGURE 2.11 Effect of cold plasma on enzymatic activity (a) amylase, (b) phytase, (c) protease of mung beans during germination

soluble protein content and better digestibility of the plasma-treated samples. An increase in soluble proteins was reported in the plasma-treated sample (Ling et al., 2014). A similar increase in soluble proteins after the plasma treatment of maize seeds was also observed (Wu et al., 2007). This may be due to change in the activity of protease enzymes, because of change in the structures of proteins. Similar kinds of result were also observed by Li et al. (2011) in lipase activity due to change in the second- ary and tertiary structures of the proteins. The germinated mung beans showed better phytase activities than the control sample. The increase in phytase activity also accounts for the better mineral bioavailability of the plasma-treated samples than that of the control samples, because it will breakdown the phytates that bind to minerals and reduce the rate of metal- enzyme complex formation. The increase in the enzyme activity is due to faster imbibition of water in the plasma-treated samples due to increased seed coat permeability. The increase in the activity of hydrolytic enzymes was also due to the disruption of the cell membrane and integrity of the cell structure.

2.3.5   EFFECT OF COLD PLASMA ON ANTINUTRITIONAL FACTORS

Anti-nutritional factors are the substance present in the food system that reduces the bioavailability of one or more nutrients. The trypsin inhibitor activity and phytic acid content decreased gradually during germination at 0, 12, and 24 h for the legume seeds. Figure 2.12 shows the phytic acid content and trypsin inhibitor activity of cold plasma-treated beans during germination. Mubarak (2005) reported that germination of mung beans for 12 h drastically reduces the trypsin inhibitor activity and phytic acid con- tent. At 12 and 24 h, all the plasma-treated samples showed reduced trypsin inhibitor activity and phytic acid content compared to that of the control. Gamma irradiation of broad bean seeds at 10 kGy showed 10% reduction in trypsin inhibitor activity and 18% reduction in phytic acid content (Al- Kaisey et al., 2003). This might be due to the breakdown of trypsin inhibi- tors through the bombardment of plasma ions on the seed surface. During germination, it is believed that protease enzymes are responsible for the breakdown of proteinaceous antinutritional factors like lectins, amylase, and trypsin inhibitors (Savelkoul et al., 1992). Enzyme activity of prote- ase was increased after the plasma treatment, which further enhanced the breakdown of trypsin inhibitors. The decrease in phytic acid content is due to the increase in phytase activity. Thus, the reduction in phytate content

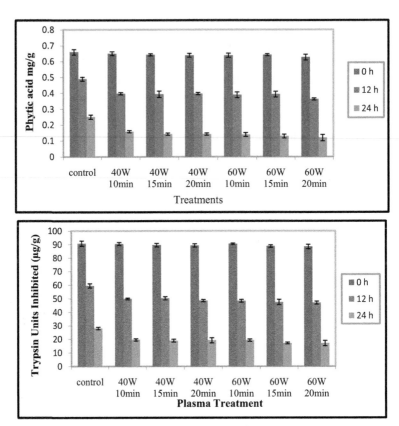

FIGURE 2.12   Effect of cold plasma on phytic acid content and trypsin inhibition

will enable better mineral bioavailability. However, the exact mechanism for the reduction of these antinutritional factors in cold plasma-treated beans is still unexplained.

The level of antinutritional factors after low-pressure plasma treatment is determined and presented in Table 2.5. A gradual decline in phytate and oxalate was observed, as a function of power. The extent of reduction increased linearly with the increase in the power of treatment. This reduction is probably due to chemical degradation of phytate to the lower inositol phosphates and inositol by the action of free radicals. Another possible mode of phytate loss could have been through cleavage of the phytate ring itself.

Tannin content of plasma-treated groundnut also increased at first at lesser power, while it was significantly decreased in groundnuts treated

TABLE 2.5 Antinutritional Factors of Cold Plasma-Treated Groundnuts

Sample		Oxalate (mg/100g DW)	Phytate (mg/100g DW)	Tannin (mg/100g DW)
Power	Min			
Control		154.31±3.31	23.46±3.5	140.07±1.39
40	5	149.12±4.52	22.15±1.6	141.01±1.90
	10	147.00±2.9	21.86±1.8	143.18±1.03
	15	144.62±3.5	21.27±2.9	146.36±1.2
60	5	147.52±4.11	18.16±3.1	143.73±0.18
	10	143.18±3.75	18.53±1.3	144.47±0.41
	15	124.062±4.75	17.66±2.7	136.37±0.81

at 60 W-15 min. The increase in tannins may be due to hydrolyzation of the condensed tannins. Plasma treatment resulted in a moderate significant reduction in antinutritional factors of peanuts compared to that in the control samples.

2.4 CONCLUDING REMARKS

The novelty of this technology primarily depends upon the type of feed gas and plasma apparatus used. The inactivation of fungal species using cold plasma seems to be feasible and gives high percentage of inactivation at 99.9% for *A. parasiticus*. The cooking time was observed to be reduced by 28% after the application of air plasma, which is an indication of energy saving. The treatments can alter cooking time and water absorption, and induce several other changes in properties and textures. Plasma etching, depolymerization, and cross-linking are the mechanisms that mainly influenced the properties of the plasma-treated starches. Cold plasma treatment of rice starch resulted in depolymerization and changes in their functional properties like an increase in gel hydration properties, syneresis, and turbidity. The increase in hydrophilic nature of the cold plasma-treated seeds enhanced the germination characteristics. The cold plasma treatment has a considerable effect on the antinutritional factors like trypsin inhibitors and phytic acid content in mung beans. Plasma applications will attract more attention in the future because this is a chemical free and eco-friendly technology with some unique properties.

KEYWORDS

- **anti-nutritional factors**
- *Aspergillus species*
- **cold plasma**
- **enzymatic activity**
- **germination**
- **hydrophilicity**
- **reactive species**
- **starch modification**

REFERENCES

Al-Kaisey, M. T., Alwan, A. K. H., Mohammad, M. H., & Saeed, A. H., (2003). Effect of gamma irradiation on antinutritional factors in broad bean. *Radiat Phys. Chem.*, *67*(3), 493–496.

Banu, M. S., Sasikala, P., Dhanapal, A., Kavitha, V., Yazhini, G., & Rajamani, L., (2012). Cold plasma as a novel food processing technology. *IJETED*, *4*(2), 803–818.

Bermudez-Aguirre, D., Wemlinger, E., Pedrow, P., Barbosa-Cánovas, G., & Garcia-Perez, M., (2013). Effect of atmospheric pressure cold plasma (APCP) on the inactivation of *Escherichia coli* in fresh produce. *Food Control*, *34*(1), 149–157.

Bertoft, E., Annor, G. A., Shen, X., Rumpagaporn, P., Seetharaman, K., & Hamaker, B. R., (2016). Small differences in amylopectin fine structure may explain large functional differences of starch. *Carbohydr Polym.*, *140*, 113–121.

Bogaerts, A., Neyts, E., Gijbels, R., & Mullen, J., (2002). Gas discharge plasmas and their applications. *Spectrochim Acta.*, *57*, 609–658.

Bormashenko, E., Grynyov, R., Bormashenko, Y., & Drori, E., (2012). Cold radiofrequency plasma treatment modifies wettability and germination speed of plant seeds. *Sci. Rep.*, *2*, 741–748.

Chen, H. H., Chang, H. C., Chen, Y. K., Hung, C. L., Lin, S. Y., & Chen, Y. S., (2016). An improved process for high nutrition of germinated brown rice production: Low-pressure plasma. *Food Chem.*, *191*, 120–127.

Chen, H. H., Chen, Y., & Chang, C. H., (2012). Evaluation of physicochemical properties of plasma treated brown rice. *Food Chem.*, *135*, 74–79.

Chen, Z., Schols, H. A., & Voragen, A. G. J., (2004). Differently sized granules from acetylated potato and sweet potato starches differ in the acetyl substitution pattern of their amylose populations. *Carbohydr Polym.*, *56*, 219–226.

De la Hera, E., Gomez, M., & Rosell, C. M., (2013). Particle size distribution of rice flour affecting the starch enzymatic hydrolysis and hydration properties. *Carbohydr Polym.*, *98*, 421–427.

Devi, Y., Thirumdas, R., Sarangapani, C., Deshmukh, R. R., & Annapure, U. S. (2017). Influence of cold plasma on fungal growth and aflatoxins production on groundnuts. *Food Control, 77*, 187–191.

Eaton, D. L., & Groopman, J. D., (1994). *Toxicology of Aflatoxins.* Academic Press.

Ellis, W. O., Smith, J. P., Simpson, B. K., Oldham, J. H., & Scott, P. M., (1991). Aflatoxins in food: occurrence, biosynthesis, effects on organisms, detection, and methods of control. *Crit Rev Food Sci Nutr., 30*(4), 403–439.

Jiang, J. F., He, X., Li, L., Li, J. G., Shao, H. L., Xu, Q. L., Ye, H. R., & Dong, Y. H., (2014). Effect of cold plasma treatment on seed germination and growth of wheat. *Plasma Sci. Technol., 16*, 54–58.

Juliano, B. O., Cartano, A. V., & Vidal, A. J., (1968). Note on a limitation of the starch-iodine blue test for milled rice amylose. *Cereal Chem., 45*, 63–65.

Kurtzman, C. P., Horn, B. W., & Hesseltine, C. W., (1987). Aspergillus nomius, a new aflatoxin-producing species related to *Aspergillus flavus* and *Aspergillus tamarii. Anton Leeuw., 53*(3), 147–158.

Laroussi, M., Mendis, D. A., & Rosenberg, M., (2003). Plasma interaction with microbes. *New J. Phys., 5*(1), 41.

Li, H. P., Wang, L. Y., Li, G., Jin, L. H., Le, P. S., Zhao, H. X., Xing, X. H., & Bao, C. Y., (2011). Manipulation of Lipase Activity by the Helium Radio-Frequency, Atmospheric-Pressure Glow Discharge Plasma Jet. *Plasma Proc Polym., 8*, 224–229.

Lii, C. Y., Liao, C. D., Stobinski, L., & Tomasik, P., (2002a) Effects of hydrogen, oxygen, and ammonia low-pressure glow plasma on granular starches. *Carbohydr polym., 49*, 449–456.

Ling, L., Jiafeng, J., Jiangang, Li.; Minchong, S., Xin, H., Hanliang, S., & Yuanhua, D., (2014). Effects of cold plasma treatment on seed germination and seedling growth of soybean. *Sci Rep., 4*, 5859.

Luo, Z., Fu, X., He, X., Luo, F., Gao, Q., & Yu, S., (2008). Effect of ultrasonic treatment on the physicochemical properties of maize starches differing in amylose content. *Starch. Starke., 60*, 646–653.

Mirabedini, S. M., Arabi, H., Salem, A., & Asiaban, S., (2007). Effect of low-pressure O_2 and Ar plasma treatments on the wettability and morphology of biaxial-oriented polypropylene (BOPP) film. *Prog Org Coat., 60*, 105–111.

Misra, N. N., Kaur, S., Tiwari, B. K., Kaur, A., Singh, N., & Cullen, P. J., (2015). Atmospheric pressure cold plasma (ACP) treatment of wheat flour. *Food Hydrocolloid., 44*, 115–121.

Misra, N. N., Pankaj, S. K., Segat, A., & Ishikawa, K., (2016). Cold plasma interactions with enzymes in foods and model systems. *Trends Food Sci Technol., 55*, 39–47.

Misra, N. N., Tiwari, B. K., Raghavarao, K. S. M. S., & Cullen, P. J., (2011). Nonthermal plasma inactivation of food-borne pathogens. *Food Eng Rev., 3*, 159–170.

Mitra, A., Yang-Fang, Li., Klämpfl, T. J., Shimizu, T., Jeon, J., Morfill, G. E., & Zimmermann, J. L., (2012). Inactivation of Surface-Borne Microorganisms and Increased Germination of Seed Specimen by Cold Atmospheric Plasma. *Food Bioproc. Technol, 7*(3), 645–653.

Mohapatra, D., & Bal, S., (2006). Cooking quality and instrumental textural attributes of cooked rice for different milling fractions. *Food Eng., 73*, 253–259.

Montie, T. C., Kelly-Wintenberg, K., & Roth, J. R., (2000). An overview of research using the one atmosphere uniform glow discharge plasma (OAUGDP) for sterilization of surfaces and materials. *IEEE Trans Plasma Sci., 28*(1), 41–50.

Morent, R., De Geyter, N., Desmet, T., Dubruel, P., & Leys, C., (2011). Plasma surface modification of biodegradable polymers: A review. *Plasma Proc Polym.*, *8*(3), 171–190.

Mubarak, A. E., (2005). Nutritional composition and antinutritional factors of mung bean seeds (*Phaseolus aureus*) as affected by some home traditional processes. *Food Chem.*, *89*, 489–495.

Pal, P., Kaur, P., Singh, N., Kaur, A. P., Misra, N. N., Tiwari, B. K., Cullen, P. J., & Virdi, A. S., (2016). Effect of nonthermal plasma on physico-chemical, amino acid composition, pasting and protein characteristics of short and long grain rice flour. *Food Res. Int.*, *81*, 50–57.

Park, B. J., Takatori, K., Sugita-Konishi, Y., Kim, I. H., Lee, M. H. et al., (2007). Degradation of mycotoxins using microwave-induced argon plasma at atmospheric pressure. *Surf. Coat. Technol.*, *201*(9), 5733–5737.

Sabularse, V. C., Liuzzo, J. A., Rao, R. M., & Grodner, R. M., (1991). Cooking quality of brown rice as influenced by gamma-irradiation, variety and storage. *J Food Sci.*, *56*, 96–98.

Sadhu, S., Thirumdas, R., Deshmukh, R. R., & Annapure, U. S., (2016). Influence of cold plasma on the enzymatic activity in germinating mung beans (Vigna radiate). *LWT – Food Sci. and Technol.*, *78*, 97–104.

Sarangapani, C., Thirumdas, R., Devi, Y., Trimukhe, A., Deshmukh, R. R., & Annapure, U. S., (2016). Effect of low-pressure plasma on physico–chemical and functional properties of parboiled rice flour. *LWT—Food Sci. and Technol.*, *69*, 482–489.

Sarangapani, C., Devi, R. Y., Thirumdas, R., Trimukhe, A. M., Deshmukh, R. R., & Annapure, U. S., (2017). Physico-chemical properties of low-pressure plasma treated black gram. *LWT—Food Sci. and Technol.*, *79*, 102–110.

Savelkoul, F. H. M. G., Van der Poel, A. F. B., & Tamminga, S., (1992). The presence and inactivation of trypsin inhibitors, tannins, lectins and amylase inhibitors in legume seeds during germination. *A review. Plant Food Hum Nutr.*, *42*(1), 71–85.

Sera, B., Stranak, V., Sery, M., Tichy, M., & Spatenka, P., (2008). Germination of Chenopodium Album in response to microwave plasma treatment. *Plasma Sci Technol.*, *10*, 506–511.

Sujka, M., & Jamroz, J., (2013). Ultrasound-treated starch: SEM and TEM imaging, and functional behaviour. *Food hydrocolloid.*, *31*, 413–419.

Thirumdas, R., Sarangapani, C., & Annapure, U. S., (2015a). Cold plasma: a novel nonthermal technology for food processing. *Food Biophys.*, *10*, 1–11.

Thirumdas, R., Deshmukh, R. R., & Annapure, U. S., (2015b). Effect of low temperature plasma processing on physicochemical properties and cooking quality of basmati rice. *Innov Food Sci. Emerg. Technol.*, *31*, 83–90.

Thirumdas, R., Saragapani, C., Ajinkya, M. T., Deshmukh, R. R., & Annapure, U. S., (2016a). Influence of low pressure cold plasma on cooking and textural properties of brown rice. *Innov. Food Sci. Emerg Technol.*, *37*, 53–60.

Thirumdas, R., Deshmukh, R. R., & Annapure, U. S., (2016b). Effect of low temperature plasma on the functional properties of basmati rice flour. *Journal of Food Sci. & Technol.*, *53*(6), 2742–2751.

Thirumdas, R., Trimukhe, A., Deshmukh, R. R., & Annapure, U. S., (2017a). Functional and rheological properties of cold plasma treated rice starch. *Carbohydrate Polymers.*, *157*, 1723–1731.

Thirumdas, R., Kadam, D., & Annapure, U. S., (2017b). Cold plasma: an alternative technology for the starch modification. *Food Biophys.*, 1–11.

Wan, J., Coventry, J., Swiergon, P., Sanguansri, P., & Versteeg, C., (2009). Advances in innovative processing technologies for microbial inactivation and enhancement of food safety–pulsed electric field and low-temperature plasma. *Trend Food Sci. Technol.*, *20*(9), 414–424.

Wu, Z. H., Chi, L. H., Bian, S. F., & Xu, K. Z., (2007). Effects of plasma treatment on maize seeding resistance. *J. Maize. Sci.*, *15*, 111–113.

Yabe, K., & Nakajima, H., (2004). Enzyme reactions and genes in aflatoxin biosynthesis. *Appl. Microbiol. Biotechnol.*, *64*(6), 745–755.

Yin, M. Q., Huang, M. J., Ma, B. Z., & Ma, T. C., (2005). Stimulating effects of seed treatment by magnetized plasma on tomato growth and yield. *Plasma Sci. Technol.*, *7*, 3143–3147.

Zhu, J., Li, L., Chen, L., & Li, X., (2012). Study on supra molecular structural changes of ultrasonic treated potato starch granules. *Food Hydrocolloid.*, *29*, 116–122.

CHAPTER 3

BIO-PROCESSING OF FOODS: CURRENT SCENARIO AND FUTURE PROSPECTS

RUPINDER KAUR,[1] PARMJIT S. PANESAR,[1] GISHA SINGLA,[2] and RAJENDER S. SANGWAN[2]

[1]*Food Biotechnology Research Laboratory, Department of Food Engineering & Technology, Sant Longowal Institute of Engineering & Technology, Longowal 148106, Punjab (India), E-mail: 29.rupinderkaur@gmail.com; pspbt@yahoo.com*

[2]*Centre for Innovative and Applied Bioprocessing, C-127, Phase-VIII, Industrial area, S.A.S. Nagar, Mohali-160071, Punjab (India), E-mail: singla.gisha@gmail.com; sangwan.lab@gmail.com*

CONTENTS

3.1   INTRODUCTION

Food processing involves the utilization of various unit operation techniques for the transformation of raw material such as crops or animals into food or other forms, thereby providing food with longer shelf-life and palatability. Further, it also contributes in improving the quality and quantity of the food by minimizing wastes and postharvest losses, besides increasing its availability and market value (Barrett et al., 1997).

Biotechnology is essentially the application of biological systems to improve or modify the process for intended use. The word biotechnology encompasses two words: "bio" and "technology." The word "bio" refers to living entities, i.e., microbes, and the word "technology" refers to the growth of microbes under controlled conditions (Arora et al., 2000; Panesar and Marwaha, 2014). Biotechnology has a great potential in the food-processing sector and has been widely used in this sector for the modification of taste, aroma, flavor, shelf-life, and texture of the food as well as for the development of new products. Biotechnological tools aim at improving process strategies, product quality, safety, yield, and process efficiency by selection and manipulation of the microorganisms used in food products. Biotechnological tools involved in food processing include both traditional tools such as fermentation techniques as well as conventional methods like recombinant DNA techniques for improvement in food processing techniques (Singh, 2010).

Biotechnological tools have been an important part of the human life since long, and with the advancements in this area, these techniques have attracted the attention of researchers and industrialists to utilize them in the food-processing sector for the development of sustainable, nutritious, and palatable food products. The role of biotechnology in various food-processing sectors is depicted in Figure 3.1 and discussed in detail in the following sections.

3.2   FERMENTED FOODS

Fermentation is one of the oldest methods of food processing. It is also reported to enhance the nutritive properties of food and acts as a natural process for preserving the food (Billings, 1998). Apart from this, fermentation processes also improve the taste and appearance of foods and reduces

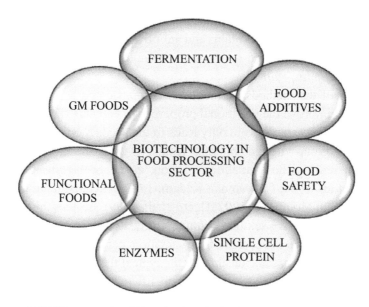

FIGURE 3.1 Role of biotechnology in the food-processing sector.

the energy required in the preparation of food, thereby providing a safer food (Simango, 1997). Fermented foods are those foods in which biochemical changes occur with the help of enzymes or microorganisms, either present naturally or added in the food (Harlander, 1992). Fermented foods have been an essential part of human life since long, and its history dates back to Southeast Asia (Osungbaro and Taiwo, 2009). Since the ancient time, the fermentation of milk, vegetables, and cereals has been carried out, and with the industrial revolution, the large-scale production of fermented foods increased tremendously. The industrial production of fermented foods includes the use of various raw materials and microorganisms to develop the product of high nutritive value (Hirahara, 1998).

3.2.1 FERMENTED CEREAL PRODUCTS

Cereals are one of the earliest sources of foods to humans (Matz, 1971). These are rich in nutrients, especially in proteins, carbohydrates, vitamins, minerals and fiber, but as compared to milk and its products, the nutritional factors in cereal grains are less owing to lysine deficiency and presence of antinutrients such as phytic acid, tannins, and polyphenols (Chavan and

Kadam, 1989). Various methods such as genetic engineering, fermentation and supplementation with proteins and amino acids, fermentation, and many more have been suggested to improve the nutritional properties of foods (Mattila-Sandholm, 1998).

Among all the methods, fermentation is considered as one of the best technique to improve the nutritional properties of cereals (Mattila and Sandholm, 1998). Fermentation not only leads to the production of certain amino acids and B group vitamins but also results in the degradation of phytate, thereby increasing the availability of iron, zinc, and magnesium cations (Gillooly et al., 1984; Chavan and Kadam, 1989; Khetarpaul and Chauhan, 1990; Nout and Motarjemi, 1997; Haard et al., 1999). Besides this, the effect of fermentation on the texture, aroma, and taste of the product has been documented. During the course of fermentation, certain volatile compounds are synthesized that are responsible for the typical flavors in the final product and improve its quality as well as the shelf-life (Chavan and Kadam, 1989).

Traditional fermented foods are prepared from common cereals such as rice, wheat, *ragi*, corn, and sorghum and are available in many parts of the world. The different fermented products from cereals include *idli, dosa, dhokla, soy sauce, kishk, tarhana, ogi, kenkey, pozol, ijera, kisra, boza,* and many more (Blandino et al., 2003; Yegin and Fernández-Lahore, 2012). The microbiology of these products is complex, and in most of these products, fermentation is carried out by the natural flora, which may include mixed cultures of bacteria, yeast, and fungi (Table 3.1).

The microbiology of many of these products is quite complex. In most of these products, the fermentation occurs naturally and is carried out by a varied species of naturally occurring microorganisms such as bacteria, yeast, and fungi. Among the bacteria, the common bacterial species capable of carrying out fermentation belong to the genera *Leuconostoc, Lactobacillus, Streptococcus, Pediococcus, Micrococcus,* and *Bacillus.* The fungal cultures belonging to the genera *Aspergillus, Paecilomyces, Cladosporium, Fusarium, Penicillium,* and *Trichothecium* are associated with the fermentation of cereal products. The common fermenting yeasts include *Saccharomyces, Candida, Kluyveromyces,* and *Trichosporon* (Steinkraus, 1998).

With the increasing trend toward the consumption and demand of the functional food as well as with the developments in the field of biotechnology, new probiotic cereal-fermented food products such as *yosa* have been developed (Wood, 1997). It is prepared from cooked oat bran, which is fermented with the lactic acid bacteria (LAB) strains and further flavored with

TABLE 3.1 List of Some Examples of Fermented Cereal Products

Fermented Cereal Products	Raw Materials	Microorganisms Involved
Ang-Kak	Rice	*Monascus purpureus*
Anarshe	Rice	*Lactic Acid Bacteria*
Boza	Wheat, Millet	*Lactobacillus* sp., *Leuconostoc* sp., *Saccharomyces cerevisiae*
Bread	Wheat	*Saccharomyces cerevisiae*
Busa	Rice or millet	*Lactobacillus* sp, *Saccharomyces* sp.
Busaa	Maize	*Lactobacillus helveticus, L. salivarus, L. casei,* *L. brevis, L. plantarum, L. buchneri, Saccharomyces cerevisiae,* *Penicillium damnosus*
Dhokla	Rice or wheat and Bengal gram	*Leuconostoc mesenteroides, Streptococcus faecalis,* *Torulopsis candida, T. pullulans*
Dosa	Rice and Bengal gram	*Leuconostoc mesenteroides, Streptococcus faecalis,* *Torulopsis candida, T. pullulans*
Idli	Rice grits and black gram	*Leuconostoc mesenteroides, Streptococcus faecalis,* *Torulopsis, Candida, Trichosporon pullulan*
Minchin	Wheat gluten	*Aspergillus* sp., *Cladosporium* sp., *Fusarium* sp., *Paecilomyces* sp., *Penicillium* sp., *Syncephalastum* sp., and *Trichothecium* sp.
Miso	Rice and soy beans or rice other cereals such as barley	*Aspergillus oryzae, Lactobacillus* sp, *Torulopsis etchellsii*

TABLE 3.1 (Continued)

Fermented Cereal Products	Raw Materials	Microorganisms Involved
Ogi	Maize, sorghum or millet	*Aerobacter* sp., *Aspergillus* sp., *Candida mycoderma, Coryne-bacterium* sp., *Cephalosporium* sp., *Fusarium* sp.,
		Lactobacillus plantarum, Penicillium sp., *Saccharomyces cere-visiae,* and *Rhodotorula* sp.,
Sake	Rice	*Saccharomyces sake*
Soy sauce	Wheat and soybeans	*Aspergillus oryzae, Aspergillus sojae, Lactobacillus* sp., *Zygosaccharomyces rouxi*
Takju	Rice, wheat	*LAB, Saccharomyces cerevisiae*
Thumba	Millet	*Endomycopsin fibuliger*
Torani	Rice	*Candida quilliermondii,*
		C. tropicalis, Candidum sp., *Geotrichum* sp., and *Hansenula anomala*
Uji	Maize Sorghum, Millet	*Lactobacillus platarum* and *Leuconostoc mesenteriodes*
Vada	Cereal/legume	*Leuconostoc* sp., *Pediococcus* sp., *Streptococcus* sp.

(Source: Modified From Chavan and Kadam, 1989; Soni and Sandhu, 1990; Adams, 1998).

sucrose or fructose and fruit jam (Salminen and Von Wright, 1998). The texture and flavor of yosa are similar to those of yogurt. Besides, some of the fermented cereal products have been modified to control some diseases, such as *Dogik*, an improved form of *ogi* and prepared using LAB (Okagbue, 1995).

3.2.2   FERMENTED DAIRY PRODUCTS

Fermented dairy products can be defined as those products that are prepared from the milk of the milch animals. Fermented milk products have been an essential part of human life since long, and these products play an important role in human nutrition. The fermented dairy products used worldwide have been classified into three categories as moderately sour (cultured buttermilk), highly sour (curd, yogurt), and acidic-and-alcoholic type (*kumis, kefir*) (Gandhi, 2000; Panesar et al., 2011). Traditionally, the fermentation of milk was carried out by the natural microflora present in milk or added from the surroundings. Currently, with the developments in the area of biotechnology and processing techniques and with the increase in health awareness among the consumers, the development of fermented dairy products has increased tremendously (Stanton et al., 2001).

LABs are the main microbial sources involved in the fermentation of dairy products. However, several factors (chemical composition of the milk, starter culture) affect the desired properties of the fermented products (Chandan, 1982; Chandan and Shahani, 1995; Chandan and Nauth, 2012).

The most common bacterial cultures employed for fermentation are *Lactobacillus delbrueckii* subsp. *bulgaricus, Streptococcus thermophilus, L. acidophilus, L. lactis* subsp. *lactis,* and *L. lactis* subsp. *cremoris.* These organisms are responsible for the development of the acidic taste in the products. The typical flavor of the sour cream is due to *Leuconostoc* sp. During cheese production, *L. lactis* subsp. *lactis* or *cremoris, L. helveticus, L. delbrueckii* subsp. *bulgaricus*, and *S. thermophilus* are responsible for acid production as well as for flavor development, while *Propionibacterium shermanii* increases the shelf-life of the product with the production of propionate (Gonzalez et al., 2011). The preparation of the fermented dairy product is depicted in Figure 3.2, and the various microorganisms employed for the production of different dairy products are listed in Table 3.2.

A number of dairy products marketed are known to have probiotic bacteria. There are many other fermented products in the world with a dairy base, which contain probiotic bacteria. An example is Yakult, which is made with a selected culture of *L. casei* (Siro et al., 2008).

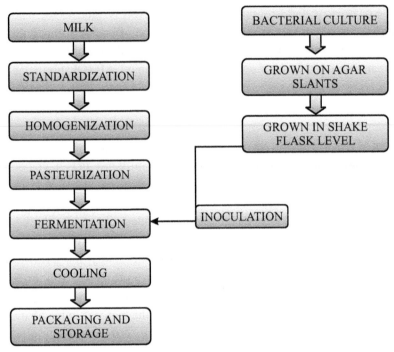

FIGURE 3.2 General scheme for the production of fermented dairy products (Source: Modified from Panesar et al., 2011).

3.2.3   FERMENTED FRUIT AND VEGETABLES PRODUCTS

Fruits and vegetables are highly nutritious as well as perishable in nature; therefore, they must be processed after harvest in order to minimize the losses (Montet et al., 1999). Lactic acid fermentation of fruits and vegetables is the best method as it not only improves the nutritional as well as the sensory properties of the product but also enhances its shelf-life (Endrizzi et al., 2009). Various LABs such as *L. plantarum, L. pentosus, L. brevis, L. fermentum, L. casei, Leuconostoc mesenteroides, L. kimchi, L. fallax, Weissella confusa, W. koreenis, W. cibaria*, and *Pediococcus pentosaceus* have been reported to carry out lactic acid fermentation as shown in Table 3.3 (Guarner and Schaafsma, 1998).

Fruits and vegetables are excellent sources of nutrients such as water-soluble vitamin B and C complex, provitamin A, dietary fibers, and minerals, which act as a suitable medium for the growth of microorganisms. Moreover, vegetables have low sugar content and possess neutral pH, thereby enhancing the microbial growth for lactic acid fermentation (Buckenhuskes, 1997).

TABLE 3.2 List of Some of the Fermented Milk Products

Fermented Milk Products	Microorganisms Involved
Curd	*Lactococcus lactis* subsp. *lactis, Lactobacillus delbrueckii* subsp. *bulgaricus, Lactobacillus plantarum, Streptococcus lactis, Streptococcus thermophilus, Streptococcus cremoris*
Yogurt	*Lactobacillus acidophilus, Streptococcus thermophilus, Lactobacillus bulgaricus*
Cultured butter milk	*Streptococcus lactis* subsp. *diacetylactis, Streptococcus cremoris*
Lassi	*Lactobacillus bulgaricus*
Shrikhand	*S. thermophilus L. bulgaricus*
Kumiss	*L. acidophilus L. bulgaricus Saccharomyces Micrococci*
Kefir	*S. lactis Leuconostoc* sp. *Saccharomyces kefir, Torula kefir, Micrococci*
Leben	*S. lactis S. thermophilus L. bulgaricus* Lactose fermenting yeast
Cheese	*L. lactis* subsp. *lactis, L. lactis* subsp. *cremoris, L. lactis* subsp. *diacetylactis, S. thermophilus, L. delbureckii* subsp. *bulgaricus Priopionibacterium shermanii, Penicillium roqueforti*

(Source: Modified From Panesar et al., 2011; Panesar and Marwaha, 2014).

Fermentation increases the organoleptic as well as nutritional properties of the food, which helps in the prevention of several diseases such as diarrhea and liver cirrhosis (Yamano et al., 2006). Besides, lactic acid fermentation helps in the retention of different pigments such as flavonoids, lycopene, anthocyanin, β-carotene, and glucosinolates (Kaur and Kapoor, 2001). The most common fruits and vegetables used for fermentation are carrots, turnips, beetroot, cucumber, olives, tomatoes, apples, pears, palms, and lemon. Among the several fermented fruits and vegetables, sauerkraut and kimchi are the most studied fermented foods (Panda et al., 2007; Panda and Ray, 2007).

Along with traditional techniques, several modern techniques such as Random Amplification of Polymorphic DNA (RAPD), Polymerase Chain Reaction (PCR), and Restriction Fragment Length Polymorphism (RFLP), and gradient gel electrophoresis have been used for the isolation and characterization of different strains for fermentation of different fruits and vegetables (Elegado et al., 2004).

Several reports have indicated that fruits and vegetables are also a suitable source for growth of probiotic microbes as these contain different

TABLE 3.3   List of Fermented Fruit and Vegetable Products

Types of fermented products	Raw Materials	Microorganisms Involved	References
Burong Mustala	Mustard Leaf	*L. brevis, Pediococcus cervisiae*	Karovicova et al. (2002)
Gundruk	Cabbage, radish, mustard, cauli-flower	*Pediococcus, Lactobacillus*	Tamang (2008)
Kimchi	Cabbage, radish	*Leuconostoc mesenteroides, L. brevis, L. plantarum,* *L. sakei*	Lee et al. (2005)
Olive	Olive	*L. plantarum, L. brevis,* *L. pentosus, P. cerevisiae,* *L. mesenteroides*	Argyri et al. (2013)
Paocai	Cabbage, celery, cucumber and radish	*L. pentosus, L. plantarum, Leuconostoc mesenteroides, L. brevis, L. lactis,* *L. fermentum*	Yan et al. (2008)
Sauerkraut	Cabbage	*L. mesenteroides,* *L. plantarum, L. brevis,* *L. rhamnosus, L. plantarum*	Viander et al. (2003)
Sinki	Radish	*L. plantarum, L. brevis,* *L. fermentum, L. fallax,* *P. pentosaceus*	Tamang (2008)
Soidon	Bamboo Shoot	*L. brevis, L. fallax, L. lactis*	Tamang et al. (2008)

prebiotic compounds. Gene technology can play an important role in determining probiotic functions (Reid, 2008).

3.2.4   FERMENTED MEAT AND MEAT PRODUCTS

Fermentation of meat not only helps in the preservation of meat but also results in the development of desired attributes in the product in terms of flavor, palatability, color, and tenderness. Furthermore, the presence of microbial sources, either cultured or wild, during the course of fermentation, decreases the pH, thereby minimizing the growth of various pathogenic and spoilage microorganisms and rendering the product as microbiologically safe (Soni and Arora, 2000).

The starter culture used for fermentation of meat and its products are facultative heterofermentative, which are responsible for the production of lactic acid through the glycolysis pathway by utilizing glucose and lactose as substrates. The potential microbial sources that carry out fermentation are the LAB strains. These microbial sources include *L. pentosus, L. casei, L. curvetus, L. plantarum, L. sakei, Pediococcus acidilactici,* and *P. pentosaceus.* Beside this, other microbial sources are used that are responsible for the development of flavors in the final product, and these include gram-positive *Staphylococcus, Kocuria,* and *Micrococcus.*

Besides fermentation, biotechnology has also played an important role in the tenderization of meat by the application of the enzymes. Different enzymes have been used for this purpose in the commercial scale. The different enzymes used are papain, protease, bromelain, and ficin (Lawrie, 1998; Grzonka et al., 2007). The enzyme protease not only finds its application in meat tenderization, but it is also widely used for bone cleaning and development of different flavors (Lantto et al., 2009). Besides protease, lipase is also used for flavor development in fermented sausages.

From the above, it can be observed that fermentation has become an integral part of the human food chain. Besides providing desirable attributes to the food product, it also plays a major role in the preservation of food. Furthermore, the traditional methods of screening of microbial cultures as starter cultures for fermentation has been replaced by the modern techniques of genetic engineering, protein engineering, and high-throughput screening methods. It can lead to the development of better, safer, and nutritious food products.

3.3 ALCOHOLIC BEVERAGES

Alcoholic fermentation, also known as ethanol fermentation, refers to the biological process where the sugars are converted into ethanol and carbon dioxide with the help of microorganisms. The alcoholic fermentation is basically carried by yeast, especially *Saccharomyces cerevisiae,* which has been used since decades both for baking as well as brewing (Panesar et al., 2000; Donalies et al., 2008). Alcoholic beverages are broadly classified into distilled and non-distilled categories.

Beer is one of the oldest fermented beverages, and the fermentation process of beer involves the extraction of malted barley followed by boiling of the extracted material with hops, cooling the extract, fermentation using yeast culture, and further clarification of the final product (Briggs et al., 1981). The hops (*Humulus lupulus*) added during the brewing, besides increasing the shelf-life of the beer, also provide bitterness and flavor to the final product (Reed, 1987). Depending upon the type of fermentation, different varieties of cultures (such as *S. uvarum*) are used during the production of bottom-fermented yeast, where the culture tends to settle down during the fermentation. Another yeast, *S. cerevisiae*, is used during the fermentation of top-fermented beer, where the yeast culture rises to the surface and is removed by skimming.

Whisky is one of the distilled alcoholic beverages, which is product of the malted barley fermented with yeast, *S. cerevisiae,* and further maturation carried out in oak barrels (Russell and Stewart, 1999). Various types of whisky differ primarily in the nature and proportion of raw materials used in addition to malted barley (Andrews, 1987).

Wine is made by complete or partial fermentation of grape or any other fruit juice using yeast (*S. cerevisiae* var. *ellipsoideus*) (Amerine et al., 1980). Beside grapes, wine can also be prepared from other fruits such as apple, plum, peach, berries, strawberry, cherries, currants, apricots, and many more (Joshi et al., 1999; Joshi et al., 2000). The type of sugars present in the grapes makes them suitable for fermentation by yeasts. The acid present in the grapes prevents the growth of spoilage microorganisms, thus making it suitable for wine production (Jackson, 2000). Although there is no accepted system for the classification of wines, wines have been classified according to the color as red, rose, or white or could be classified as table, sparkling, or fortified wines. Another product prepared from grapes and by the distillation of wine is brandy. Similar to wine, brandies can be prepared from different fruits such as plum, peach, berries, and many other fruits.

3.4 FUNCTIONAL FOODS

Health awareness among people has evoked demand for natural food supplement or additive that can not only ease the risk of disease but also provide nutrition and better health. Therefore, the focus has been shifted toward functional foods, and their demands have increased tremendously. These foods, basically probiotics, prebiotics, and synbiotics, have been developed in order to provide nutrition and health by stimulating the activity of the beneficial gut microflora. The microbiota of the human gastrointestinal tract plays a key role in efficient nutrient absorption and thus maintaining human health (Szakaly et al., 2012). Under certain conditions such as changes in diet, medication, stress, age, and general living conditions, the microbial balance may be changed, and pathogenic microorganisms may outnumber the beneficial microflora, which may result in different gut disorders like inflammatory bowel disease and colon cancer. Functional foods are resistant to the digestive enzymes, and they pass into the colon in the intact form, where they act as substrates for fermentation. The end products of fermentation, such as organic acids, vitamins, and short chain fatty acids, stimulate the activity and/or the growth of the beneficial bacteria, which reduces the risk of various diseases (Venema, 2012).

The term "probiotic" was coined by Parker (1974) and can be defined as "a live microbial food ingredient that is beneficial to health." The most common bacterial genera used as probiotics include *Lactobacilli* and *Bifidobacteria*. Currently, probiotics are consumed in the form of fermented dairy products such as yogurt or freeze-dried cultures.

Prebiotics are defined as "non-digestible food ingredients that beneficially affect the host by selectively stimulating the growth and/or activity of one or a limited number of bacteria in the colon" (Gibson and Roberfroid, 1995). The digestive enzymes are unable to breakdown the prebiotics, and therefore, these prebiotics reach the colon in an intact form. Once these non-digestible carbohydrates pass into the intestine, they serve as a feast for the probiotic bacteria that live there (Panesar et al., 2011). A food ingredient can only be classified as prebiotics if it is not hydrolyzed or absorbed in the stomach or small intestine. Moreover, it stimulates the growth and/or activity of the gut microflora by inducing beneficial luminal/systemic effects within the host (Gibson and Roberfroid, 1995; Panesar and Bali, 2015). The most common prebiotics used are inulin, fructo-oligosaccharides (FOS), galacto-oligosaccharides lactulose, dietary fiber, and gums (Panesar et al., 2013; Bandyophadyay and Mandal, 2014).

Synbiotics are the combination of probiotics as well as prebiotics (Panesar et al., 2009). These act as a substrate of fermentation for the gut microflora and support the growth of the probiotic organisms (Farnworth, 2001).

The commercial synbiotic supplements include combinations of *Bifidobacteria* and FOS; *Lactobacillus* GG and inulins; and *Bifidobacteria, Lactobacilli*, and FOS or inulins (Sekhon and Jairath, 2010). The list of important probiotics, prebiotics, and synbiotics is depicted in Table 3.4.

3.5   ENZYMES

Enzymes are proteins that are responsible for carrying out all the biochemical reactions as catalysts as they speed the rate of reaction by lowering the activation energy. These are very specific in nature and are able to carry out the reactions under mild conditions. Although enzymes have been exploited by humans since the traditional times for the processing of various food prod-

TABLE 3.4   List of Important Probiotics, Prebiotics, and Synbiotics

Probiotics	Prebiotics	Synbiotics
Lactobacillus acidophulis	Fructooligosaccharides (FOS)	*Lactobacilli* + Lactitol
L. amylovorus	Lactitol	*Lactobacilli* + Inulin
L. brevis	Lactulose	*Lactobacilli* + FOS
L. casei	Galactooligosaccharides (GOS)	*Lactobacillus rhamnosus* GG+ Inulin
L. bulgaricus	Inulin	*Bifidobacteria* +FOS
L. fermentum	Isomaltooligosaccharides	*Bifidobacteria* + GOS
L. plantarum	Xylooligosaccharides	*Bifidobacteria* and *Lactobacilli* +FOS or Inulin
L. reutri	Soyoligosaccahrides	
L. lactis	Raffinose	
L. rhamnosus	Lactosucrose	
Bifidobacterium animalis		
B. bifidum		
B. breve		
B. infantum		
B. longum		
Streptococcus thermophilus		
S. salivarius		
Saccharomyces cerevisae		
S. boulardii		

(Source: Modified from Sekhon and Jairath 2010).

ucts such as beer, wine, cheese, yogurt, *kefir*, and other fermented beverages, owing to their multiple applications, their demand in various food sectors has increased tremendously (Panesar et al., 2010). The traditional sources of food enzymes have been the tissues of plants and animals. However, the trend has shifted toward the use of microbial sources as an alternative for enzyme production. In contrast to the plant and animal sources, microbial sources are the preferred due to economic as well other technical benefits such as higher yields obtained within the shortest fermentation time. Moreover, these can also be genetically manipulated to improve the productivity of the enzyme (Trevan, 1988). Different microbial sources such as bacteria, yeast, and fungus have been utilized for enzyme production by utilizing different substrates (Aunstrup, 1974). The enzymes mostly used in the food industry, such as amylase, protease, pectinase, β-galactosidase, invertase, tannase, laccase, and lipases, are efficiently produced by *Bacillus subtilis, Trametes versicolor, Candida, Rhizopus, Rhizomucor, Geotrichum, Penicillium, Aspergillus, Bacillus, Pseudomonas, Staphylococcus, Bacillus circulans*, *Candida pseudotropicalis, Kluyveromyces marxianus, K. fragilis*, and *K. lactis.*

Moreover, biotechnological strategies have been applied for the utilization of agro-industrial wastes for enzyme production, as these are rich in macro- as well as micronutrients and can serve as a substrate for production of different value-added products (Panesar et al., 2016).

3.6 GENETICALLY MODIFIED FOODS

The term genetically modified (GM) food generally refers to those crops that have been created for human or animal consumption using molecular technology. According to the definition of FAO and European Commission, GM foods are the products produced from GM plants or animals. These plants have been modified in the laboratory either to improve the desired traits or to enhance the nutritional properties. Traditionally, the methods of plant breeding had been carried to improve the traits within the plants, but these techniques have been replaced by the genetic engineering methods, which give accurate results at the rapid speed as compared to the plant breeding techniques (FAO, 2005).

The process of genetic modification can be carried by transferring the foreign DNA into the host through vectors, which includes different steps from isolation of suitable gene to its verification in the transferred material and finally regeneration of the modified plant (Van den Eedea et al., 2004; Querci et al., 2007; 2010). The target genes are expressed in the vector with the help of promoters and terminators. The most common promoters used include cauliflower mosaic virus, *Agrobacterium tumefaciens nos*-promoter (dicotyledons), and promoters of the maize alcohol dehydrogenase gene and the rice actin gene (monocotyledons) (Finnegan and McElroy, 1994; Van den Eedea et al., 2004).

Various gene transfer techniques have been proposed for the efficient modification of the plants, either by electroporation, microinjection, and fusion of bacterial spheroplasts or by chemically induced endocytosis. Other alternative techniques have also been used, which include the transformation of organelles (Sanford, 1990; Kapila et al., 1997; Van den Eedea et al., 2004). The latter method has limited applications, and therefore, new techniques have been proposed. These techniques include agrobacterial and ballistic methods of modification. The agrobacterial technique utilizes *Agrobacterium* spp. (*A. tumefaciens*, *A. rhizogenes*), but as compared to the ballistic one, this is not an efficient method (Singer and Berg, 1991; Van den Eedea et al., 2004). Ballistic method of genetic modification involves bombarding the plant cells with gold and tungsten or other metals with high molecular weight, which further acts as a carrier (Russian Federation National Methodical Recommendations, 2006).

The recent technological developments in genetic engineering have proved to be beneficial in improving the yield as well as the quality of the grain. Besides, the nutritional properties of the foods are also improved. Certain reports are available where the foods have been modified to increase the vitamin as well as protein content; oilseeds have been modified to lower the content of saturated fatty acids, and fruits have been modified to delay ripening (Skryabin and Tutelyan, 2013).

Soybean has been genetically modified to increase its productivity by making them resistant to pests. Herbicide-resistant soybeans have also been developed by transferring 5-enolpyruvylshikimate-3-phosphate synthase from *Agrobacterium* sp. Further, research includes the production of soybeans having high nutritional value. One such example includes Vistive Gold Soybeans that have monounsaturated fat levels similar to olive oil and saturated fat content as low as that of canola oil.

Similar to soybeans, potatoes have also been modified to make them resistant to pests. Several varieties of Bt potato have been developed by Monsanto. Moreover, other GM potatoes have also been developed, which make them resistant to PVX virus, phytophthora, and Colorado beetle. Further studies have been conducted for the development of potato with improved nutritional value, such as the starch potato "Amflora." The genes responsible for synthesizing amylose was suppressed to develop the potato "Amflora" having higher amylopectin content (over 98%) (GMO Compass, 2010).

GM maize has also been developed to make the crop resistant to stem moth. Moreover, herbicide-resistant and Bt-resistant maize has also been developed. Future research includes the development of modified maize having higher levels of lysine and tryptophan, structural changes in the maize starch, and improved technological parameters of maize (James, 2011).

3.6.1   SAFETY CONCERNS OF GM FOODS

As accepted by FAO, the approach to determine the safety aspects of GM foods is based on "substantial equivalence," or "composite equivalence." It is the assessment test where the equivalence of the GM food and its traditional analogue is compared in terms of major nutrients, allergens, antialimentary, and toxic substances. These also form the basis of safety classes as first, second, and third. First class of safety is considered if the GM product is found to be chemically equivalent to its traditional analogue and therefore does not need any further study. If there are some difference among the GM product and its analogue, i.e., presence or absence of any of the components, then the product is considered under the second class of safety. Generally, most of the GM products developed belong to the second class of safety (Skryabin and Tutelyan, 2013).

3.7   FOOD ADDITIVES

Due to the increase in the consumer awareness about health, the use of chemical additives in food has been replaced by those synthesized by microbial fermentation. The various microbial sources and their respective food additives are depicted in Table 3.5.

3.7.1   BACTERIOCINS

Bacteriocins are the antimicrobial peptides or proteins synthesized by the members of LAB, which have the ability to kill closely related species (Jack

TABLE 3.5 Microbial Sources and their Synthesized Food Additives

Food Additives	Types of additives	Microbial Sources
Bacteriocins	Nisin	*Lactococcus lactis*
	Lacticin 3147	*Lactococcus lactis*
	Pediocin	*Pediococcus acidilactici*
Biopigments	Astaxanthin	*Agrobacterium aurantiacum, Paracoccus carotinifaciens, Xanthophyllomyces dendrorhous, Haematococcus Pluvialis, Phaffia rhodozyma*
	β-carotene	*Dunaliella salina, Blakeslea trispora, Mucor circinelloides, Neurospora crassa* and *Phycomyces blakesleeanus*
	Lycopene	*Blakeslea trispora, Fusarium sporotrichioides*
	Ankaflavin	*Monascus* sp.
	Zeaxanthin	*Flavobacterium* sp., *Paracoccus zeaxanthinifaciens, Achromobacter, Bacillus, Brevibacterium, Corneybacterium michigannise*
	Canthaxanthin	*Bradyrhizobium* sp. *Haloferax alexandrines, Hematococcus,Monascus roseus*
	Anthraquinone	*Pacilomyces farinosus, Penicillium oxalicum*
	Lutein	*Chlorococcum*
	Riboflavin	Ashbya gossypi
Biofavors	Monosodium glutamate	*Micrococcus glutamicus*
	Glutamic acid	*Corynebacterium glutamicum*
	Terpenes	*Ceratocystis* sp., *Ascoidea hylecoeti, Cronartium fusiforme, Phellinus* sp., *Pleurotus euosmus*
	Esters	*Saccharomyces* sp., *Hansenula* sp., *Candida utilis*
	Lactones	*Sporobolomyces odorus, Trichoderma viride*
	Pyrazines	*Bacillus subtilis, Corynebacterium glutamicum*
	Vanillin	*Aspergillus niger*
Gums	Xanthan	*Xanthomonas campestris*
	Pullulan	*Aureobasidium pullulans*
	Dextran	*Leuconostoc mesenteroides, L. dextranicum, Pseudomonas* sp., *Azotobacter* sp.

TABLE 3.5 (Continued)

Food Additives	Types of additives	Microbial Sources
	Alginates	*Sphingomonas paucimobilis, Pseudomonas aeruginosa*
	Gellan gum	*Sphingomonas paucimobilis, Sphingomonas elodea*
	Welan gum	*Alcaligenes* sp.

(Source: Malik et al. 2012; Bali et al. 2014).

et al., 1995). Bacteriocins, as identified, are the cationic molecules that constitute up to 60 amino acid residues. Their various characteristics such as proteinaceous nature, thermo-resistance, and broad bactericidal activity have led to their use as biopreservative agents.

Bacteriocins have been classified into four classes, namely lantibiotics, thermostable nonlantibiotic, heat labile bacteriocins, and circular antibacterial peptide (Bali et al., 2014). Although extensive research is being carried for the biopreservative potential of all the bacteriocins, the most extensively studied bacteriocin is nisin, which is also used as a biopreservative in the food industry. Nisin is effective against *Listeria monocytogenes* and other gram-positive spoilage microorganisms.

The application of bacteriocin in food systems can be carried out by different ways such as inoculating it with strains of LAB where bacteriocins can be produced within the product, use of the food ingredient that has been previously fermented with bacteriocin-producing strains, and addition of bacteriocins in the food (Jeevaratnam et al., 2005). The potential of bacteriocins in various food products such as cheese, sausages, fish, and many other products has been tested. Several studies have revealed the effect of nisin against *Clostridium botulinum* and *Listeria monocytogenes* in cheese such as camembert, ricotta, and manchego. Moreover, other bacteriocins such as pediocin AcH also have shown a profound effect against *L. monocytogenes*, *S. aureus*, and *E. coli* O157:H7. The most extensively studied bacteriocins in meat and meat products include nisin, enterocin AS-48, enterocins A and B, sakacin, leucocin A, and pediocin PA-l/AcH. These can either be used alone or in combination with other treatments such as modified atmosphere packaging, high hydrostatic pressure (HHP), heat, and chemical preservatives to control the growth of *L. monocytogenes* as well as other spoilage organisms (Bali et al., 2014).

The use of bacteriocins or bacteriocin-producing microorganisms in food as biopreservative tools has been of keen interest to the researchers. However, the effect of the environmental factors on the growth of the bacteriocin-producing strains as well as activity of the bacteriocin to ensure food safety is the major focus of research.

3.7.2 BIOPIGMENTS

Pigments have been an essential part of the food industry and are associated with the sensory as well as with the quality of the food product (Joshi et al., 2003). The colors used in the food industry have been the synthetic ones; however, these synthetic colors are now being replaced with the natural ones owing to the increase in consumer awareness about the health benefits (Malik et al., 2012). Therefore, the focus has been shifted to explore the natural sources for the extraction of these pigments as well as on the industrial processes for the large-scale production of biopigments (Cho et al., 2002). Biopigments, in contrast to the synthetic ones, possess anticancer activity as well as pro-vitamin A, which makes them suitable to be used as food additives.

Although the natural sources of biopigments include plants, animals as well as microorganisms, the preferred source for the industrial production of these natural colors include microorganisms owing to their various technical advantages (Babitha, 2009). Microorganisms are capable of producing wide variety of colors such as riboflavin, carotenoids, xanthohphylls, and many more. Different species of bacteria, yeast, and fungus have been reported to produce wide variety of biopigments (Mohankumari et al., 2009; Panesar et al., 2014). Among the fungal species, the microbial sources belonging to the genus *Monsascus, Penicillium, Rhodotorula*, and *Aspergillus* have the potential to produce higher yields of biopigments (Mendez et al., 2011). Bacterial sources such as *Serratia, Pseudomonas, Bacillus, Vibrio*, and *Streptomyces* have been reported as the potential pigment producers (Gupta et al., 2011). Among the yeast, *Cryptococcus, Phaffia, Yarrowia*, and *Rhodotorula* are the natural sources of pigment (Duffose, 2006).

These biopigments have been used for the preparation of flavored milk and flavored yoghurt. Moreover, pigments like carotenoids have been used to color the shell or yolk of the egg, to color the flesh of the fish, and to color the shell of the crustaceans (Khanafari et al., 2008). Further, riboflavin has been used in baby foods, breakfast cereals, cheese, sauces, and energy drinks (Malik et al., 2012).

Thus, biotechnological tools have played a major role in the large-scale production of biopigments using different microbial sources as the demand for these natural colorants has increased.

3.7.3 BIOFLAVORS

Flavors have wide range of applications in the food, pharmaceutical, feed as well as cosmetic industries. These are an important part of the food industry as they determine the product quality and acceptance. Various aromatic compounds such as hydrocarbons, aldehydes, ketones, esters, or lactones provide desirable aroma to the foods. The synthesis of these aromatic compounds could be carried either by chemical means or by extraction from plants or animals or by biotechnological means, i.e., by microbial fermentation (Cheetnam, 1993; Franco, 2004). Various microbial sources are responsible for carrying out the fermentation of flavors, such as microbial species *Ascoidea hylecoeti, Cronartium fusiforme, Phellinus* sp., and *Pleurotus euosmus* are responsible for the production of terpenes (Francke et al., 1978). Similarly, the production of esters is efficiently carried out by the microorganisms of the genus *Saccharomyces* sp., *Hansenula* sp., and *Candida utilis* (Armstrong et al., 1984). *Trichoderma viridae* and *Sporobolomyces odorus* have been reported to synthesize lactones, and *B. subtilis* as well as *Corynebacterium glutamicum* have been studied for pyrazine synthesis (Kosuge and Kamiya, 1962; Demain et al., 1967; Welsh et al., 1989).

Recently, biotechnological process of bioflavor production includes the utilization of agro-industrial residues as the substrates for microbial fermentation in order to reduce the cost of bioflavor production. Various agro-industrial residues such as citric pulp (Rossi et al., 2009), coffee husk (Pandey et al., 2000a), sugarcane bagasse (Pandey et al., 2000b), and cassava bagasse (Pandey et al., 2000c) have been reported as potential substrates for flavor production from *Ceratocystis fimbriata* (Soares et al., 2000) and *Rhizopus oryzae* (Bramorski et al., 1998).

The biotechnological tools for flavor production in contrast to the other techniques have gained much importance not only because of increasing demand of the consumers toward the use for natural products but also because of higher yield and better economic benefits. Furthermore, the recent developments in biochemistry, recombinant technology, and molecular biological techniques have led to the production of cost-effective bioflavors (Bicas et al., 2010).

3.7.4 POLYSACCHARIDES/GUMS

Microbial polysaccharides constitute a wide variety of biopolymers produced by diverse groups of microorganisms. The application of polysaccharides is diverse, and they have been used since ages in the food industry. These are used as viscofiers, gelling agents, thickeners, stabilizers, texturizers, and emulsifiers. Besides, these also offer various health benefits as they have antitumor, anti-inflammatory, and antimicrobial properties, thereby making them suitable as functional foods (Giavasis and Biliaderis, 2006). Although various microbial polysaccharides such as xanthan, gellan, pullulan, curdlan, and levan have been produced by the microorganisms and have vast applications in the food industry, xanthan gum, produced by *Xanthomonas campestris* is the most utilized bacterial polysaccharide (Kennedy and Bradshaw, 1984).

Among the bacteria, the most common species capable of producing polysaccharides are *Streptococcus salivarius, L. sanfranciscensis, B. subtilis and B. polymyxa, Acetobacter xylinum, Gluconoacetobacter xylinus, Microbacterium levaniformans, Zymomonas mobilis, Lactobacillus, Lactococcus, Pediococcus*, and many more (Newbrun and Baker, 1967; De Vuyst and Degeest, 1999; Notararigo et al., 2012). The most common fungal polysaccharide is pullulan, produced by *Aureobasidium pullulans. Elsinoe leucospila*, another fungal species is capable of producing elsinan, extracellular, linear α-d-glucan (Misaki et al., 1978). Glycan, another polysaccharide, is produced by the yeast *Saccharomyces cerevisae* (Xu et al., 2009).

Fermentation of polysaccharides can be carried by batch, continuous as well as fed-batch culture (Cheng et al., 2011). However, to minimize the cost of production, synthetic media are being replaced by cheap agro-industrial wastes.

3.8 SINGLE CELL PROTEIN

The term single cell protein (SCP) was coined in 1966 by Carol L. Wilson. SCP can be defined as the dried microbial cells, which are used as protein source (Litchfield, 1983). Besides having high protein content of approximately about 60.82% of dry cell weight, SCP is also rich in other nutritive factors such as fats, carbohydrates, nucleic acids, vitamins, minerals, and essential amino acids like lysine and methionine (Suman et al., 2015).

Microorganisms like bacteria, yeast, fungi, and algae have been used as a protein source by providing them the necessary environmental conditions and substrates for their growth. However, with the continuing research, the synthetic ones have been replaced by various cheap agro-industrial by-products, which not only help in minimizing the pollution caused by these wastes but also reduce the cost of production (Litchfield, 1983). These wastes are composed of cellulose as well as hemicelluloses, and therefore, pretreatment of these wastes is carried by physical, chemical, or enzymatic methods to convert them into simple sugars, thereby making them suitable substrates (Nasseri et al., 2011). Some studies have revealed the production of SCP from *S. cerevisiae* by utilizing different wastes such as banana skin, rind of pomegranate, apple waste, mango waste, and sweet orange peel. Among all the wastes, banana peel was found to be the suitable substrate followed by rind of pomegranate (Adedayo et al., 2011). Similarly, cucumber wastes have also found to be suitable substrates for biomass production from *S. cerevisiae*. Besides yeasts, fungal strains are also found to be efficient in utilizing these wastes. *Aspergillus* sp. have been reported to grow efficiently on cheap substrates like eichornia and banana peel (Jaganmohan et al., 2013).

Whey, a by-product of the dairy industry, has also been used for SCP production. Whey can be utilized for biomass production in three different ways. It can either be used in its native form for the utilization of lactose by microorganisms or the lactose in the whey can be enzymatically or chemically hydrolyzed into its respective monomers. The other technique involves the prior fermentation of whey by LAB, thereby resulting in the production of a mixture of lactic acid and galactose (Boze et al., 1995). Most of the studies reveal the potential of *Kluyveromyces* (*Saccharomyces*) *fragilis* strains for high SCP production by utilizing whey (Powell and Robe, 1964). Besides, concentrated whey permeate has also been used as substrates for fermentation under both aerobic and anaerobic conditions using *Kluyveromyces* sp. strain (Mahmoud and Kosikowski, 1982).

3.9 BIOSENSORS

"A biosensor is a self-contained integrated device which is capable of providing specific quantitative or semi-quantitative analytical information using a biological recognition element (biochemical receptor) which is in direct spatial contact with a transducer element (Cammann, 1977). A biosensor should be clearly distinguished from a bioanalytical system, which requires additional processing steps, such as reagent addition. Furthermore, a biosensor should be distinguished from a bioprobe which is either disposable after one measurement, i.e., single use, or unable to continuously monitor the analyte concentration," a definition proposed by IUPAC (Thevenot et al., 1999). The first biosensors, used to determine glucose concentration in the sample contained immobilized glucose oxidase on an amperometric oxygen electrode surface, were described by Clark and Lyons in 1962 (Sassolas et al., 2011; Nambair et al., 2011).

Biosensors can be classified as electrochemical, optical, piezoelectric, and thermal sensors. Electrochemical sensors can further be classified into potentiometric, amperometric, and conductometric sensors. Electrochemical sensors have the advantage over the other sensors in terms of enzyme utilization as these have high specificity and biocatalytic activity (Thevenot et al., 1999). Other elements used can be nucleic acids, antibodies, and microorganisms that have to be immobilized on the surface of the electrode. Different techniques have been reported for the immobilization of these biorecognition elements, such as adsorption, microencapsulation, entrapment, covalent binding, and cross-linking techniques.

Biosensors in the food industry can either be used for the determination of carbohydrates, amino acids, and phenol by using enzyme sensors in the liquor and beverage industry, or they can be used for the detection of pathogenic microorganisms in the food material (Kaur et al., 2011; Murugaboopathi et al., 2013). Electrochemical sensors have been reported to detect *Salmonella* and *E. coli* O157:H7 in the food samples in less than 90 min (Arora et al., 2011).

Biosensors are the fast-growing areas that encompass several fields such as environment, agriculture, medicine, and food sector. In the food industry, quality control is the major area and employs the need for faster and accurate monitoring methods. In contrast to the conventional techniques, which are time consuming, expensive, and labor intensive, biosensors are cost effective as well as rapid in the determination of food quality. Future developments must include the provision for multianalyte detection combined with signal transmitters for remote sensing (Murugaboopathi et al., 2013).

3.10 CONCLUDING REMARKS

Biotechnological tools have a great influence in the food-processing sector as they can improve the quality and nutritional properties of the food products. Recent advancements in the field of biotechnology have played a significant role in improving the traits of the microorganisms to obtain quality product at higher yields. Further, other biotechnological strategies have been employed for the utilization of various agro-industrial wastes for the production of value-added products such as fermented foods, enzymes, functional foods, and food additives; thereby minimizing the cost of production. Moreover, biotechnological tools have also played a major role in the food quality and safety in order to achieve customer satisfaction, which has been attained by the development of biosensors that can detect the presence of pathogenic microorganisms or adulteration in the food products. Therefore, biotechnological tools have significantly affected the food-processing sector in terms of new product development, cost reduction, and novel processing techniques.

KEYWORDS

- biotechnology
- food processing
- microbes
- safety
- techniques

REFERENCES

Adedayo, M. R., Ajiboye, E. A., Akintunde, J. K., & Odaibo A, (2011) SCP as nutritional enhancer. *J. Microbiol.*, *2*(5), 396–409.

Amerine, M. A., Kunkee, K. E., Ough, C. S., Singleton, V. L., & Webb, A. D., (1980). *The Technology of Wine Making*, AVI Publishing Co. Inc.

Argyri, A. A., Zoumpopoulou, G., Karatzas, K. A. G., Tsakalidou, E., Nychas, G. J., Panagou, E. Z., & Tassou, C. C., (2013). Selection of potential probiotic lactic acid bacteria from fermented olives by in vitro tests. *Food Microbiol.*, *33*, 282–291.

Armstrong, D. W., Martin, S. M., & Yamazaki, H., (1984). Production of ethyl acetate from dilute ethanol solutions by *Candida utilis. Biotechnol. Bioeng.*, *26*, 1038–1041.

Arora, J. K., Marwaha, S. S., & Bakshi, A., (2000). *Biotechnological advancement in food processing*. In *Food Processing: Biotechnological Applications*; Marwaha, S. S., Arora, J. K., Eds., Asiatech Publishers Inc., New Delhi, pp. 1–24.

Arora, P., Sindhu, A., Dilbaghi, N., & Chaudhury, A., (2011). Biosensors as Innovative tools for the detection of food borne pathogens. *Biosens. Bioelectron.*, *28*, 1–12.

Aunstrup, K., (1974). Industrial production of proteolytic enzymes. *Ind. Aspects Biochem.*, *30*(1), 23–46.

Babitha, S., (2009). *Microbial Pigments In Biotechnology for Agro-Industrial Utilization*; Singh nee' Nigam, P., Pandey, A., Eds., Springer, Netherlands, pp. 147–162.

Bagwan, D. J., Patil, J. S., Mane, S. A., Kadam, V. V., & Vichare, S., (2010). Genetically modified crops: Food of the future (review). *Int. J. Adv. Biotechnol. Res.*, *1*, 21–30.

Bali, V., Panesar, P. S., Bera, M. B., & Kennedy, J., (2014). Bacteriocins: Recent trends and potential applications. *Crit. Rev. Food Sci. Nutr.*, *56*, 817–834.

Bandyopadhyay, B., & Mandal, N. C., (2014). Probiotics, prebiotics and synbiotics – In health improvement by modulating gut microbiota: The concept revisited. *Int. J. Curr. Microbiol. App. Sci.*, *3*(3), 410–420.

Barrett, T., Fang, P., & Swaminathan, B., (1997). Amplification methods for detection of food-borne pathogens. In *Nucleic Acid Amplification Techniques: Application to Disease Diagnosis.* Lee H, Morse S, Slovak, O., (eds), Eaton Publishing, Boston, USA, pp. 171–181.

Billings, T., (1998). On fermented foods. Available: http://www.livingfoods.com.

Bicas, J. L., Silva, J. C., Dionisio, A. P., & Pastore, G. M., (2010). Biotechnological production of bioflavors and functional sugars. *Ciênc Tecnol Aliment Campinas.*, *30*(1), 7–18.

Blandino, A., Al-Aseeri, M. E., Pandiella, S. S., Cantero, D., & Webb, C., (2003). Cereal-based fermented foods and beverages. *Food Res. Intl.*, *36*, 527–543.

Boze, H., Moulin, G., Galzy, P., (1995). Production of microbial biomass, In: *Enzymes, Biomass, Food and Feed Biotechnology;* Reed, G., Nagodawithana, T. W., Eds., Weinheim, DEU:VCH, pp. 167–220

Bramorski, A., Christen, P., Ramirez, M., Soccol, C. R., & Revah, S., (1998). Production of volatile compounds by the edible fungus *Rhizopus oryzae* during solid-state cultivation on tropical agro-industrial substrates. *Biotechnol. Lett.*, *20*, 359–362.

Briggs, D. E., Wadison, A., Statham, R., & Taylor, J. F., (1981). The use of extruded barley, wheat and maize as adjunct in mashing. *J. Inst. Brew.*, *92*, 468–474.

Buckenhuskes, H. J., (1997). *Fermented Vegetables.* ASM Press.

Cammann, K., (1977). Biosensors based on ion-selective electrode. *Fresenius Z. Anal. Chem.*, *287*, 1–9.

Chandan, R. C., (1982). Other fermented dairy products. In *Industrial Microbiology*; Reed, G., Ed., 4th edn, AVI Publishing: Westport, CT, pp. 113–184.

Chandan, R. C., & Nauth, K. R., (2012). Yogurt In *Handbook of Animal Based Fermented Food and Beverage Technology;* Hui, Y. H., Evranuz, E. O., Chandan, R. C., Cocolin, L., Drosinos, E. H., Goddik, L., Rodriguez, A., Toldra, F., Eds., 2nd edn, CRC Press, Boca Raton, USA, pp. 213–233.

Chandan, R. C., & Shahani, K. M., (1995). *Other Fermented Dairy Products In Biotechnology;* Reed, G., Nagodawithana, T. W., Eds., 2nd edn, VCH Publishing: Weinheim, Germany, pp. 386–418.

Chavan, J. K., Kadam, S. S., & Beuchat, L. R., (1989). Nutritional improvement of cereals by fermentation. *Crit. Rev. Food Sci. Nutr.*, *28*, 349–400.

Cheetham, P. S. J., (1993). The use of biotransformation for the production of flavours and fragrances. *Trends Biotechnol.*, *11*, 478–488.

Cheng, K. C., Demirci, A., & Catchmark, Pullulan, J. M., & (2011). Biosynthesis, production, and applications. *Appl. Microbiol. Biotechnol.*, *92*, 29–44.

Cho, Y. J., Park, J. P., Hwang, H. J., Kim, S. W., Choi, J. W., & Yun, J. W., (2002). Production of red pigment by submerged culture of *Paecilomyces sinclairii*. *Lett. Appl. Microbiol.*, *3*, 195–202.

De Vuyst, L., & Degeest, B., (1999). Heteropolysaccharides from lactic acid bacteria. *FEMS Microbiol. Rev.*, *23*, 153–177.

Demain, A. L., Jackson, M., & Trenner, N. R., (1967). Thiamine-dependent accumulation of tetramethyl pyrazine accompanying a mutation in the isoleucine-valine pathway. *J. Bacteriol.*, *94*, 323–326.

Donalies, U. E.,Nguyen, H. T.,Stahl, U., & Nevoigt, E., (2008). Improvement of *Saccharomyces* yeast strains used in brewing, wine making and baking. *Adv. Biochem. Eng. Biotechnol.*, *111*, 67–98.

Dufosse, L., (2006). Microbial production of food grade pigments. *Food Technol. Biotechnol.*, *44*, 313–321.

Elegado, F. B., Guerra, M. A. R. V., Macayan, R. A., Mendoza, H. A., & Lirazan, M. B., (2004). Spectrum of bacteriocin activity of *Lactobacillus plantarum* BS and fingerprinting by RAPDPCR. *Int. J. Food Microbiol.*, *95*, 11–18.

Endrizzi, I, Pirretti, G., Calo, D. G., & Gasperi, F., (2009). A consumer study of fresh juices containing berry fruits. *J. Sci. Food Agri.*, *89*, 1227–1235.

Farnworth, E. R., (2001). Probiotics and prebiotics. In *Handbook of Nutraceuticals and Functional Foods*, Wildman, R. E. C., Ed., CRC Press: Boca Raton, USA, pp. 407–422.

FAO.Annex 3, (2005) Special note from the experts who participated in the consultation. In Genetically modified organisms in crop production and their effects on the environment: methodologies for monitoring and the way ahead. Report of the Expert Consultation, 18–20 January. Food and Agriculture Organization of the United Nations, Rome. http://www.fao.org/docrep/008/ae738e/ae738e14.htm,

Finnegan, H., & McElroy, D., (1994). Transgene inactivation: Plants fight back! *Biotechnol.*, *12*, 883–888.

Francke, W., & Brummer, G., (1978). Terpene aus *Ascoidea hylecoeti*. *Planta. Medica.*, *34*, 426–429.

Franco, M. R. B., (2004). *Aroma and Flavor of Food: Current Issues,* Livraria Varela: Sao Paulo, p. 246.

Gandhi, D. N., (2000). Fermented dairy products and their role in controlling food borne diseases. In: *Food Processing: Biotechnological Applications*; Marwaha, S. S., Arora, J. K., Eds.; Asiatech Publishers Inc., New Delhi, pp. 209–220.

Giavasis, I., & Biliaderis, C., (2006). *Microbial polysaccharides* In *Functional food carbohydrates*; Biliaderis, C., Izydorczyk, M., Eds.; CRC Press: New York, pp. 167–214.

Gibson, G. R., & Roberfroid, M. B., (1995). Dietary modulation of the human colonic microflora: introducing the concept of prebiotics. *J. Nutr.*, *125*, 1401–1412.

Gillooly, M., Bothwell, T. H., Charlton, R. W., Torrance, J. D., Bezwoda, W. R., Macphail, A. P., Deman, D. P., Novelli, L., Morrau, D., & Mayet, F., (1984). Factors affecting the adsorption of iron from cereals. *Br. J. Nutr.*, *51*,37–46.

GMO Compass EU., (2010). *The First Harvest of the Amflora Potato.*, http://www.gmocompass.org/eng/news/534.docu.html (accessed on March 25, 2013).

Grzonka, Z., Kasprzykowski, F., & Wiczk, W., (2007). Cysteine proteases. In: *Industrial Enzymes: Structure Function and Applications*; Ploaina, J., MacCabe, A. P., Eds., Springer: The Netherlands, pp. 181–195.

Guarner, F., & Schaafsma, G. J., (1998). Probiotics. *Int. J. Food Microbiol.*, *39*, 237–238.

Gupta, C., Garg, A. P., Prakash, D., Goyal, S., & Gupta, S., (2011). Microbes as potential source of biocolours. *Pharmacology Online*, *2*, 1309–1318.

Haard, N. F., Odunfa, S. A., Lee, C. H., Quintero-Ramirez, R., Lorence-Quinones, A., & Wacher-Radarte, C., (1999). Fermented cereals: A global perspective. *FAO Agricultural Services Bulletin.*, *138*.

Harlander, S., (1992). Food biotechnology. In: *Encyclopaedia of Microbiology*; Lederberg, J., Ed., Academic Press: New York, pp. 191–207.

Hirahara, T., (1998). Functional food science in Japan. In: *Functional Food Research in Europe*; Mattila-Sandholm, T., Kauppila, K., Eds.; Julkaisija-Utgivare: Finland., pp. 19–20.

Hofvander, P., Persson, T. P., Tallberg, A., & Wikstrom, O., (2004). *Genetically Engineered Modification of Potato to Form Amylopectin-Type Starch.* U.S. Patent, 6,784,338, B1.

Jack, R. W., Bierbaum, G., Hiedrich, C., & Sahl, H. G., (1995). The genetics of lantibiotic biosynthesis. *Bioessays.*, *17*, 793–802.

Jackson, R. S., (2000). Wine science-principles, *Practices, Perception.* 2nd Edition, Academic Press, San Diego.

Jaganmohan, P., Purushottam, B., & Prasad, S. V., (2013). Production of SCP with *Aspergillus terrus* using solid state fermentation. *Eur. J. Biol. Sci.*, *5*(2), 38–45.

James, C., (2011). Global status of commercialized biotech/GM crops. ISAAA brief no.43, ISAAA, Ithaca, NY.

Jeevaratnam, K., Jamuna, M., & Bawa, A. S., (2005). Biological preservation of foods: Bacteriocins of lactic acid bacteria. *Indian J. Biotechnol.*, *4*, 446–454.

Joshi, V. K., Attri, D., Baja, A., & Bhushan, S., (2003). Microbial Pigments. *Indian J. Biotechnol.*, *2*, 362–369.

Joshi, V. K., Chauhan, S. K., & Bhushan, S., (2000). Technology of fruit based alcoholic beverages. In: *Postharvest technology of fruits and vegetables;* Verma, L. R., Joshi, V. K., Eds., Indus Publishing: New Delhi, pp. 1019.

Joshi, V. K., Sandhu, D. K., & Thakur, N. S., (1999). Fruit based alcoholic beverage. In: *Biotechnology: Food Fermentation (Microbiology, Biochemistry and Technology)*, Joshi, V. K., Pandey, A., Eds., Education Publishers and Distributors: New Delhi, pp. 647.

Kapila, J., De Rycke, R., Van Montagu, M., & Angenon, G., (1997). An *Agrobacterium* mediated transient gene expression system for intact leaves. *Plant Sci.*, *122*, 101–108.

Karovicova, J., Kohajdova, Z., & Greif, G., (2002). The use of PCA, CA, FA for evaluation of vegetable juices processed by lactic acid fermentation. *Czech J. Food Sci.*, *20*, 135–143.

Kaur, C., & Kapoor, H. C., (2001). Antioxidants in fruits and vegetables-the millennium's health. *Int. J. Food Sci. Technol.*, *36*, 703–725.

Kaur, S., Panesar, P. S., Bera, M. B., & Panesar, R., (2011). Biotechnological tools: potentials in food quality and safety. In: *Bio-processing of Foods.* Panesar, P. S., Sharma, H. K., Sarkar, B. C., Eds., Asiatech Publishers: New Delhi.

Kennedy, J. F., & Bradshaw, I. J., (2001). Production, properties and applications of xanthan. *Prog. Ind. Microbiol.*, *19*, 319–371.

Khanafari, A., Tayari, K., & Emami, M., (2008). Light requirement for the carotenoids production by *Mucor hiemalis*. *Iranian J. Basic Med. Sci.*, *11*, 25–32.

Khetarpaul, N., & Chauhan, B. M., (1990). Effect of fermentation by pure cultures of yeasts and lactobacilli on the available carbohydrate content of pearl millet. *Tropical Sci.*, *31*, 131–139.

Kosuge, T., & Kamiya, H., (1962). Discovery of a pyrazine in a natural product: Tetramethyl pyrazine from cultures of a strain of *Bacillus subtilis*. *Nature.*, *193*, 776.

Kumar, S., Prasanna, L. P. A., & Wankhade, S., (2011). Potential benefits of *Bt* brinjal in India-An economic assessment. *Agr. Econom. Res. Rev.*, *24*, 83–90.

Lantto, R., Kruus, K., Puolanne, E., Honkapää, K., Roininen, K., & Buchert, J., (2009). Enzymes in meat processing In *Enzymes in food technology;* Whitehurst, R. J., van Oort, M., Eds.; Wiley-Blackwell: Oxford, U.K, pp. 264–289.

Lardizabal, K., Effertz, R., Levering, C., Mai, J., Pedroso, M. C., Jury, T., Aasen, E., Gruys, K., & Bennett, K., (2008). Expression of *Umbelopsis ramanniana* DGAT2A in seed increases oil in soybean. *Plant Physiol.*, *148*, 89–96.

Lawrie, R. A., (1998). *Lawrie's Meat Science*. Woodhead Publishing Ltd., Cambridge, U.K.

Lee, J. S., Heo, G. Y., Lee, J. W., Oh, Y. J., Park, J. A., Park, Y. H., Pyun, Y. R., & Ahn, J. S., (2005). Analysis of kimchi microflora using denaturing gradient gel electrophoresis. *Int. J. Food Microbiol.*, *102*, 143–150.

Litchfield, J. H., (1983). Single cell proteins. *Science*, *219*, 740–746.

Mahmoud, M. M., & Kosikowski, F. W., (1982). Alcohol and single-cell protein production by *Kluyveromyces* in concentrated whey permeates with reduced ash. *J. Dairy Sci.*, *65*, 2082–2087.

Malik, K., Tokkas, J., & Goyal, S., (2012). Microbial Pigments: A review. *Int. J. Microb. Res. Technol.*, *1*, 361–365.

Mattila-Sandholm, T., (1998). VTT on lactic acid bacteria. *VTT Symposium*, *156*, 1–10.

Matz, S. A., (1971). *Cereals Science*. AVI publishing Co. Westport Connecticut.

Mendez, A., Perez, C., Montanez, C. J., Mrtinez, G., & Aguilar, N. C., (2011). Red pigment production by *Pencillium purpurogenum* GH2 is influenced pH and temperature. *Biomed. Biotechnol.*, *12*, 961–968.

Misaki, A., Tsumuraya, Y., & Takaya, S., (1978). A new fungal α-d-glucan, elsinan, elaborated by *Elsinoe leucospila Agric. Biol. Chem.*, *42*, 491–493.

Moellenbeck, D. J., Peters, M. L., Bing, J. W., et al., (2001) Insecticidal proteins from *Bacillus thuringiensis* protect corn from corn rootworms. *Nature Biotechnol.*, *19*, 668–672.

Mohankumari, H. P., Naidu, A., Vishwanatha, S., Narasimhamurthy, K., & Vijayalakshmi, G., (2009). Safety evaluation of *Monascus purpureus* red mould rice in albino rats. *Food Chem. Toxicol.*, *47*, 1739–1746.

Montet, D., Loiseau, G., Zakhia, N., & Mouquet, C., (1999). Fermented fruits and vegetables In *Biotechnology: food fermentation*, Joshi, V. K., Pandey, A., Eds.; Educational Publishers & Distributors: New Delhi., pp. 951–969.

Murugaboopathi, G., Parthasarathy, V., Chellaram, C., Prem Anand, T., & Vinurajkumar, S., (2013). Applications of biosensors in food industry. *Biosci. Biotechnol. Res. Asia.*, *10*(2), 711–714.

Nambiar, S., & Yeow, J. T. W., (2011). Conductive polymer-based sensors for biomedical applications. *Biosens. Bioelectron.*, *26*, 1825–1832.

Nasseri, A. T., Rasoul-Amini, S., Morowvat, M. H., & Ghasemi, Y., (2011). Single cell protein: production and process. *Am. J. Food. Technol.*, *6*, 103–116.

National Research Council (2010). The Impact of Genetically Engineered Crops on Farm Sustainability in the United States. National Academies Press, Washington, DC.

Newbrun, E., & Baker, S., (1967). Physico-chemical characteristics of the levan produced by *Sreptococcus salivarious*. *Carbohydr. Res.*, *6*, 165–170.

Newell, C. A., Rozman, R., Hinchee, M. A., Lawson E. C., Haley, L., Sanders, P., Kaniewski, W., Tumer, N. E., Horsch, R. B., & Fraley, R. T., (1990). Agrobacterium-mediated transformation of *Solanum tuberosum* L. cv. "Russet Burbank." Plant Cell Rep *10*, 30–34.

Nochi, T., Takagi, H., Yuki, Y., Yang, L., et al., (2007). Rice-based mucosal vaccine as a global strategy for cold-chain- and needle-free vaccination. *Proc. Natl. Acad. Sci. USA*, *104*, 10986–10991.

Notararigo, S., Nácher-Vázquez, M., Ibarburu, I., Werning, M. L., de Palencia, P. F., Duenas, M. T., Aznar, R., López, P., & Prieto, A., (2012). Comparative analysis of production and purification of homo- and hetero-polysaccharides produced by lactic acid bacteria. *Carbohydr. Polym.*, *93*, 57–64.

Nout, M. J. R., & Motarjemi, Y., (1997). Assessment of fermentation as a household technology for improving food safety: a joint FAO/WHO workshop. *Food Control, 8*, 221–226.

Okagbue, R. N., (1995). Microbial biotechnology in Zimbabwe: current status and proposals for research and development. *J. Appl. Sci. South Afr.*, *1*, 148–158.

Osungbaro, T. O., (2009). Physical and nutritive properties of fermented cereal foods. *Afr. J. Food Sci.*, *3*, 023–027.

Osusky, M., Osuska, L., Kay, W., & Misra, S., (2005). Genetic modification of potato against microbial diseases: In vitro and in planta activity of a dermaseptin B1 derivative, MsrA2. *Theor. Appl. Genet.*, *111*, 711–722.

Panda, S. H., Parmanick, M., & Ray, R. C., (2007). Lactic acid fermentation of sweet potato (*Ipomoea Batatas* L.) into pickles. *J. Food Process Preserv.*, *31*, 83–101.

Panda, S. H., & Ray, R. C., (2007). Lactic acid fermentation of β-carotene rich sweet potato (*Ipomoea batatas* L.) into lacto-juice. *Plant Food Hum. Nutr.*, *62*, 65–70.

Pandey, A., Soccol, C. R., Nigam, P., Brand, D., Mohan, R., & Roussos, S., (2000a). Biotechnological potential of coffee pulp and coffee husk for bioprocesses. *Biochem. Engg. J.*, *6*, 153–162.

Pandey, A., Soccol, C. R., Nigam, P., & Soccol, V. T., (2000b) Biotechnological potential of agro-industrial residues. I: sugarcane bagasse. *Bioresour. Technol.*, *74*, 69–80.

Pandey, A., Soccol, C. R., Nigam, P., Soccol, V. T., Vandenberghe, L. P. S., & Mohan, R., (2000c) Biotechnological potential of agro-industrial residues. II: cassava bagasse. *Bioresour. Technol.*, *74*, 81–87.

Panesar, P. S., (2011). Fermented dairy products: starter cultures and potential nutritional benefits. *Food Nutr. Sci.*, *2*, 47–51.

Panesar, P. S., & Bali, V., (2015). Prebiotics. In *Biotransformation of Waste Biomass into High Value Biochemicals;* Brar, S., Dhillon, G. S., Marcelo, F., Eds., Springer Science + Business Media: New York, pp. 237–259.

Panesar, P. S., Chopra, H., Marwaha, S. S., & Joshi, V. K., (2000). *Technologies for the Production of Alcoholic Beverages. In Food Processing: Biotechnological Applications*, Marwaha, S. S., Arora, J. K., Eds., Asiatech Publishers, New Delhi., pp. 191–208.

Panesar, P. S., Kaur, R., Singla, G., & Sangwan, R. S., (2016). Bio-processing of agro-industrial wastes for production of food-grade enzymes: Progress and prospects. *Appl. Food Biotechnol.*, *3*(4), 208–227.

Panesar, P. S., Kumari, S., & Panesar, R., (2013). Biotechnological approaches for the production of prebiotics and their potential applications. *Crit. Rev. Biotechnol.*, *33*(4), 345–364.

Panesar, P.S., Kaur, G., Panesar, R., & Bera, M. B., (2009). *Synbiotics: Potential Dietary Supplements in Functional Foods.* http://www.foodsciencecentral.com/fsc/ixid15649.

Panesar, P. S., & Marwaha, S. S., (2014). Biotechnology and its role in agriculture and food processing. In: *Biotechnology in Agriculture and Food Processing: Opportunities and Challenges,* Panesar, P. S., Marwaha, S. S., Eds., CRC Press: Boca Raton, USA., pp. 3–44.

Panesar, P. S., Marwaha, S. S., & Chopra, H. K., (2010). *Enzymes in Food Processing: Fundamentals & Potential Applications,* IK International Pvt. Ltd: New Delhi,.

Panesar, P.S., Kumari, S., & Panesar, R., (2011). Prebiotics: Current status and perspectives. *Int. J. Food Ferment. Technol.*, *1*(2),149–159.

Parker, R. B., (1974). Probiotics, the other half of the antibiotic story. *Animal Nutr. Health.*, *29*, 4–8.

Powell, M., & Robe, K., (1964). High protein feed production by lactose fermenting yeast on whey. *Food Proc.*, *25*, 80.

Querci, M., Van Den Bulcke, M., Zel, J., Van den Eedea, G., & Broll, H., (2010). New approaches in GMO detection. *Anal. Bioanal. Chem.*, *396*(6),1991–2002.

Querci, M. C., Paoletti, C., & Van den Eedea, G. (2007) From sampling to quantification: Developments and harmonization of procedures for GMO testing in the European Union. In: *Collection of Biosafety Reviews;* Craig, W., (ed.), International Centre for Genetic Engineering and Biotechnology, Trieste, Italy, 8–41.

Reed, G., (1987). *Prescott and Dunn's Industrial Microbiology, 4th edn.,* CBS Publishers, New Delhi.

Reid, G., (2008). Probiotics and prebiotics-progress and challenges. *Int. Dairy J.*, *18*, 969–975.

Rossi, S. C., Vandenberghe, L. P. S., Pereira, B. M. P., Gago, F. D., Rizzolo, J. A., Pandey, A., Soccol, C. R., & Medeiros, A. B. P., (2009). Improving fruity aroma production by fungi in SSF using citric pulp. *Food Res. Int.*, *42*, 484–486.

Russell, I., & Stewart, R., (1999). Cereal based alcoholic beverages. In: *Biotechnology: Food Fermentation,* Joshi, V. K., Pandey, A., Educational Publishers and Distributors, New Delhi., pp. 745–780.

Russian Federation National Methodical Recommendations (Guidelines) (2006). For GM plants (soybean RR40–3-2, maize MON810 NK603, GA21, T25) quantitative identification in raw material and food with application of test kits.

Salminen, S., & Von Wright, A., (1998). Safety of probiotic bacteria: Current perspectives. In: *Functional Food Research in Europe,* Mattila-Sandholm, T., Kauppila, T., Eds., Julkaisija-Utgivare, Finland, pp. 105–106.

Sanford, J. C., (1990). Biolistic plant transformation. *Physiol. Plant.*, *79*, 206–209.

Sassolas, A., Blum, L. J., & Leca-Bouvier, B. D., (2011). Immobilization strategies to develop enzymatic biosensors. *Biotechnol. Adv.*, *30*(3), 489–571.

Sekhon, B. S., & Jairath, S., (2010). Probiotics, prebiotics and synbiotics: An overview. *J. Pharm. Edu. Res.*, *1*, 13–36.

Simango, C., (1997). Potential use of traditional fermented foods for weaning in Zimbabwe. *J. Soc. Sci. Med.*, *44*, 1065–1068.

Singer, M., & Berg, P., (1991). *Genes & Genomes.* University Science Books, Mill Valley, CA.

Singh, B. D., (2010). *Biotechnology: Expanding Horizons*. Kalyani Publishers, New Delhi.

Siro, I., Kapolna, E., Kapolna, B., & Lugasi, A., (2008). Functional food: Product development, marketing and consumer acceptance: A review. *Appetite.*, *51*, 456–467.

Skryabin, K., & Tutelyan, V., Genetically Modified Foods. In: *Biotechnology in Agriculture and Food Processing: Opportunities and Challenges,* Panesar, P. S., Marwaha, S. S., Eds., CRC Press: Boca Raton, USA, pp. 480–505.

Soares, M., Christen, P., & Pandey, A., (2000). Fruit flavor production by *Ceratocystis fimbriata* grown on coffee husk in solid-state fermentation. *Process Biochem.*, *35*, 857–861.

Soni, S. K., & Arora, J. K., Indian fermented foods: Biotechnological approaches. In: *Food Processing: Biotechnological Applications.* Marwaha, S. S., Arora, J. K., Eds.; Asiatech Publishers, Inc., New Delhi, pp. 143–190.

Stanton, C., Gardiner, G., Meehan, H., Collins, K., Fitzgerald, G., Lynch, P. B., & Ross, P., (2001). Market potential for probiotics. *Am. J. Clin. Nutr.*, *73*, 476S–483S.

Steinkraus, K. H., (1998). Bio-enrichment: production of vitamins in fermented foods. In: *Microbiology of Fermented Foods.* Wood, J. B. (ed.). Blackie Academic and Professional, London, pp. 603–619.

Suman, G., Nupur, M., Anuradha, S., & Pradeep, B., (2015). Single cell protein production: A review. *Int. J. Curr. Microbiol. App. Sci.*, *4*(9), 251–262.

Tamang, B., Tamang, J. P., Schillinger, U., Franz, C. M. A. P., Gores, M., & Holzapfel, W. H., (2008). Phenotypic and genotypic identification of lactic acid bacteria isolated from ethnic fermented bamboo tender shoots of North East India. *Int. J. Food Microbiol.*, *121*, 35–40.

Tamang, J. P., (2009). *Himalayan Fermented Foods: Microbiology, Nutrition and Ethnic Values*. CRC Press, New Delhi.

Tengel, C., Schubler, P., Setzke, E., Balles, J., & Haubels, S. M., (2001). PCR based detection of genetically modified soybean and maize in raw and highly processed foodstuffs. *Biotechniques.*, *31*, 426–429.

Thevenot, D. R., Toth, K., Durst, R. A., & Wilson, G. S., (1999). Electrochemical biosensors: Recommended definitions and classification. *Pure Appl. Chem.*, *7*, 2333–2348.

Trevan, M. D. (1988). *Enzyme Production in Biotechnology: The Biological Principles*; Trevan, M. D., Boffey, S., Goulding, K. H., Stanbury, P. Eds.; Tata McGraw Hill Publishing Company Ltd: New Delhi, pp. 155–177.

Van den Eedea, G., Aarts, H., Buhk, H. J., Corthier, G., Flint, H. J., Hammes, W., Jacobsen, B., Midtvedt, T., van der Vossen, J., von Wright, A., Wackernagel, W., & Wilcks, A., (2004). The relevance of gene transfer to the safety of food and feed derived from genetically modified (GM) plants. *Food Chem. Toxicol.*, *42*, 1127–1156.

Venema, K., (2012). Intestinal fermentation of lactose and prebiotic lactose derivatives, including human milk oligosaccharides. *Int. Dairy J.*, *22*, 123–140.

Viander, B., Maki, M., & Palva, A., (2003). Impact of low salt concentration, salt quality on natural large-scale sauerkraut fermentation. *Food Microbiol.*, *20*, 91–395.

Welsh, F. W., Murray, W. D., & Williams, R. E., (1989). Microbiological and enzymatic production of flavor and fragrance chemicals. *Crit. Rev. Biotechnol.*, *9*, 105–169.

Whitman, B. D., (2000). Genetically modified foods: Harmful or helpful? CSA Discovery Guides, http://www.csa.com/discoveryguides/gmfood/overview.php.

Wood, P. J., (1997). Functional foods for health. In: *Cereals;* Campbell, G. M., Webb, C., McKee, S. L., Eds., Springer US: New York, 233–238.

Xu, X., Pu, Q., He, L., Na, Y., Wu, F., & Jin, Z., (2009). Rheological and SEM studies on the interaction between spent brewer's yeast b-glucans and k-carrageenan. *J. Texture Stud.*, *40*, 482–496.

Yamano, T., Iino, H., Takada, M., Blum, S., Rochat, F., & Fukushima, Y., (2006). Improvement of the human intestinal flora by ingestion of the probiotic strain *Lactobacillus johnsonii* La1. *Br. J. Nutr.*, *95*, 303–312.

Yan, P. M., Xue, W. T., Tan, S. S., Zhang, H., & Chang, X. H., (2008). Effect of inoculating lactic acid bacteria starter cultures on the nitrite concentration of fermenting Chinese paocai. *Food Control.*, *19*, 50–55.

Yarasi, B., Sadumpati, V., Immanni, P. C., Vudem, R. D., & Khareedu, R. V., (2008). Transgenic rice expressing *Allium sativum* leaf agglutinin (ASAL) exhibit high level resistance against major sap sucking pests. *BMC Plant Biol.*, *8*, 102–106.

Yegin, S., & Fernandez-Lahore, M., (2012). Boza: a traditional cereal-based, fermented Turkish beverage. In: *Handbook of Plant-Based Fermented Food and Beverage Technology;* Hui, Y. H., Evranuz, O. E., Eds., CRC Press: Boca Raton, US, pp. 533–542.

CHAPTER 4

TRENDS IN ENZYMATIC SYNTHESIS OF HIGH FRUCTOSE SYRUP

R. S. SINGH, K. CHAUHAN, and R. P. SINGH

Carbohydrate and Protein Biotechnology Laboratory, Department of Biotechnology, Punjabi University, Patiala–147 002, Punjab, India, E-mail: rssingh11@lycos.com; rssbt@pbi.ac.in

CONTENTS

4.1   INTRODUCTION

Sweeteners have been the part of our diet for over thousands of years. Among them, honey and sucrose are chiefly used to enhance the taste of the food. Both comprise almost similar content of fructose and dextrose. Later, the discovery of depolymerization of starch into sweet monomer subunits of dextrose provided an attractive alternative to sucrose. However, its lower relative sweetness left it as an inadequate replacement for conventional sugar in various food and beverage products. Concurrent advances in refining and separation technologies in mid-1960s led to the production of high fructose syrup (HFS) with sweetness equivalent to sucrose. Later in 1980s, HFS became an important component of various beverages like Coca-Cola and

Pepsi. The large-scale production of HFS was first carried out by A.E Staley Manufacturing Company in 1987. HFS can be prepared by the conventional multienzymatic method from starch or single-step hydrolysis of inulin using inulinases. In the conventional method, three enzymes, namely α-amylase, glucoamylase, and glucose isomerase, act simultaneously on starch to produce maximum 45% of HFS, whereas in single-step hydrolysis of inulin, exoinulinase (EC 3.2.1.80; β-2-1-D-fructan fructohydrolase) sequentially degrades β, 2-1 linkages of inulin to give as high as 95% of fructose yield.

Fructose is a monosaccharide and a simple sugar found in various fruits, vegetables, and honey. An important difference between sucrose and fructose is that fructose is twice sweeter than sucrose. Hence, it is used as an idyllic sweetener alternative to conventional sugar by many food industries. It is GRAS food ingredient. It has also been reported to play various physiological and beneficial roles in human body (Table 4.1).

Some of its significant physiological role includes bypassing the glucose metabolic pathway, thereby not requiring hormonal (insulin) regulation of glucose; provide sensitivity to insulin in noninsulin dependent diabetes mellitus (Wolfgang and Südzucker, 2004); enhance ethanol metabolism (Ylikahari et al., 1972); increase iron and calcium intestinal absorption (Pawan, 1973); and stimulate *Bifidobacteria* growth in human gut. Moreover, due to low glycemic index, it does not abruptly affect the blood glucose level. It is also free from common health problems caused by sucrose like cariogenecity, corpulence, and artherosclerosis. Fructose also possesses innumerable technically superior and functional properties (Table 4.2) which has further broadened the possibility of its use as an important food ingredient by many food industries. High solubility at low temperature, large freezing point depression, and lesser crystallization as compared to sucrose are some of its important properties that strongly influence taste and texture of dairy products. Hence, these properties are beneficial

TABLE 4.1   Beneficial Roles of High Fructose Syrup Consumption

- Insulin independent metabolism
- Safe alternate sweetener (non-cariogenic, non-corpulence, non-artherosclerosis, non-diabetic)
- Increases absorption of iron and zinc
- Speeds up ethanol metabolism
- Low calorie sweetener & desirable organoleptic properties
- Reduced intake of calories

TABLE 4.2   Superior Technical Properties of Fructose Over Other Sweeteners

- More sweeter
- Humectancy
- Flavor enhancer
- High osmotic pressure
- Causes no problem in acidic foods
- High solubility and low viscosity
- High freezing point depression
- No crystal formation and smooth consistency
- Masks the bitter after-taste of some intense sweeteners
- Antioxidant properties

for the formulation of ice-creams and other frozen desserts. Fructose has higher osmotic pressure and lower water activity than sucrose, sorbitol, mannitol, dextrose, etc. Both these properties make it a good preservative as high osmotic pressure prevents microbial contamination, while lower water activity is important for sustaining microbial stability in food items without affecting their total moisture content. Fructose is also a good humectant as it easily retains water in a frozen system, which is further significant in providing longer shelf-life to food products (Hanover and White, 1993). It is also advantageous for its hygroscopic and antioxidant properties and for retaining anions (Pawan, 1973). Moreover, it is preferable over sucrose in acidic foods as it does not breakdown easily at high temperature and acidic conditions. These functional and technical superiorities of fructose have extended its necessity from food industries to health and pharmaceutical sectors also. In pharmaceutical industry, it is used in formulation of tablets or solution for injection and infusion (Hanover and White, 1993) and conversion of its derived furans into eco-friendly chemicals like pyrroles, pyridazines, and diazepinones (Lichtenthaler et al., 2001).

4.2   SUBSTRATES FOR THE PRODUCTION OF HIGH FRUCTOSE SYRUP (HFS)

Starch and inulin are the two important substrates for the production of HFS. Starch is hydrolyzed to HFS by the action of different amylolytic enzymes, while inulinases carry out inulin hydrolysis to produce fructose with high productivity.

4.2.1   STARCH

Starch is a plant reserve polysaccharide that consists of a chain of glucose molecules. This chain of glucose molecules can be straight or branched. The branchless form is called amylose in which glucose molecules are linked by α-(1, 4) bonds with the ring of oxygen atoms on all same sides, whereas amylopectin is the name given to the branched form containing branches linked to linear backbone by α-(1, 6) linkages.

The relative proportion of amylose to amylopectin solely depends upon the source of starch (Hanover and White, 1993). Because starch is entirely made up of glucose molecules, it is termed as homosaccharide which stands for the chain of sugars made from a single type of molecule. The basic chemical formula of starch is $(C_6H_{10}O_5)n$, where n is the number of glucose molecules in the chain. Wheat, potato, tapioca, rice, arrowroot, sago, etc. are some common sources of starch, while corn is considered its most abundant source. Corn wet-milling is carried out to separate starch from other constituents like proteins, fiber, and oil. The use of corn by refining industry has increased significantly since the mid-1970s, exclusively due to its abundant supply, low cost, and ease of storage that further ensures its continuous supply as a raw material for HFS production. The separated starch can be acid or enzyme hydrolyzed for HFS production. Dent corn variety is commonly used for the production of high fructose corn syrup (HFCS). Other than syrup production, corn starch can also be used in various food and non-food applications. It is generally used as thickener, water binder, stabilizer, and gelling agent in food industries. Moreover, approximately 6–7% of corn starch produced is utilized for baking, confectionary, and canning purposes (Orthoefer, 1987). More than 90% of starch is consumed by the paper industry to improve paper strength and appearance, which is one of its important nonfood applications.

4.2.2   INULIN

Inulin is a natural reserve polysaccharide with a wide variety of food and pharmaceutical applications. It is widely distributed in plants representing more than 30,000 plant species, which commonly include the tubers of *Helianthus tuberosus, Cichorium intybus, Dahlia pinnata, Polymnia sonchifolia* etc. (Table 4.3). Roots of *Pombalia calceolaria*, a Brazilian plant, have been recently determined to be a rich source of inulin (Pontes et al., 2016). Mostly, inulin-containing plants are dicotyledonous belonging to *Asteraceae* and *Campanulaceae* families, but some inulin-rich monocotyledonous plants from *Poaceae, Liliaceae,* and *Amaryllidaceae* families have also been reported (Singh and Chauhan, 2016). Inulin was first discovered by Valentine Rose in the early 1800s from the roots of *Inula helenium* and named

TABLE 4.3 Inulin Content of Some Plants (Singh and Singh, 2010)

Botanical Name	Common Name	Plant Part	Inulin*
Agave Americana	Agave	Lobes	7–10
Allium ampeloprasum var. *porrum*	Leek	Bulbs	3–10
Allium cepa	Onion	Bulbs	2–6
Allium sativum	Garlic	Bulbs	9–16
Arctium sp.	Burdock	Roots	3.5–4.0
Asparagus officinalis	Shatwaar	Root tubers	10–15
Asparagus racemosus	Safed musli/Shatwaar	Root tubers	10–15
Camassia sp.	Camas	Bulbs	12–22
Cichorium intybus	Chicory	Roots	15–20
Cynara cardunculus	Artichoke	Leaves-heart	3–10
Dahlia sp.	Dahlia	Root tubers	15–20
Helianthus tuberosus	Jerusalem artichoke	Root tubers	14–19
Hordeum vulgare	Barley	Grains	0.5–1.5
Microseris lanceolata	Murnong	Roots	8–13
Musa acuminate	Banana	Fruit	0.3–0.7
Pombalia calceolaria	-	Roots	13
Secale cereale	Rye	Grains	0.5–1.0
Smallanthus sonchifolius	Yacon	Roots	3–19
Taraxacum officinale	Dandelion	Leaves	12–15
Tragopogon sp.	Salsify	Roots	15–20
Scorzonera hispanica	Spanish salsify	Roots	8.15–10.75
Saussurea lappa	Kuth	Roots	18–20

*Percentage of fresh mass.

'inulin' by Thomson in 1817. In 1864, the plant physiologist Julius Sachs detected spherocrystals of inulin in the tubers of *D. pinnata*, *H. tuberosus*, and *I. helenium* after ethanol precipitation by using a microscope. Edelman and Jefford (1968) first proposed a model for the biosynthesis of inulin in *H. tuberosus*. They reported inulin as a mixture of oligo- and polysaccharides containing fructose units with β-configuration at anomeric C_2. The presence of this β-configuration makes inulin nondigestible by human intestinal enzymes as these enzymes are specific for α-glycosidic bonds only (Roberfroid, 2007).

Five types of fructans, namely inulin-type fructans (1-kestose), levan-type fructans (6-kestose), fructans of the inulin neoseries (neokestose), mixed-type levans (bifurcose), and fructans of the levan neoseries (mixed-type F_3 fructan) have been reported in different plant species. The presence of specific type of fructans depends upon environmental conditions and developmental stage of plants as well as plant species. Inulin is a inulin-type fructan that constitutes variable number of fructose units (2 to 100) linked through β-(2→1) D-fructosyl-fructose bonds terminating with one glucose unit linked through an α-(1→2) D-glucopyranosyl bond (Figure 4.1) as present in sucrose (Bruyn et al., 1992).

It is a heterodisperse molecule with varied degree of polymerization (DPn) and polydispersity (Mensink et al., 2015). Polydispersity and DPn of plant inulin are influenced by plant species, phase in the plant life cycle, harvesting period, and extraction and postextraction methods (Ronkart et al., 2007; Barclay et al., 2010). DPn affects physicochemical as well as functional properties of inulin. For instance, short-chain fractions of oligofructose are more soluble in water and sweeter than native and long-chain fractions, while long-chain fraction is less soluble, more viscous, and thermostable. Long-chain fractions also provide rheological and sensory properties to dairy products as a fat substitute in low-fat products (Apolinário et al., 2014). High DPn also represents good prebiotic potential of inulin (van de Wiele et al., 2007). The physicochemical properties of chicory inulin are summarized in Table 4.4.

Inulin has various food and medical applications. Some of its important health applications include improving the gastrointestinal system and kidney function, providing relief from constipation (Mensink et al., 2015), having low calorific value (Roberfroid et al., 1993), having prebiotic potential (Singh and Singh, 2010), reducing the chances of colon and breast cancer occurrence (Taper and Roberfroid, 1999; Kato, 2000), and reducing blood

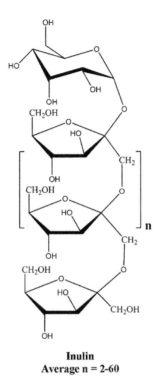

Inulin
Average n = 2-60

FIGURE 4.1 Structure of inulin.

sugar and cholesterol level (Hofer and Jenewein, 1999). In the food industry, it is mainly used as a dietary fiber, low calorie sweetener, and sugar
and fat substitute in many dairy products. Inulin has also been reported to
improve organoleptic properties of various food products and is also used
to form gels with increased viscosity (Kaur and Gupta, 2002). Currently,
inulin has gained tremendous interest as a potent substrate for HFS, fructo-
oligosaccharide, and inulinase production (Singh and Singh, 2010; Singh
and Chauhan, 2016).

4.3 ENZYMATIC SYNTHESIS OF HFS

HFS can be produced by conventional multienzymatic hydrolysis of starch,
single-step enzymmatic hydrolysis of inulin using inulinases, or chemical
hydrolysis of starch/inulin. The chemical approach is generally not used due

TABLE 4.4   Physico-Chemical Properties of Chicory Inulin (Franck and Levecke, 2003)

Properties	Standard Inulin	Long-Chain Inulin
Chemical structure	GF_n $(2 \leq n \leq 60)$	GF_n $(10 \leq n \leq 60)$
Average degree of polymerization	12	25
Dry matter (%)	> 95	> 95
Inulin content (% of d.m.)	92	99.5
Sugar content (% of d.m.)	8	0.5
pH (10 wt %)	5-7	5-7
Sulfated ash (% of d.m.)	< 0.2	<0.2
Heavy metals (ppm of d.m.)	< 0.2	<0.2
Appearance	White powder	White powder
Taste	Neutral	Neutral
Sweetness (vs sucrose = 100%)	10%	None
Solubility in water at 25°C (g/L)	120	10
Viscosity in water (5%) at 10°C (mPa.s)	1.6	2.4
Heat stability	Good	Good
Acid stability	Fair	Good

G: Glucosyl unit; F: Fructosyl unit; d.m.: Dry matter.

to some drawbacks like the formation of difructose anhydrides and other colored inulin hydrolysates that require additional energy expense and improved downstream processing. Hence, the above-mentioned two enzymatic approaches are generally used for the synthesis of HFS. Out of the two methods, the conventional multienzymatic method is commercially well established, while single-step hydrolysis of inulin is still at the laboratory scale.

4.3.1   CONVENTIONAL MULTIENZYMATIC METHOD FOR HFS PRODUCTION FROM STARCH

Concurrent availability of commercial-scale enzymes and development in fractionation techniques has made HFS production possible in the late 1960s. Conventionally, HFS is produced by the depolymerization of starch into monosaccharide dextrose by α-amylase and glucoamylase, which is then isomerized into fructose using glucose isomerase. The conventional production of HFS from starch is completed in three steps: (a) liquefaction and saccharification of starch into monomer units of dextrose, (b) isomerization of dextrose into fructose, and (c) fractionation to increase final fructose yield by ion-exchange techniques. For commercial purpose, later crystallization of fructose is also carried out to obtain it in a dry and crystalline form.

4.3.1.1 Starch Liquefaction and Saccharification

The first step in HFS production from starch is enzymatic liquefaction (partial hydrolysis) and saccharification of liquefied starch into soluble, low viscosity, and dextrose equivalents for their subsequent conversion into different valuable products. Liquefaction is conducted using thermostable α-amylases (EC 3.2.1.1) derived from several bacterial sources such as *Bacillus stearothermophilus*, *B. licheniformis*, and *B. subtilis* (Teague and Brumm, 1992). Amylases from *B. stearothermophilus* and *B. licheniformis* are very thermostable and can withstand temperature higher than 100°C, while amylase from *B. subtilis* is less thermostable and generally lose activity above 90°C. α-Amylase randomly catalyzes the hydrolysis of starch backbone comprising α-1,4 glycosidic linkages to produce soluble oligosaccharides and relatively small amount of medium- and low-molecular-weight saccharides (Hebeda, 1993). Parameters like starch solids, pH, calcium level, time, and temperature are important factors that should be carefully controlled to carry out successful enzymatic liquefaction. Starch solids level is generally maintained at 30–40% to achieve proper starch gelatinization without increasing the cost for removing extra water content. The pH of starch slurry is optimally controlled at 5.8–6.5. Calcium in chloride or oxide form is significant in enhancing α-amylase thermostability. A calcium level higher than 100–200 ppm and 300 ppm is sufficient for α-amylases from *B. stearothermophilus*, *B. licheniformis*, and *B. subtilis*, respectively. Reaction time and temperature are generally balanced to uphold the optimal conditions for enzyme activity and prompt starch gelatinization. Normally, high temperature is used to achieve accurate degree of hydrolysis to prevent any microbial contamination. However, maintaining high temperature can increase by-product formation resulting in loss of final product yield, while low temperature can inactivate the enzyme, which correspondingly affects the reaction between the reactant and the biocatalyst. The Enzyme-Heat-Enzyme (EHE) process was the first developed method used for enzy-

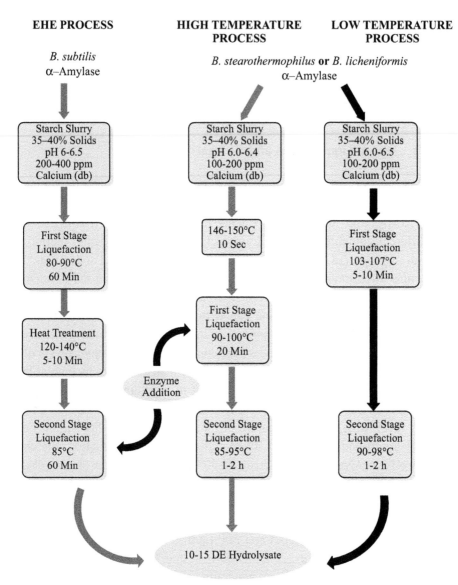

FIGURE 4.2 Starch liquefaction process (EHE: Enzyme-Heat-Enzyme; DE: Dextrose Equivalents; db: dry basis).

matic liquefaction at high temperature (85–90°C) employing α-amylase from *B. subtilis*. Later, high temperature (146–150°C) and low temperature (103–107°C) methods were developed (Figure 4.2), taking the advantage of thermostable property of α-amylases of *B. stearothermophilus* and *B. licheniformis* for HFS production (Hebeda, 1993).

After liquefaction, saccharification of di/oligomeric products produced by amylases is carried out using glucoamylase (EC 3.2.1.3). Glucoamylase breaks α-1,4 and α-1,6 glycosidic linkages that join consecutive dextrose units. *Aspergillus* and *Rhizopus* fungal sources are generally used for the commercial production of glucoamylase due to their adequate thermostability. Rate of hydrolysis by glucoamylase solely depends upon the type and number of linkages, size of the molecule, and the arrangement of α-1,4 and α-1,6 linkages (Abdullah et al., 1963). For example, the hydrolysis rate increases if α-1,6 bonded adjacently to α-1,4 linkage rather than α-1,6 bond. Approximately 95% of dextrose is obtained after saccharification along with other sugars like maltose (1%), isomaltose (1.5%), and maltulose (0.2%), which are formed by the reverse reaction catalyzed by glucoamylase (Shiraishi et al., 1985). For commercial purpose, minimizing this reversion is important to maximize the total dextrose yield. Overdosing glucoamylase and shortening the reaction time can be helpful in preventing excessive reversion. However, overdosing and extending reaction time beyond the maximum dextrose level will result in low dextrose yield due to continuous formation of isomaltose. Dextrose yield can also be increased on combining glucoamylase with a debranching enzyme derived from *B. acidopullulyticus* and an enzyme exhibiting both amylase and transferase properties isolated from *B. megaterium* (Hebeda et al., 1988). The debranching enzyme specifically hydrolyses α-1,6 glycosidic bonds providing additional substrates for glucoamylase, whereas *B. megaterium* amylase degrades oligosaccharides to panosyl units that form maltotetraose after combining with dextrose. This maltotetraose saccharide can be easily hydrolyzed to dextrose. The saccharified hydrolysate is then clarified using vacuum filtration, and the resulting product is further refined by combining carbon and ion-exchange techniques. The refined hydrolysate can be used as feed for HFS production or processed to high dextrose corn syrup and crystalline or liquid dextrose.

4.3.1.2   Isomerization

The next major step in producing HFS is the isomerization of dextrose into fructose. Various chemical methods have been attempted using carbonates and diluted hydroxides to isomerize dextrose into fructose. But, these chemicals were reported to produce unacceptably high color and flavor with low fructose yield (Hanover and White, 1993). Moreover, fructose molecules become more susceptible to degradation under such conditions. Efforts have also been made to enzymatically isomerize dextrose into fructose, but were hampered due to the complexity of the biochemical pathway linking the two sugars and the expense in regenerating essential co-factors. Later, Akabori et al. (1952) discovered an enzyme named glucose isomerase that bears the ability to catalyze isomerization of dextrose into fructose without requirement of any co-factor regeneration. Takasaki and Tanabe (1971) further upgraded economics of enzyme catalysis by using an immobilized glucose isomerase system. Today, glucose isomerase from various sources like *B. coagulans*, *Flavobacterium arborescens*, *Actinoplanes missouriensis*, and *Streptomyces* sp. is commercially used for immobilization technology. Several parameters such as feed purity, solids, temperature, pH, oxygen level, magnesium content, and reaction time must be regulated carefully to maximize the product yield from glucose isomerase (Table 4.5). Ion-exchange treatment and efficient refining are required to lower calcium levels and to remove any insoluble material that could cause inactivation of glucose isomerase. Feed solids level should be controlled between 40–50% range. Low solids level may cause microbial contamination, while high solids level increase the chances of carrier pore clogging and further reduction in product yield. Temperature must be controlled at 55–61°C to prevent enzyme deactivation and microbial cross-contamination. Glucose isomerase stability is highly pH dependent. Generally, the pH range of 7.5–8.2 is recommended for isomerase action. Maintaining the oxygen level is also an important factor to maintain enzyme activity. Deaeration of dextrose feed to remove

TABLE 4.5 Commercially Available Glucose Isomerase Enzymes and Their Recommended Operating Conditions

Manufacturer	Source	Operating Conditions		
		pH	Temperature (°C)	Solids (%)
Enzyme Bio-Systems	*Streptomyces olivochromogenes*	7.8–8.6	55–60	35–50
Genencor International	*Streptomyces rubiginosus*	7.6	54–62	40–50
Gist-Brocades	*Actinoplanes missouriensis*	7.5	58–60	40–45
Solvay Enzyme Products	*Microbacterium arborescens*	7.5	< 60	40–50
Novo Nordisk	*Bacillus coagulans*	7.8–8.3	59–61	40–45
Novo Nordisk	*Streptomyces murinus*	7.5	50–60	35–45

excess oxygen or addition of SO_2 can prove helpful in maintaining enzyme conformation (Volkin and Klibanov, 1989). Magnesium is an important co-factor for glucose isomerase activation and stabilization. Generally, 0.5–5 mM magnesium level is recommended to counteract the adverse effects of added calcium. Other co-factors like cobalt and iron also enhance glucose isomerase activity and stability. Moreover, long reaction time is also desirable to achieve fructose yield between 42–45%. Refining by carbon and ion-exchange treatment of 45% fructose product containing 51–54% dextrose is carried out to remove traces of salts and other impurities, color, and off-flavors. A storage temperature of 27–32°C is sustained to prevent dextrose crystallization (Hebeda, 1993).

4.3.1.3   Fractionation

A higher content of fructose in HFS is required to match its sweetness equivalent to sucrose at amounts typically used in food and beverages. Chromatographic separation technologies using activated carbon or iron oxide have been employed to increase fructose concentration, but dextrose: fructose efficiency is very poor and costly. Later, Mitsubishi Chemical Industries separated fructose from dextrose and other products by passing 45% HFS through a cationic adsorbent containing calcium groups on the basis of greater affinity of fructose for the calcium salt form. Fructose was retained in the column, while the stream predominantly comprising nonfructose saccharide (80–90% dextrose and 5–10% fructose) can be collected. The higher saccharide fraction is again recycled to the saccharification and isomerization process. This high fructose fraction with water yields HFS product with 80–90% fructose and 7–19% dextrose. Using this technology, Japan and US industries also started manufacturing HFS by the late 1970s. By late 1984, HFS became predominant sweetener in colas.

For various commercial purposes, the crystalline form of fructose is required. The 90% of fructose obtained after fractionation is used as a substrate to crystallize fructose. Several solvent and aqueous-based systems have been applied for fructose crystallization. But today, the aqueous system is most successfully used for the crystallization process due to its several advantageous features like low solvent expense and no requirement of solvent disposal. Batch and continuous systems have been devised with pre-programmed cooling, concentrating, warming, and re-cooling to attain an efficient crystallization. The resultant crystals are then harvested and washed to remove traces of saturated fructose solution from the surface of the crystals (Hanover and White, 1993).

4.3.2 SINGLE-STEP ENZYMATIC HYDROLYSIS OF INULIN FOR HFS PRODUCTION USING INULINASES

Enzymatic hydrolysis of inulin using inulinases is becoming a prompt process for HFS production because as high as 95% of fructose yield can be achieved by this method in a batch or continuous system. Inulinases are fructofuranosyl hydrolases produced by a wide array of microbial sources, viz. bacteria, yeast, and fungi (Table 4.6). Among various sources, *Bacillus* sp., *Streptomyces* sp., *Kluyveromyces* sp., *Pichia* sp., *Aspergillus* sp., and *Penicillium* sp. are some of the common potent inulinase producers (Singh and Chauhan, 2016).

4.3.2.1   HFS Production in a Batch System

Exoinulinase (EC 3.2.1.80), a member of glycoside hydrolase family 32 (GH 32), is a fructofuranosyl hydrolase that degrades β-(2, 1) linkages of inulin from its non-reducing end to produce fructose units. Most of the inulolytic enzymes from various microbial sources are thermophilic in nature that produce fructose at elevated temperature such as above 50°C (Tomotani and Vitolo, 2007). Inulin is an inexpensive and abundant substrate for single-step enzymatic hydrolysis for HFS production using inulinases. Various reports are available in literature on the hydrolysis of pure inulin or inulin-rich plant materials for HFS production using whole cells (Table 4.7), soluble enzyme (Table 4.8), and immobilized inulinase in a batch system (Table 4.9) under a wide range of temperature (28–60°C) and pH (4.5–6.0) conditions. The batch system can be run for a number of cycles that exclusively depends upon enzyme residual activity and structural conformation. A maximum of 55 cycles have been reported in a batch system for HFS production using inulinases. De Andrade et al. (1992) constantly obtained approximately 70% of fructose recovery after hydrolyzing Jerusalem artichoke tubers extract by using whole cells of *Cladosporium cladosporioides* for 11 cycles. Whole cells of a recombinant inulinase-secreting *Saccharomyces cerevisiae* strain have also been reported to give high fructose productivity after hydrolyz-

TABLE 4.6 Inulinase-Producing Microorganisms

Moulds	Yeasts	Bacteria
Aspergillus aureus	*Candida macedoniensis*	*Alternaria alternata*
Aspergillus awamori	*Candida kefyr*	*Arthrobacter globiformis*
Aspergillus candidus	*Candida kutaonensis*	*Arthrobacter ureafaciens*
Aspergillus clavatus	*Candida pseudotropicalis*	*Bacillus polymyxa*
Aspergillus ficuum	*Debaryomyces cantarellii*	*Bacillus stearothermophilus*
Aspergillus fischeri	*Debaryomyces phaffii*	*Bacillus subtilis*
Aspergillus flavus	*Hansenula beijerinckii*	*Bifidobacterium sp.*
Aspergillus foetidus	*Hansenula polymorpha*	*Brevibacillus centrosporus*
Aspergillus fumigatus	*Kluyveromyces cicerisporus*	*Clostridium acetobutylicum*
Aspergillus fumigatus	*Kluyveromyces fragilis*	*Criptococcus aureus*
Aspergillus glaucus	*Kluyveromyces lactis*	*Clostridium thermoautotrophicum*
Aspergillus nidulans	*Kluyveromyces marxianus*	*Clostridium thermosuccinogenes*
Aspergillus niger	*Kluyveromyces marxianus var. bulgaricus*	*Escherichia coli*
Aspergillus nivius	*Pichia fermentan*	*Flavobacterium multivorum*
Aspergillus oryzae	*Pichia guilliermondii*	*Geobacillus stearothermophilus*
Aspergillus tamari	*Pichia polymorpha*	*Marinimicrobium sp.*
Aspergillus terreus	*Rhodotorulaa glutinis*	*Nocardiopsis sp.*
Aspergillus tubingensis	*Saccharomyces fermentati*	*Paenibacillus sp.*
Aspergillus wentti	*Saccharomyces cerevisiae*	*Pseudomonas sp.*
Botryosphaeria sp.	*Saccharomyces flozenzani*	*Sphingomonas sp.*
Chrysosporium pannorum	*Saccharomyces rosei*	*Staphylococcus sp.*
Cladosporium cladosporioides	*Schizosaccharomyces pombe*	*Streptomyces griseus*
Cladosporium phoenicis	*Schwanniomyces allucius*	*Streptomyces rochei*
Fusarium oxysporum	*Torulopsis colliculosa*	*Xanthomonas campestris* pv. *phaseoli*
Fusarium roseum	*Zygosaccharomyces bailii*	*Xanthomonas oryzae*
Geotrichum candidum	-	-
Monascus sp.	-	-
Panaeolus papillonaceus	-	-
Penicillium citrinum	-	-

TABLE 4.6   (Continued)

Moulds	Yeasts	Bacteria
Penicillium expansum	-	.
Penicillium glaucum	-	-
Penicillium italicum	-	-
Penicillium jancze-wskii	-	-
Penicillium luteum	-	-
Penicillium purpuro-genum	-	-
Penicillium rubrum	-	-
Penicillium rugulo-sum	-	-
Penicillium subrube-scens	-	-
Penicillium trzebin-skii	-	-
Rhizopus delemar	-	-
Scytalidium acidophi-lum	-	-
Sterigmatocystis nigra	-	-
Thielavia terrestris	-	-
Ulocladium atrum	-	-

ing Jerusalem artichoke tuber extract (Yu et al., 2011). Synergistic effect of exo- and endoinulinase has also been reported for HFS production. Recently, the exoinulinase gene (rEXINU) from *Kluyveromyces marxianus* and the endoinulinase gene (rENINU) from *Aspergillus niger* cloned and expressed in *Yarrowia lipolytica* have been observed to show synergistic effect for hydrolyzing inulin in a tuber meal of Jerusalem artichoke. The maximum degree of synergistic effect for fructose production was obtained with a molar ratio of rEXINU to rENINU at 1:1 (Liu et al., 2016). This kind of synergistic effect of enzymes can be considered as a potential driving force for various industrial applications. Soluble inulinase from *A. niger* has also been reported to give higher fructose productivity from pure chicory inulin hydrolysis (Mutanda et al., 2009). Inulinase from other sources like *Candida guilliermondii* (Sirisansaneeyakul et al.,

TABLE 4.7   HFS Preparation from Inulin Using Whole Cells

Inulinase Source	Substrate	Hydroly-sis Time (h)	Number of cycles	Fructose Productiv-ity (g/L)	Reference
Clado-sporium cladospo-rioides	Jerusalem artichoke tubers extract	16	11	5	De Andrade et al. (1992)
	Jerusalem artichoke tubers extract	2.5	12	N/A	Ferreira et al. (1991)
Saccha-romyces cerevisiae[a]	Jerusalem artichoke tubers extract	96	1	≈83	Yu et al. (2011)
	Inulin	50	1	≈19	Brevnova et al. (1998)
	Inulin	50	1	≈55	Brevnova et al. (1998)
	Inulin	40	1	≈85	Brevnova et al. (1998)

N/A: Not available.

[a]Recombinant cells of *Saccharomyces cerevisiae* were used for hydrolysis.

2007), *K. marxianus* (Singh et al., 2007a; García-Aguirre et al., 2009), and *Fusarium oxysporium* (Gupta et al., 1989) have also been reported for HFS production in a batch system.

Currently, a lot of work deals with the development of immobilization technology for HFS production. In the modern food industry, enzymes immobilized on insoluble matrices offer a number of technological advantages over soluble precursors, such as they are heterogeneous molecules and can be easily separated from the reaction medium; the reaction can be stopped any time, and hence, the catalyst can be re-used repeatedly; and controlled catalyzed reaction is possible, specifically in fermenters.

Moreover, the product does not get contaminated by the enzyme fraction. Covalent binding, encapsulation, cross-linking, and adsorption methods have been used for the immobilization of biocatalyst on various support/ matrices like ion-exchange resin KU-2 and ion-exchange fiber VION KN-1 (Holyavka et al., 2016), octadecyl substituted nanoporous silica (Karimi et al., 2016), chitin (Gill et al., 2006), chitosan beads (Yewale et al., 2013;

TABLE 4.8 HFS Preparation from Inulin Using Soluble Inulinase

Inulinase Source	Substrate	Hydrolysis Time (h)	Number of Cycles	Fructose Productivity (g/L)	Reference
Aspergillus ficuum	Pure inulin (Chicory roots)	48	1	106.4	Mutanda et al. (2009)
A. niger	Artichoke tubers extract	3	1	N/A	Cruz et al. (1998)
	Chicory tubers extract	3	1	N/A	Cruz et al. (1998)
	Dahlia inulin extract	3	1	N/A	Cruz et al. (1998)
	Pure inulin (Chicory roots)	25	1	37.5	Sirisansaneeyakul et al. (2007)
	Kuth (Saussurea lappa) root powder extract	3	1	35	Viswanathan and Kulkarni (1995)
A. niger and Kluyveromyces marxianus (mixed culture)	Jerusalem artichoke tubers extract	4	1	N/A	Öngen-Baysal and Sukan (1996)
Candida guilliermondii	Pure inulin (Chicory roots)	25	1	35.3	Sirisansaneeyakul et al. (2007)
Fusarium oxysporium	Chicory root tubers extract	8	1	N/A	Gupta et al. (1989)
K. marxianus	Agave juice	48	1	N/A	García-Aguirre et al. (2009)

TABLE 4.8 (Continued)

Inulinase Source	Substrate	Hydrolysis Time (h)	Number of Cycles	Fructose Productivity (g/L)	Reference
	Pure inulin	4	1	43.6	Singh et al. (2007a)
	Asparagus racemosus root extract	4	1	41.3	Singh et al. (2007a)
Kluyveromyces marxianus var. bulgaricus	Pure inulin	24	1	653	Manzoni and Cavazzoni (1992)
		24	1	691	
	Fresh Jerusalem artichoke tubers extract	24	1	773	Manzoni and Cavazzoni (1992)
		24		754	
	Dried Jerusalem artichoke tubers extract	24		839	Manzoni and Cavazzoni (1992)
		24		857	

N/A: Not available.

TABLE 4.9 HFS Preparation from Inulin Using Immobilized Inulinase in a Batch System

Inulinase Source	Substrate	Hydrolysis Time (h)	Immobilization Matrix/Suport	Number of Cycles	Fructose Productivity (g/L/h)	Reference
Aspergillus ficuum	Jerusalem artichoke tubers extract	10	Chitin	N/A	77.5	Kim and Rhee (1989)
Aspergillus sp.	Pure inulin (Chicory roots)	28	Sepabeads	N/A	9.6	Ricca et al. (2010)
A. niger	Pure inulin (Chicory roots)	24	Sol-gel	29	47±4	Santa et al. (2011)
	Pure inulin (Chicory roots)	24	Magnetic sol-gel particles	29	46±2	Santa et al. (2011)
Fusarium oxysporum	Chicory root tubers extract	NS	DEAE-Cellulose	3	N/A	Gupta et al. (1992)
Kluyveromyces fragilis	Jerusalem artichoke tubers extract	20	2-Aminoethyl-cellulose	N/A	34	Kim and Byun (1982)
Kluyveromyces marxianus	Jerusalem artichoke tubers extract	10	Agar gel	9	≈ 40	Bajpai and Margaritis (1985a)
Kluyveromyces marxianus[a]	Jerusalem artichoke tubers extract	7	Open pore gelatin matrix	10	42	Bajpai and Margaritis (1985b)
	Jerusalem artichoke tubers extract	20	Alginate beads	16	24	Parekh and Margaritis (1986)
	Pure inulin	4	Duolite A568	55	40.2	Singh et al. (2007b)
	Asparagus racemosus root extract	4	Duolite A568	55	39.2	Singh et al. (2007b)

N/A: Not available.

[a]Immobilized whole cells were used for hydrolysis.

Missau et al., 2014), and montmorillonite (Coghetto et al., 2012). Some low-cost supports like activated carbon, diatomite, hen eggshell, amberlite, and porous silica have also been reported for inulinase immobilization for HFS production (Paula et al., 2007). Singh et al. (2007a) immobilized inulinase from *K. marxianus* on Duolite A568 for HFS production. For 55 batch cycles, immobilized inulinase was able to give high fructose productivity (40.2 g/L) from *Asparagus racemosus* root extract hydrolysis, which validates the good candidature of immobilized biocatalyst for large-scale applications.

4.3.2.2 HFS Production in a Continuous System

Continuous reactors also termed as flow reactors carry the reactant material as a flowing stream at one point, allow the reaction to take place, and remove the products from another point, thus maintaining the equal flow rate of reactants and products. At the industrial level, several types of continuous reactors like tubular reactors, fixed bed reactors, fluid bed reactors, and continuous stirred tank reactors are used. Continuous system possesses various advantages over batch system, such as possibility of using high reactant concentration due to superior heat transfer capability, coherent quality of product and its concentration and lesser energy requirement for raw material and product storage. A number of continuous systems have been reported for the preparation of HFS (Table 4.10). Packed-bed reactors are the most commonly studied continuous system reactors, while the use of a tubular and miniaturized reactor system is also gaining momentum due to several advantages like low requirement for consumables and man-power, high mass and heat transfer, improved parallelization, automation, and the possibility of scaling-out instead scaling-up. Holyavka et al. (2016) designed a tubular continuous reactor system for the effective splitting of pure inulin or inulin content of *H. tuberosus* by inulinase from *K. marxianus* immobilized on an ion-exchange fiber VION KN-1. They observed almost 2 times increase in hydrolysis efficiency. The inner walls of a silicone tube coated with amino groups and inulinase and subsequently immobilized using the cross-linking method has also been reported as an efficient continuous system for HFS production (Ribeiro and Fernandes, 2013). Thus, from these observations, it can be stated that a miniaturized reactor and immobilized enzyme system has been devised to provide better platform for bioconversion systems. High

TABLE 4.10 HFS Preparation from Inulin Using Immobilized Inulinase in a Continuous System

Inulinase Source	Substrate	Immobilization Matrix/ Support	System Operation (days)	Fructose Productivity (g/L/h)	Reference
Aspergillus sp.	Pure inulin (Chicory roots)	Polyvinyl alcohol particles	20	146	Anes and Fernandes (2014)
	Pure inulin (Chicory roots)	Sepabeads	7[a]	≈ 5.7	Ricca et al. (2010)
A. niger	Pure inulin (Dahlia tubers)	Amino-Cellulofine beads	45	397.7	Nakamura et al. (1995)
	Pure inulin (Chicory roots)	Silicone tube activated by APTES	5	N/A	Ribeiro and Fernandes (2013)
	Kuth (*Saussurea lappa*) root powder	Chitosan beads	6	68	Yewale et al. (2013)
A. ficuum	Jerusalem artichoke tubers juice	Chitin	14	≈ 61	Kim and Rhee (1989)
A. fumigatus	Pure inulin	Chitin	50	2.7	Gill et al. (2006)
	Pure inulin	ConA-linked silica beads	50	3.4	Gill et al. (2006)
	Pure inulin	QAE-Sephadex	50	2.3	Gill et al. (2006)
Debaryomyces phaffii	Pure inulin	DEAE-Cellulose	> 21	N/A	Guiraud et al. (1981)
Kluyveromyces sp.	Jerusalem artichoke tubers extract	Ionic polystyrene beads	35	199.66	Wenling et al. (1999)
Kluyveromyces fragilis	Jerusalem artichoke tubers extract	2-Aminoethyl-cellulose	12	N/A	Kim and Byun (1982)
Kluyveromyces marxianus	Jerusalem artichoke tubers extract	Agar gel	15	31	Bajpai and Margaritis (1985a)

TABLE 4.10   (Continued)

Inulinase Source	Substrate	Immobilization Matrix/ Support	System Operation (days)	Fructose Productivity (g/L/h)	Reference
Kluyvero- myces marx- ianus[b]	Jerusalem ar- tichoke tubers extract	Open pore gelatin matrix	10	90	Bajpai and Margaritis (1985b)
	Pure inulin	Duolite A568	71	27.4	Singh et al. (2008)
	Asparagus racemosus root extract	Duolite A568	11	49.7	Singh et al. (2008)
Saccha- romyces cerevisiae[c]	Pure inulin (Dahlia roots)	Amino- Cel- lulofine	46	614	Kim et al. (1997)
	Jerusalem ar- tichoke tubers extract	Amino- Cel- lulofine	46	N/A	Kim et al. (1997)

N/A: Not available.

APTES: (3-aminopropyl) triethoxysilasne.

[a]Reactor was operated only for 7 hours.

[b]Immobilized whole cells were used for hydrolysis.

[c]Recombinant exoinulinase from *S. cerevisiae* was used for hydrolysis.

fructose productivity ranging from 2.3–614 g/L/h has been reported from continuous systems. Some of these systems have been observed to run continuously for more than 70 days.

4.4   CONCLUDING REMARKS

Fructose is the sweetest natural sugar commonly found in many fruits and vegetables. Due to its various functional properties and technical superiorities over conventional sucrose, it has become an important component in various food products and beverages. Moreover, it is generally recognized as safe (GRAS) food ingredient. In the past few years, the consumption of HFS has increased tremendously. HFS can be produced enzymatically or by acid hydrolysis of starch/inulin. The acid hydrolysis method is generally not recommended due to the formation of colored hydrolysates that affect the final product quality. Commercially, fructose is synthesized by multienzymatic hydrolysis of starch based on the action of three enzymes, namely α-amylase, glucoamylase, and glucose isomerase. The first two enzymes hydrolyze starch into dextrose, while glucose isomerase catalyzes the isomerization of dextrose into fructose. A maximum of 45% (approx.) of fructose yield can be obtained by this method. HFS can also be prepared by single-step hydrolysis of inulin using inulinase, which is coming up as more promising technology. In this method, exoinulolytic degradation of β-(2,1) linkages of inulin releases fructose units and delivers as high as 95% of fructose yield. The present chapter describes the enzymatic methods for the synthesis of HFS from starch and inulin.

KEYWORDS

- glucoamylase
- high fructose syrup
- inulinase
- inuulin
- isomerase
- starch
- α-amylase

REFERENCES

Abdullah, M., Fleming, I. D., Taylor, M., & Whelan, W. J., (1963). Substrate specificity of the amyloglucosidase of *Aspergillus niger*. *Biochem. J.*, *89*, 35–36.

Akabori, S., Nehara, K., & Muramatsu, I., (1952). Biochemical formation of tetrose, pentose and hexose. *J. Chem. Soc. Jpn.*, *73*, 311.

Anes, J., & Fernandes, P., (2014). Towards the continuous production of fructose syrups from inulin using inulinase entrapped in PVA-based particles. *Biocatal. Agric. Biotechnol.*, *3*, 296–302.

Apolinário, A. C., de Lima Damasceno, B. P. G., de Macêdo Beltrão, N. E., Pessoa, A., Converti, A., & da Silva, J. A., (2014). Inulin-type fructans: a review on different aspects of biochemical and pharmaceutical technology. *Carbohydr. Polym.*, *101*, 368–378.

Bajpai, P., & Margaritis, A., (1985a) Production of high fructose syrup from Jerusalem artichoke tubers using *Kluyveromyces marxianus* cells immobilized in agar gel. *J. Gen. Appl. Microbiol.*, *31*, 305–311.

Bajpai, P., & Margaritis. A., (1985b) Immobilization of *Kluyveromyces marxianus* cells containing inulinase activity in open pore gelatine matrix: 1. Preparation and enzymatic properties. *Enzyme Microb. Technol.*, *7*, 373–376.

Barclay, T., Ginic-Markovic, M., Cooper, P., & Petrovsky, N., (2010). Inulin: a versatile polysaccharide with multiple pharmaceutical and food chemical uses. *J. Excip. Food Chem.*, *1*, 27–50.

Brevnova, E. E., Kozlov, D. G., Efremov, B. D., & Benevolensky, S. V., (1998). Inulinase-secreting strain of *Saccharomyces cerevisiae* produces fructose. *Biotechnol. Bioeng.*, *60*, 492–497.

Bruyn, A., Alvarez, A. P., Sandra, P., & De Leenheer, L., (1992). Isolation and identification of O-β-D-fructofuranosyl-(2→1)-O-β-D-fructofuranosyl-(2→1)-D-fructose, a product of the enzymatic hydrolysis of the inulin from *Cichorium intybus*. *Carbohydr. Res.*, *235*, 303–308.

Coghetto, C. C., Scherer, R. P., Silva, M. F., Golunski, S., Pergher, S. B.C., de Oliveira, D., Oliveira, J. V., & Treichel, H., (2012). Natural montmorillonite as support for the immobilization of inulinase from *Kluyveromyces marxianus* NRRL Y-7571. *Biocatal. Agric. Biotechnol.*, *1*, 284–289.

Cruz, V. D., Belote, J. G., Belline, M. Z., & Cruz, R., (1998). Production and action pattern of inulinase from *Aspergillus niger*-245: hydrolysis of inulin from several sources. *Rev. Microbiol.*, *29*, 301–306.

De Andrade, A. V. M., Ferreira, M. S. S., & Kennedy, J. F., (1992). Selective fructose production by utilization of glucose liberated during the growth of *Cladosporium cladosporioides* on inulin or sucrose. *Carbohydr. Polym.*, *18*, 59–62.

Edelman, J., & Jefford, T., (1968). The mechanism of fructosan metabolism in higher plants as exemplified in *Helianthus tuberosus*. *New Phytol.*, *67*, 517–531.

Ferreira, M. S. S., Andrade, A. V. M. D., & Kennedy, J. F., (1991). Properties of a thermostable nonspecific fructofuranosidase produced by *Cladosporium cladosporioides* cells for hydrolysis of Jerusalem artichoke extract. *Appl. Biochem. Biotechnol.*, *31*, 1–9.

Franck, A., & Levecke, B., (2003). Inulin. In *Ullmann's Encyclopedia of Industrial Chemistry*, Elvers, B., & Hawkins, S., Eds., Wiley-VCH Verlag GmbH & Co. KGaA, Germany, pp. 1–12.

García-Aguirre, M., Saenz-Alvaro, V. A., Rodri´Guez-Soto, M. A., Vicente-Magueyal, F. J., Botello-Alvarez, E., Jimenez-Islas, H., Cardenas-Manriquez, M., Rico-Martinez, R., & Navarrete-Bolanos, J. L., (2009). Strategy for biotechnological process design applied to the enzymatic hydrolysis of agave fructo-oligosaccharides to obtain fructose-rich syrups. *J. Agric. Food Chem.*, *57*, 10205–10210.

Gill, P. K., Manhas, R. K., & Singh, P., (2006). Hydrolysis of inulin by immobilized thermostable extracellular exoinulinase from *Aspergillus fumigatus*. *J. Food Eng.*, *76*, 369–375.

Guiraud, J. P., Demeulle, S., & Galzy, P., (1981). Inulin hydrolysis by the *Debaryomyces phaffii* inulinase immobilized on DEAE cellulose. *Biotechnol. Lett.*, *3*, 683–688.

Gupta, A. K., Kaur, M., Kaur, N., & Singh, R., (1992). A comparison of properties of inulinases of *Fusarium oxysporum* immobilised on various supports. *J. Chem. Technol. Biotechnol.*, *53*, 293–296.

Gupta, A. K., Kaur, N., & Singh, R., (1989). Fructose and inulinase production from waste *Cichorium intybus* roots. *Biol. Wastes*, *29*, 73–77.

Hanover, L. M., & White, J. S., (1993). Manufacturing, composition and applications of fructose. *Am. J. Clin. Nutr.*, *58*, 724S–732S.

Hebeda, R. E., (1993). Starches, sugars, and syrups. In *Enzymes in food processing*; Nagodawithana, T., & Reed, G., Eds.; Elsevier, London, England, pp. 321–346.

Hebeda, R. E., Styrlund, C. R., & Teague, W. M., (1988). A kinetic model of *Bacillus stearothermophilus* α-amylase under process conditions. *Starch/Stärke, 40*, 412–418.

Hofer, K., & Jenewein, D., (1999). Enzymatic determination of inulin in food and dietary supplements. *Eur. Food Res. Technol., 209*, 423–427.

Holyavka, M. G., Evstigneev, M. P., Artyukhov, A. G., & Savin, V. V., (2016). Development of heterogenous preparation with inulinase for tubular reactor systems. *J. Mol. Catal. B: Enzym., 129*, 1–5.

Karimi, M., Habibi-Rezaei, M., Rezaei, K., Moosavi-Movahedi, A. A., & Kokini, J., (2016). Immobilization of inulinase from *Aspergillus niger* on octadecyl substituted nanoporous silica: inulin hydrolysis in a continuous mode operation. *Biocatal. Agric. Biotechnol., 7*, 174–180.

Kato, I., (2000). Antitumour activity of lactic acid bacteria. In *Probiotics*; Fuller, R., & Perdigon, G. Eds.; Kluwer Academic Publishers, Dordrecht, Netherlands, pp. 115–138.

Kaur, N., & Gupta, A. K., (2002). Applications of inulin and oligofructose in health and nutrition. *J. Biosci., 27*, 703–714.

Kim, B. W., Kim, H. W., & Nam, S. W., (1997). Continuous production of fructose-syrups from inulin by immobilized inulinase from recombinant *Saccharomyces cerevisiae*. *Biotechnol. Bioprocess Eng., 2*, 90–93.

Kim, C. H., & Rhee, S. K., (1989). Fructose production from Jerusalem artichoke by inulinase immobilized on chitin. *Biotechnol. Lett., 11*, 201–206.

Kim, W. Y., & Byun, S. M., (1982). Hydrolysis of inulin from Jerusalem artichoke by inulinase immobilized on aminoethylcellulose. *Enzyme Microb. Technol., 4*, 239–244.

Lichtenthaler, F. W., Brust, A., & Cuny, E., (2001). Sugar-derived building blocks. Part 26. Hydrophilic pyrroles, pyridazines and diazepinones from D-fructose and isomaltulose. *Green Chem., 3*, 201–209.

Liu, Y., Zhou, S. H., Cheng, Y. R., Chi, Z., Chi, Z. M., & Liu, G. L., (2016). Synergistic effect between the recombinant exo-inulinase and endo-inulinase on inulin hydrolysis. *J. Mol. Catal. B: Enzym., 128*, 27–38.

Manzoni, M., & Cavazzoni, V., (1992). Hydrolysis of topinambur (Jerusalem artichoke) fructans by extracellular inulinase of *Kluyveromyces marxianus* var. *bulgaricus*. *J. Chem. Technol. Biotechnol., 54*, 311–315.

Mensink, M. A., Frijlink. H. W., van der Voort, M., & Hinrichs, W. L. J., (2015). Inulin, a flexible oligosaccharide I: review of its physicochemical characteristics. *Carbohydr. Polym., 130*, 405–419.

Missau, J., Scheid, A. J., Foletto, E. L., Jahn, S. L., Mazutti, M. A., & Kuhn, R. C., (2014). Immobilization of commercial inulinase on alginate-chitosan beads. *Sustainable Chem. Process., 2*, 13

Mutanda, T., Wilhelmi, B., & Whiteley, C. G., (2009). Controlled production of fructose by an exoinulinase from *Aspergillus ficuum*. *Appl. Biochem. Biotechnol., 159*, 65–77.

Nakamura, T., Ogata, Y., Shitara, A., Nakamura, A., & Ohta, K., (1995). Continuous production of fructose syrups from inulin by immobilized inulinase from *Aspergillus niger* mutant 817. *J. Ferment. Bioeng., 80*, 164–169.

Öngen-Baysal, G., & Sukan, S. S., (1996). Production of inulinase by mixed culture of *Aspergillus niger* and *Kluyveromyces marxianus*. *Biotechnol. Lett., 18*, 1431–1434.

Orthoefer, F. T., (1987). Corn starch modification and uses. In: *Corn: chemistry and technology*, Watson, S. A., & Ramstad, P. E., Eds.; American Association of Cereal Chemists, St. Paul Minnesota, pp. 479–499.

Parekh, S. R., & Margaritis, A., (1986). Application of immobilized cells of *Kluyveromyces marxianus* for continuous hydrolysis to fructose of fructans in Jerusalem artichoke extracts. *Int. J. Food Sci. Technol.*, *21*, 509–515.

Paula, F. C., Cazetta, M. L., Monti, R., & Contiero, J., (2007). Screening of supports for *Kluyveromyces marxianus* var. *bulgaricus* inulinase immobilization. *Curr. Trends Biotechnol. Pharm.*, *1*, 34–40.

Pawan, G. L. S., (1973). Fructose. In: *Molecular Structure and Function of Food Carbohydrates*, Birch, G. G., & Green, L. F., Eds.; Applied Science Publishers, London, pp. 65–80.

Pontes, A. G. O., Silva, K. L., da Cruz Fonseca, S. G., Soares, A. A., de Andrade, Feitosa, J. P., Braz-Filho, R., Romero, N. R., & Bandeira, M. A. M., (2016). Identification and determination of the inulin content in the roots of the Northeast Brazilian species *Pombalia calceolaria* L. *Carbohydr. Polym.*, *149*, 391–398.

Ribeiro, E. M., & Fernanades, P., (2013). Coated-wall mini reactor for inulin hydrolysis. *Curr. Biotechnol.*, *2*, 47–52.

Ricca, E., Calabrò, V., Curcio, S., Basso, A., Gardossi, L., & Iorio, G., (2010). Fructose production by inulinase covalently immobilized on sepabeads in batch and fluidized bed bioreactor. *Int. J. Mol. Sci.*, *11*, 1180–1189.

Roberfroid, M., Gibson, G. R., & Delzenne, N., (1993). Biochemistry of oligofructose, a non-digestible fructo-oligosaccharide: an approach to estimate its caloric value. *Nutr. Rev.*, *151*, 137–146.

Roberfroid, M. B., (2007). Inulin-type fructans: functional food ingredients. *J. Nutr.*, *137*, 2493S–2502S.

Ronkart, S. N., Blecker, C. S., Fourmanoir, H., Fougnies, C., Deroanne, C., Van Herck, J. C., & Paquot, M., (2007). Isolation and identification of inulooligosaccharides resulting from inulin hydrolysis. *Analytica Chimica Acta*, *604*, 81–87.

Rose, V., (1804). About a peculiar substance. *Neues Algem. Chem.*, *3*, 217–219 (in German).

Santa, G. L. M., Bernardino, S. M. S. A., Magalhães, S., Mendes, V., Marques, M. P. C., Fonseca, L. P., & Fernandes, P., (2011). From inulin to fructose syrups using sol-gel immobilized inulinase. *Appl. Biochem. Biotechnol.*, *165*, 1–12.

Shiraishi, F., Kawakami, K., & Kusunoki, K., (1985). Kinetics of condensation of glucose into maltose and isomaltose in hydrolysis of starch by glucoamylase. *Biotech. Bioeng.*, *27*, 498–502.

Singh, R. S., & Chauhan, K., (2016). Production, purification, characterization and applications of fungal inulinases, *Curr. Biotechnol.*, *5*, doi: 10.2174/221155010566616051214 2330 (E-pub ahead of print).

Singh, R. S., Dhaliwal, R., & Puri, M., (2007a) Partial purification and characterization of exoinulinase from *Kluyveromyces marxianus* YS-1 for preparation of high-fructose syrup. *J. Microbiol. Biotechnol.*, *17*, 733–738.

Singh, R. S., Dhaliwal, R., & Puri, M., (2007b) Production of high fructose syrup from *Asparagus* inulin using immobilized exoinulinase from *Kluyveromyces marxianus* YS-1. *J. Ind. Microbiol. Biotechnol.*, *34*, 649–655.

Singh, R. S., Dhaliwal, R., & Puri, M., (2008). Development of a stable continuous flow immobilized enzyme reactor for the hydrolysis of inulin. *J. Ind. Microbiol. Biotechnol.*, *35*, 777–782.

Singh, R. S., & Singh, R. P., (2010). Production of fructooligosaccharides from inulin by endoinulinases and their prebiotic potential. *Food Technol. Biotechnol.*, *48*, 435–450.

Sirisansaneeyakul, S., Worawuthiyanan, N., Vanichsriratana, W., Srinophakun, P., & Chisti, Y., (2007). Production of fructose from inulin using mixed inulinases from *Aspergillus niger* and *Candida guilliermondii*. *World J. Microbiol. Biotechnol.*, *23*, 543–552.

Takasaki, Y., & Tanabe, O., (2007). Enzyme method for converting glucose in glucose syrups to fructose. U. S. Patent 3616221.

Taper, H. S., & Roberfroid, M., (1999). Influence of inulin and oligofructose on breast cancer and tumor growth. *J. Nutr.* (Suppl), *129*, 1488–1491.

Teague, W. M., & Brumm, P. J., (1992). Commercial enzymes for starch hydrolysis products. In: *Starch Hydrolysis Products: Worldwide Technology, Production and Applications*, Schenck, F. W., & Hebeda, R. E., Eds., VCH Publishers, New York, pp. 45–77.

Tomotani, E. J., & Vitolo, M., (2007). Production of high-fructose syrup using immobilized invertase in a membrane reactor. *J. Food Eng.*, *80*, 662–667.

van de Wiele, T., Boon, N., Possemiers, S., Jacobs, H., & Verstraete, W., (2007). Inulin-type fructans of longer degree of polymerization exert more pronounced *in vitro* prebiotic effects. *J. Appl. Microbiol.*, *102*, 452–460.

Viswanathan, P., & Kulkarni, P. R., (1995). Properties and application of inulinase obtained by fermentation of costus (*Saussurea lappa*) root powder with *Aspergillus niger*. *Die Nahrung*, *39*, 288–294.

Volkin, D. B., & Klibanov, A. M., (1989). Mechanism of thermos inactivation of immobilized glucose isomerase. *Biotech. Bioeng.*, *33*, 1104–1111.

Wenling, W., Le Huiying, W. W., & Shiyuan, W., (1999). Continuous preparation of fructose syrups from Jerusalem artichoke tuber using immobilized intracellular inulinase from *Kluyveromyces* sp. Y-85. *Process Biochem.*, *34*, 643–646.

Wolfgang, W., & Südzucker, A. G. M., (2004). Fructose. In *Electronic Ullmann's Encyclopedia of Industrial Chemistry*, Hubert, P., Ed.; Wiley-VCH GmbH & Co. KGaA, Germany, doi: 10.1002/14356007.a12_047.pub2.

Yewale, T., Singhal, R. S., & Vaidya, A. A., (2013). Immobilization of inulinase from *Aspergillus niger* NCIM 945 on chitosan and its application in continuous inulin hydrolysis. *Biocatal. Agric. Biotechnol.*, *2*, 96–101.

Ylikahari, R. H., Kahonen, M. T., & Hassinen, I., (2013). 1972 Modification of metabolic effects of ethanol by fructose. *Acta Med. Scand. Suppl.*, *542*, 141–150.

Yu, J., Jiang, J., Ji, W., Li, Y., & Liu, J., (2011). Glucose-free fructose production from Jerusalem artichoke using a recombinant inulinase-secreting *Saccharomyces cerevisiae* strain. *Biotechnol. Lett.*, *33*, 147–152.

CHAPTER 5

GAMMA IRRADIATION OF FOOD PROTEINS: RECENT DEVELOPMENTS IN MODIFICATION OF PROTEINS TO IMPROVE THEIR FUNCTIONALITY

MUDASIR AHMAD MALIK, HARISH KUMAR SHARMA, and CHARANJIV SINGH SAINI

Department of Food Engineering and Technology, Sant Longowal Institute of Engineering and Technology Longowal, Sangrur, Punjab, 148106, India, E-mail: malikfet@gmail.com

CONTENTS

5.1   INTRODUCTION

Functional properties of proteins affect their behavior during food preparation, processing, storage, and consumption and contribute to the quality and sensory attributes of food system (Liu et al., 2012). These properties are influenced by the physicochemical characteristics of proteins, interaction with protein and nonprotein components, and environmental conditions of the food system. Proteins differ in physiochemical properties (molecular weight, net surface charge, conformation, amino acid composition, and sequence) and hence differ in functional properties. Solubility, swelling power, water retention capacity, gelling capacity, foaming properties, emulsifying, and fat binding properties are the important functional properties of proteins in food applications (Zayas, 1997). Therefore, the main functional properties of proteins are to form and/or stabilize network (gels and films), emulsion, foams, and sols. The different mechanisms underlying functional properties are hydration, protein surface activity, and alteration in protein structure. Hydration characteristic of protein affects its water/oil solubility, wettability, and thickening characteristics. Protein surface activity (hydrophilicity, hydrophobicity, and net charge) affects protein-lipid film formation, foaming, and emulsion activities. The changes in protein structure such as shape, size, amino acid sequence, and composition can affect viscosity, adhesiveness, elasticity, aggregation, and gelation (Speroni et al., 2009).

Proteins show broad spectrum of functional properties due to their heterogeneous structure and interaction with other food components (Table 5.1). The unique structure of proteins and their ability to interact with other food

TABLE 5.1   Functional Properties Performed by Proteins in Food Systems

Property	Mechanism	Applications in food system
Solubility	Protein solvation	Beverages
Absorption and binding of water	Hydrogen bonding and entrapment of water	Meat, sausages
		Cakes breads
Viscosity	Water binding and thickening	Gravies, soups
Gelation	Formation of protein matrix and setting	Meats, cheese, curd
Elasticity	Hydrophobic bonding and disulfide bonds	Meats, bakery
Emulsification	Fat emulsion formation and stabilization	Sausages, bologna, soups, cakes
Fat absorption	Binding of free fat	Meats, doughnuts, sausages
Flavor binding	Adsorption and entrapment	Simulates meats, bakery
Foaming	Formation of stable film	Whipped toppings, angle cakes, chiffon dessert

components make their molecular basis for protein functionality. Protein molecular weight and shape, diversity in structure, conformation and charge distribution, along with the primary structure, i.e., the number of amino acids and their disposition in the polypeptide chains, affect their functional properties. Apolar amino acids in protein influence their conformation, solubility, hydration, and gelation properties. Polar amino acid content in protein enhances the electrostatic interactions that stabilize the globular proteins. Globular protein having more charged amino acids toward the surface can increase the solubility, swelling, and hydration properties of proteins. However, proteins with compact globular structure such as soy protein have poor emulsifying capability than other protein such as milk protein (Molina et al., 2001; Roesch and Corredig, 2003). Some specific proteins have been exploited for many years for their unique functional properties, e.g., wheat gluten in leavened breads for their viscoelastic properties, egg albumin in cakes for aerating properties, and papain for tenderization of meat and manufacturing of cheese. In addition, proteins perform many functions in foods, and they need to be recognized and manipulated to cater the demand of the food industry. Protein functional properties are affected by and vary with the source of protein (zein, gluten, casein, egg albumin, whey), method(s) of extraction (precipitation and drying), concentration, modification (chemical, enzymatic, alkaline, or acid hydrolysis) and various environmental conditions such as temperature, ionic strength, and pH.

5.2 MODIFICATION OF FOOD PROTEINS

Deliberate modification of food proteins to change their properties has been increasingly popular area for researchers from the last couple of decades (Banach et al., 2013; Huang et al., 2013). Modification is an important tool that eliminates the undesirable elements (toxin) and undesirable properties, removes the antinutritional factors, and improves the nutritional properties and protects the proteins against the process-induced modifications such as glycation, phosphorylation, enzymatic modification, and acylation (Li et al., 2010). Such changes are related to the alteration to protein structure, such as partial unfolding, protein dissociation, decreased enthalpy of denaturation, altered isoelectric point, and increased hydrophobicity (Gruener and Ismond, 1997a, 1997b). Functional properties of protein are associated with their secondary and tertiary structure. Hence, alteration in the protein structure during food processing by chemical, enzymatic, or physical means have direct consequences on protein functionality.

In chemical modification, the chemical reagent of interest is covalently crosslinked with the protein molecule. The behavior of protein is largely changed by these molecules and imparts the characteristics to the proteins that are little present in the unmodified proteins such as changing the solubility of protein as required, improving foaming properties, and inhibiting the aggregation of proteins. Noncovalent forces that determine the protein conformation are altered during chemical modification in a manner that results in desirable functional and structural changes (Kester and Richardson, 1984). Deamination of proteins is also a chemical modification that has a strong effect on protein charge and isoelectric point (Mirmoghtadaie et al., 2009). The chemical modification is not used at a large scale due to its adverse health effects such as toxicity, food allergy, and impaired nutrition. An alternative approach to chemical crosslinking is enzymatic hydrolysis of protein to smaller peptides with protease enzymes that results in improvement of nutritional and textural characteristics of proteins (MacLeod et al., 1988; Periago et al., 1998). Amino acids are not destroyed during enzymatic hydrolysis of proteins, as it is considered as a mild transformation and is specific and allows for controlled processing (Ribotta et al., 2012). An enzymatic treatment to change the protein functionality is not often cost-effective (Beilen and Li, 2002).

Physical protein modification may involve thermal treatment, irradiation, high pressure, ultrasound treatment, or a texturization procedure. Thermal treatment changes the degree of protein denaturation and aggregation and hence contributes to change in protein functionality, which can lead to alterations in the properties of the finished product (Livney et al., 2003; Nicorescu et al., 2008; Raikos, 2010). High pressure disrupts the intermolecular, hydrophobic, and electrostatic interactions with important consequences for tertiary and quaternary structures (Ker and Toledo, 1992; Subirade et al., 1998). Ultrasound involves the usage of mechanical waves at a frequency above the threshold of human hearing (>16 kHz) and can be divided into high frequency (100 kHz–1 MHz) and low frequency (16–100 kHz) (Soria and Villamiel, 2010). In extrusion, the combined effect of high pressure and shear and high temperature affects the integrity of the protein molecule and makes the molecule accessible to the enzymes and hence increases the digestibility (Surówka and Zmudzinski, 2004; MacLean et al., 1983).

5.3 IRRADIATION

Food irradiation is a process where food is exposed to ionizing radiations such as gamma rays emitted from the radio isotopes ^{60}Co and ^{137}Cs or high-energy electrons and X-rays produced by machine sources (Farkas, 2006). Radiation processing improves the microbiological safety and/or enhances shelf-life of raw and processed food materials without affecting the nutritional quality (WHO, 1999). Irradiation has been acknowledged as a valuable technique to improve the nutritional value of foods (Diehl, 2002). Microbial decontamination of foods by irradiation has already become legally accepted in several countries (Anon, 1991). In 1980s, irradiation of foods started on the industrial scale after the joint FAO/IAEA/WHO expert committee approved the use of 10 kGy as an overall average dose for food (WHO, 1981). US-Food and Drug Administration (FDA) in 1981 came up with the recommendation that foods irradiated at 50 kGy or less is considered safe for human (FDA, 1981), and animal consumption. The application of irradiation in foods could substantially improve food safety and benefit the food availability, inhibit the spread of pathogenic microbes in human food and animal feed, control foodborne illnesses, and make safe and nutritious foods available (WHO, 1999). In addition, irradiation finds application in sprout inhibition, increase in juice yield, delay in ripening, improvement in rehydration, reduction in antinutritional components, improvement in functional and nutritional properties, and improvement in the overall quality of plant produce (Bhat et al., 2007; Bhat and Sridhar, 2008; Bhat et al., 2011). While carrying out irradiation on foods, food components especially carbohydrates, lipids, proteins, and vitamins are subsequently changed. Therefore, the structure of these components can be modified by gamma irradiation for better utilization in the food industry.

Modification of food protein by gamma irradiation is not a common practice in food processing, but it is recognized as an essential tool to change the microstructure and physical performance of proteins. Gamma irradiation of proteins is carried out to alter the protein structure to obtain desired functional properties at the industrial level and expand the area of fabricated foods. The effects of radiation on food proteins are either due to the direct action of radiation on proteins or indirect action through the reactive intermediates formed by radiolysis of water (Diehl, 1995). In direct action, the gamma rays lose their energy while passing through material. At the same time, this energy is absorbed by proteins, which leads to ionization or excitation of atoms or molecules, which subsequently results in the chemical changes in proteins (Stewart, 2001). The direct action is usually observed when the protein is in solid state. However, when the protein is in liquid state, indirect action dominates in which proteins are mainly modified by free radicals generated by hydrolysis of water (Figure 5.1) due to irradiation. Exposure of protein to these generated radicals results in the random and nonrandom fragmentation of the protein (Kempner, 1993). Gamma irradiation affects proteins by promoting reactions such as protein-protein association, cleavage of peptide and disulfide bonds, deamination, and association of heterocyclic and aromatic residues. All these changes are associated with the production of free radicals (Urbain, 1986; Cho et al., 1999). The modifications of primary structure of proteins by gamma irradiation can result in distortions of secondary and tertiary structures of the protein by generation of hydroxyl and superoxide anion radicals (Figure 5.1) (Davies and Delsignore, 1987). During irradiation, changes depend upon various factors

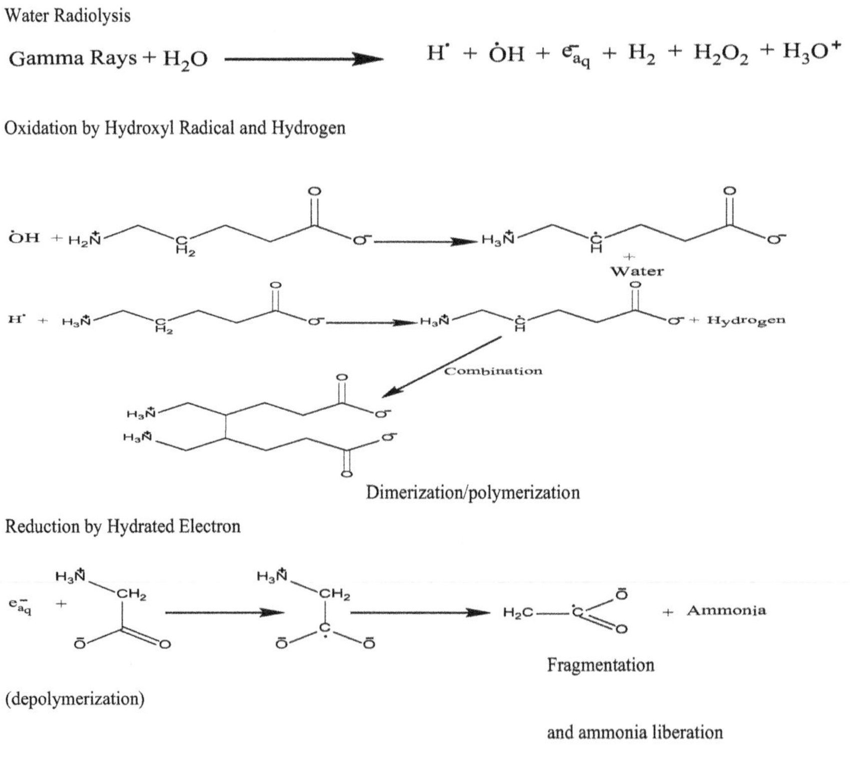

Water Radiolysis

$$\text{Gamma Rays} + H_2O \longrightarrow H^{\cdot} + \overset{\cdot}{O}H + e^-_{aq} + H_2 + H_2O_2 + H_3O^+$$

Oxidation by Hydroxyl Radical and Hydrogen

Reduction by Hydrated Electron

(depolymerization)

and ammonia liberation

FIGURE 5.1   Radiolysis of water due to gamma irradiation and its effect on protein structure.

such as dose, hydration state, pH, and temperature (Simic, 1978) and the presence or absence of oxygen (Davies and Delsignore, 1987; Giroux and Lacroix, 1998). A number of food proteins have been modified by irradiation to change the protein functionality, protein structure, protein digestibility, and protein reactivity, such as red kidney bean proteins (Dogbevi et al., 2000), myoglobin (Lee and Song, 2002), bovine serum albumin (BSA) (Gaber, 2005), cowpea protein (Abu et al., 2006), whey protein (Chawla et al., 2009), egg white (Song et al., 2009), and fish myofibrillar proteins (Shi et al., 2015).

5.4 EFFECT OF GAMMA IRRADIATION ON FUNCTIONAL PROPERTIES OF PROTEINS

Effect of gamma irradiation on functional properties of proteins depends on the irradiation dose and more on the state of protein in which it is subjected to irradiation. Irradiation carried out on protein in the solid form has lesser effect than that in the solution form due to the generation of free radicals by hydrolysis of water molecules. Bovine and porcine plasma protein solutions and powders were irradiated by Lee et al. (2003) with 1, 5, 7, and 10 kGy at room temperature using ^{60}Co. The study reported significant changes in the molecular weight of irradiated protein solutions due to formation of oxygen radicals generated by the radiolysis of water that led to the aggregation and degradation of proteins. However, in the solid form, no significant changes were observed because of the absence of oxygen radicals by the radiolysis of water, which led to minimal damage to the protein structure. A similar observation was reported by Clark et al. (1992) on egg white. Egg white sample that were irradiated in an aqueous solution showed major changes in viscosity, but minor changes were observed in sample that were irradiated in freeze-dried powder form.

Functional properties like solubility depend upon the molecular weight profile of proteins. At low dose of irradiation, reduction in the molecular weight of fragmented proteins can increase the solubility of proteins. However, solubility can decrease at higher doses due to the aggregation of proteins (Lee et al., 2003). Protein-related functional properties are affected by different factors such as protein denaturation, conformation, structure, size, charge, and amino acid sequence and composition of the protein molecules (Zayas, 1997). Dogbevi et al. (2000) irradiated red kidney beans with 2,

4, and 8 kGy and studied the effect of irradiation on its proteins. Irradiation increased the solubility of extracted proteins due to the deamination (removal of the amide group) of proteins by irradiation. Improvement in functional properties of proteins after deamination has also been reported (Shih and Kalmar, 1987).

Spray dried egg white powder was irradiated in the range of 2–16 kGy (Clark et al., 1992). Irradiation caused a marked increase in foamability and foam stability for all the irradiated samples (2–16 kGy). The minimum thickness and drainage rate of air suspended thin liquid films that were stabilized by egg white protein was determined interferometrically. With the increase in the irradiation dose, there was a decrease in rate of film thinning. Improvement in foaming ability of egg white protein has been reported (Song et al., 2009). Irradiation lowered the surface tension by reducing the viscosity of protein solution, created a large surface area that is essential for foaming. Viscosity of egg white is positively related with foaming ability as incorporation of air is obstructed at high viscosity (Yang and Baldwin, 1995).

Camillo and Sabato (2004) studied the viscosity behavior of whey protein dispersion by combining thermal treatment with irradiation. A dispersion of whey protein in water (5 and 8% w/v) and a dispersion containing protein and glycerol at the ratio of 1:1 and 2:1 (protein:glycerol) were subjected to thermal plus irradiation treatment and thermal plus vacuum and N_2 plus irradiation treatment. The dispersions were heated to 60 °C for 20 min and irradiated at 0, 5, 15, and 25 kGy. For both 5 and 8% protein dispersion, thermal treatment combined with gamma irradiation resulted in significant ($p<0.05$) increase in viscosity with an increase in the irradiation dose. However, for protein and glycerol dispersion, the irradiation treatment did not enhance the viscosity to a large extent.

Proteins isolated from gamma-irradiated flour and pastes of cowpea were analyzed for functional properties (Abu et al., 2006). Nitrogen solubility index (NSI) of the protein extracted from both flour and paste decreased with an increase in the irradiation dose. Reduction in solubility was observed, which may be due to protein denaturation followed by production of free radicals during irradiation. Denaturation of the globular protein leads to the exposure of nonpolar sites, which were previously buried inside, with the resultant increase in hydrophobicity (Zayas, 1997) and subsequent decrease in solubility. Gamma irradiation promotes the protein-protein crosslinking that leads to the increase in molecular weight and hence could be a reason for decreased solubility of both flour and paste proteins. The other functional properties like emulsion capacity and oil-binding capacity were increased with an increase in the dose of irradiation. Increase in emulsion capacity and oil-binding capacity of irradiated samples was observed, which may be due to an increase in hydrophobicity of proteins mainly due to denaturation. Foaming capacity of flour proteins decreased at 2 kGy but not at higher doses, while the foaming capacity of paste protein decreased only at 50 kGy. The gels prepared from the irradiated proteins showed an improvement in gel strength at 10- and 50-kGy dose. The increase in gel strength was mainly due to the crosslinking of proteins after irradiation.

Choi et al. (2015) irradiated the chicken salt soluble proteins with 0, 3, 7, and 10 kGy. No significant difference in fat, protein, ash content and sarcoplasmic protein solubility was observed. A significant reduction in water binding capacity was observed with an increase in the irradiation dose. The color of gels showed a significant difference upon treatment with gamma irradiation. The lightness and yellowness were higher in the irradiated sample than in the unirradiated sample and increased with an increase in the gamma irradiation level. Gels prepared from the treated samples showed higher total protein solubility and myofibrillar protein solubility than control. The textural properties of the gel prepared from the treated sample showed lower values than the treated samples. The decrease in textural properties was attributed to the denaturation effect of gamma irradiation on the chicken salt soluble protein.

5.5 EFFECT OF GAMMA IRRADIATION ON STRUCTURAL PROPERTIES OF PROTEINS

Gamma irradiation causes irreversible changes in the protein structure at the molecular level, involving breakage of covalent bonds of polypeptide chains (Table 5.2). The structural modification of proteins by gamma irradiation involves crosslinking, fragmentation, aggregation, and oxidation by oxygen radicals generated by the hydrolysis of water (Filali-Mouhim et al., 1997; Cho and Song, 2000). The primary structure of proteins could be modified by hydroxyl and superoxide anion radicals generated by radiation, which would result in the distortion of secondary and tertiary structure of proteins (Davies and Delsignore, 1987). Protein exposed to oxygen radicals show both random and nonrandom fragmentation (Kempner, 1993). Several factors affect the fragmentation of protein in an aqueous solution, such as local conformation of amino acids in the protein, its accessibility to the product of

TABLE 5.2   Effect of Gamma Irradiation on the Structure of Proteins

Protein	Irradiation dose	Effect on protein structure	Reference
Ovalbumin and ovomu-coid	0.5, 1, 5 and 10 kGy	Disruption in ordered structure of proteins	Moon and Song (2001)
		Caused initial fragmentation and subsequent aggregation	
Myoglobin	0.5, 1, 5 and 10 kGy	Decrease in α-helix content and concurrent increase in aperiodic structure	Lee and Song (2002)
		Increases the emission intensity that was excited at 280 nm	
		Crosslinking and aggregation of protein molecules	
Porcine and bovine blood plasma protein	1, 5, 7 and 10 kGy	Decrease in ordered structure of protein	Lee et al. (2003)
		Caused initial fragmentation and subsequent aggregation of proteins	
Whey protein	0, 5, 15 and 25 kGy	Increase viscosity of protein dispersion	Camillo and Sabato (2004)
Bovine serum albumin	0.5, 1 and 5 kGy	Disruption in ordered structure	Gaber (2005)
		Transformation of β-turn into β-sheets	
		Fragmentation and aggreagation of protein molecules	
Cow pea Pro-tein isolate	2, 10 and 50 kGy	Decrease denaturation temperature and enthalpy	Abu et al. (2006)
		Increase in protein molecular weight	
		and protein-protein crosslinking	
Fish myo-fibrillar proteins	2, 4, 6, 8 and 10 kGy	Increase in surface hydrophobicity	Shi et al. (2015)
		Decrease in sulphydryl content	

water radiolysis, and the primary amino acid sequence (Filali-Mouhim et al., 1997). The production of free radicals during irradiation affects the protein structure by promoting different reactions such as cleavage of peptide and disulfide bonds and protein-protein association (Cho et al., 1999).

Moon and Song (2001) studied the effect of gamma irradiation (0, 0.5, 1, 5, and 10 kGy) on the molecular properties of ovalbumin and ovomucoid by examining the effect on secondary and tertiary structure and molecular weights of proteins. The secondary structure of protein determined by circular dichroism (CD) spectra was disrupted after irradiation. Both α-helix and β-sheet contents were decreased with a parallel increase in unordered structural contents. Irradiation caused changes in the CD spectra of proteins mainly due to breakdown of covalent bonds and formation of aggregates, indicating the disruption of ordered structural content and production of unnatural products. The effect of irradiation on molecular properties was further examined by the measurement of fluorescence emission intensity after irradiation. Emission intensity of both ovalbumin and ovamucoid was quenched after irradiation. It was also observed that irradiation at low dose caused breakdown of polypeptide chains and as a result led to the formation of low-molecular-weight molecules. However, at higher doses, high-molecular-weight aggregates were formed that increased significantly with an increase in dose. Proline residues are believed to be the targets for chain scission during irradiation (Schuessler and Schilling, 1984). Wolff et al. (1986) reported that direct oxidation of proline residues could result in cleavage of peptide bonds. During irradiation, the generation of inter-protein crosslinking reactions, electrostatic and hydrophobic interactions, and formation of disulfide bonds are responsible for the formation of high-molecular-weight aggregates (Cho and Song, 1999).

Gaber (2005) irradiated (0, 0.5, 1, and 5 kGy) BSA and studied the effect on it molecular properties. The ordered structure of protein molecules was disrupted due to irradiation, along with degradation, crosslinking, and aggregation of polypeptide chains. The emission intensity of BSA, which reflects its tertiary structure, decreased with an increase in the irradiation dose. The decrease in emission intensity was due to the change in local environment around the tryptophan and tyrosine residues (Moon and Song, 2001). Irradiation caused transformation of β-turns into β-sheets as indicated by Fourier transform infrared spectroscopy (FTIR). The effect of irradiation on the molecular weight of protein was determined by light scattering analysis, and there was a marked decrease in the molecular weight when exposed to 1 and 5 kGy dose. A similar study earlier reported by Lee and Song (2002) on irradiation (0, 0.5, 1, 5, and 10 kGy) of myoglobin showed similar results. However, the emission intensity of myoglobin increased, when excited at 280 nm, with an increase in the irradiation dose, unlike BSA. This difference in emission intensity could be attributed to the difference in tertiary structure of myoglobin and BSA.

The presence of water during gamma irradiation has a significant effect on the structure of proteins. The proteins from gamma-irradiated (0, 2, 5, 10, and 50 kGy) cowpea flour and paste showed that peak denaturation temperatures and enthalpies of both flour protein and paste protein decreased progressively with an increase in the irradiation dose, except for paste protein at 50 kGy (Abu et al., 2006). This general decrease indicates that gamma irradiation denatured proteins partially in a dose-dependent manner. However, at 50-kGy dose, the increase in peak denaturation temperature and comparatively stable enthalpy of the paste protein might be due to the extensive protein crosslinking at this dose. Irradiation had more pronounced effect on the paste protein than on the flour protein. This effect was attributed to the generation of free radicals by radiolysis of water, which facilitates more protein denaturation. Size exclusion HPLC revealed that with the increase in the gamma irradiation dose, protein molecular weight was increased, probably due to crosslinking of proteins. Reducing SDS-PAGE of both flour and paste proteins revealed that disulfide bonds contribute less in the radiation-induced protein-protein crosslinking. Improvement in protein-protein crosslinking for increasing the molecular weight has also been reported.

Fish myofibrillar proteins were irradiated at the dose of 0, 2, 4, 6, 8, and 10 kGy at the dose rate of 0.5 kGy/h (Shi et al., 2015). The hydrophobicity of myofibrillar proteins increased initially up to 6 kGy and decreased after 8-kGy dose. The initial increase in hydrophobicity was attributed to the exposure of some buried hydrophobic groups by irradiation. However, with an increase in the irradiation dose, the exposed hydrophobic groups were again covered by protein crosslinking with the resultant decrease in hydrophobicity. The protein structure was further influenced by the formation of S-S bonds by the oxidation of SH groups after irradiation. The sulfhydryl content decreased with an increase in irradiation dose, which reflects the structural changes in myofibrillar protein after irradiation. High molecular myosin heavy chain content of irradiated samples decreased in the SDS-PAGE patterns.

The irradiation of sugar-amino acid solution resulted in the formation of Maillard reaction products (MRPs) (Chawla et al., 2009). The decrease in the amino group and lactose in a dose-dependent manner suggested the formation of glycated proteins. The development of fluorescence property and an increase in UV absorbance in a dose-dependent manner was observed. The increase in absorbance (A_{420}) and hunter color values upon irradiation validated the formation of brown pigments (MRPs). Irradiation increased the iron chelating abilities and reducing power of MRPs and was able to scavenge superoxide anion radicals and hydroxyl under in vitro condition.

5.6   EFFECT OF GAMMA IRRADIATION ON PROTEIN FILMS

The application of biopolymers in the food industry has witnessed increased interest in recent years. One of the possible applications is edible and bio-degradable packaging materials. Protein is one of the biopolymers used for the preparation of edible and biodegradable films. The films prepared from proteins as a biodegradable packaging material have environmental compat-ibility and offer longer shelf-life of food materials (Vachon et al., 2000). Mechanical properties of foods can be improved by the use of protein films, which will also minimize the loss of volatile components (McHugh and Krochta, 1994). These films act as a selective barrier for moisture, solute, and gas migration and as well as the carrier of food additives. Protein films have low carbon dioxide and oxygen permeability and higher water vapor perme-ability than plastic films (Krochta and Mulder-Johnston, 1997), which is due to the hydrophilic nature of proteins that have a tendency to absorb water (Ouattara et al., 2002). Therefore, it is necessary to search for a new process and composition that permit to obtain better products. Crosslinking agents or ionizing radiation have been tried to improve functional properties of protein films (Vachon et al., 2000). Application of gamma irradiation to proteins to induce crosslinking was found as an efficient method for enhancement of both mechanical and barrier properties of the edible films and coatings pre-pared from sodium caseinates and calcium alone or pooled with some globu-lar proteins (Le-Tien et al., 2000; Sabato et al., 2001; Lacroix et al., 2002). Molecular properties of proteins are modified by superoxide and hydroxyl anion radical generated by the irradiation of the film-forming solution, and hence, the structure of protein films can be altered due to the formation of linkages during irradiation (Garrison, 1987). The effect of gamma irradia-tion on different physicochemical properties of protein films is summarized in Table 5.3.

The ordered structure of gluten molecules was disrupted along with aggregation and degradation of polypeptide chains due to application of gamma irradiation. Viscosity of the film-forming solution was decreased below 16 kGy due to the breakage of polypeptide chains and increased above 32 kGy because of aggregation of proteins molecules (Lee et al.,

TABLE 5.3   Effect of Gamma Irradiation on the Properties of Protein Films

Protein	Irradiation dose	Effect on films	Reference
Milk Protein (sodium casein-ate and calcium caseimate)	4, 8, 12, 15 and 20 kGy	Improved the mechanical strength by incorporating the crosslinks between the protein molecules	Brault et al. (1997)
Milk (caseine)	4 and 64 kGy	Decreased solubility	Mezgheni et al. (2000)
Soy protein isolate	32 kGy	Improved puncture strength and puncture deformation	Sabato et al. (2001)
Soy protein isolate and whey protein isolate	4-128 kGy	Improved the puncture strength	Lacroix et al. (2002)
		Improved the water vapour permeability	
		Slowdown the biodegradation of materials used	
Soy protein isolate	0, 4, 16, 32 and 50 kGy	Decreases viscosity of film forming solution.	Lee et al. (2005b)
		Decreases water vapour permeability	
		Increases color value, b	
Wheat gluten	10, 20 and 40 kGy	Increases tensile strength	Lee et al. (2005a)
		Decreases elongation at break	
Cacium Caseinate and whey protein isolates	32 kGy	Improved mechanical and barrier properties	Ciesla et al. (2006)
Corn Zein	0, 10, 20, 30 and 40 kGY	Decreases viscosity of film forming solution	Soliman et al. (2009)
		Improved color and surface density	
		Improved mechanical properties and water vapour permeability	
Calcium caseinate	32 kGy	Increases molecular weight of proteins and decreases solubility	Vachon et al. (2000)
		Puncture strength was improved significantly	
		Pore size was denser	

2005a). Tensile strength of the irradiated films was higher than that of the nonirradiated films. It was found that tensile strength increased by 1.5-fold at 50 kGy. The elongation of the irradiated films was lower than that of the non-irradiated films. After irradiation, water vapor permeability was reduced by 29%. Gamma irradiation can be used as a useful tool to improve the functional properties of protein films by inducing the crosslinking between the protein molecules.

In another study on soy protein, a decrease in viscosity was observed because of cleavage of the polypeptide chains. Water vapor permeability was decreased by 13%, while tensile strength of the film was increased by two times after gamma irradiation. The irradiated film solutions showed more yellow films than the nonirradiated ones. Smooth and glossy surfaced films were obtained from the irradiated solutions (Lee et al., 2005b)

Gamma irradiation was also tried to enhance the structure and functional characteristics of films prepared from mixture of proteins. Lacroix et al. (2002) applied gamma irradiation to produce sterilized crosslinked films from inconsistent variable concentrations of whey protein in the form of whey protein concentrate (WPC) and whey protein isolate (WPI) and calcium caseinate (CAS) or mixture of WPI with soy protein isolate (SPI). Gamma irradiation improved the mechanical properties (tensile strength) of all crosslinked films. X-ray diffraction analysis revealed that gamma irradiation modified the conformation of protein to a certain extent where they are adopted to more ordered and stable structure. Gamma-irradiated films had improved water vapor permeability and microstructure. It was also observed that gamma irradiation slowed down the biodegradation of crosslinked films.

Gamma irradiation has been used to impart crosslinking between the protein polymers and improve the chemical stability and water vapor permeability of milk protein films (Ouattara et al., 2002). The films were prepared using calcium caseinate (CAS) with various proportions of WPC and WPI. A significant reduction in water vapor permeability was observed. Caseinate and whey protein exerted a synergistic effect on water vapor permeability. A higher combined effect was obtained for CAS:WPI (25:75) formulation with permeability values of 1.38 and 2.07 gmm/m^2 d mm Hg for irradiated and unirradiated samples, respectively. A substantial increase in the molecular weight of protein components in the film-forming solutions was observed after gamma irradiation. The major fraction was $\geq 10 \times 10^6$ Da for the irradiated film-forming solutions, compared to $<0.2 \times 10^6$ Da for native unirradiated solutions. The improvement in physiochemical properties of the edible film prepared from CAS, WPI, and glycerol were also reported by Vu et al. (2012). The creation of crosslinked proteins after gamma irradiation enhanced the physiochemical properties of films. Further, it was reported that there was an increase in β-sheet structural content and decrease in α-helix and unordered fraction in gamma-irradiated crosslinked proteins. Puncture strength and water vapor permeability were also improved after gamma irradiation. In the study of Micard et al. (2000) on gamma irradiation of gluten films, an increase in water vapour permeability (WVP), tensile strength and decrease in elongation was observed at 10 kGy dose; however, at a higher dose (20 and 40 kGy), a reduced effect was observed. The increase in mechanical properties may be due to crosslinking of polypeptide chains during gamma irradiation. Reduction in insoluble glutenin polymers was responsible for decrease in elongation.

The film-forming properties of corn zein protein were evaluated after exposure to gamma irradiation in different doses (0, 10, 20, 30, and 40 kGy at the rate of 10.5 kGy/h) (Soliman et al., 2009). It was observed that α-helical content was decreased with a concomitant increase in β-sheet content with an increase in the gamma irradiation dose. Viscosity of the film-forming solution was decreased significantly at 10 kGy and increased at 20 kGy and then decreased again with a further increase in the irradiation dose at 40 kGy. The variation of viscosity with gamma irradiation was attributed to the unfolding, aggregate formation, and crosslinking or scission of the polypeptide chains by gamma irradiation. It was further revealed that water barrier, color, and appearance of the zein films can be improved by the application of gamma irradiation to the film-forming solution.

5.7 CONCLUDING REMARKS

Protein functionality is dependent on the condition prevailing during food processing and preparation. The literature available clearly indicates that the protein functionality could be modified by varying the environment around the protein molecule as well as by the application of some external forces. It is evident that gamma irradiation is a promising method of protein modification and changes protein functionality. The change in structure due to gamma irradiation has direct consequences on protein functionality. The improvements in protein functionality like solubility, foaming, and emulsion properties after gamma irradiation could be used to prepare number of products having industrial significance. The gamma irradiation technique could also be applied to other food systems to achieve the desired functionality.

KEYWORDS

- edible films
- functional properties
- gamma irradiation
- modification
- protein isolate
- structural properties

REFERENCES

Abu, J. O., Muller, K., Duodu, K. G., & Minnaar, A., (2006). Gamma irradiation of cowpea (*Vigna unguiculata* L. Walp) flours and pastes: Effects on functional, thermal and molecular properties of isolated proteins. *Food Chem.*, *95*, 138–147.

Anon. Regulations in the Field of Food Irradiation., (1991) *IAEA-TECDOC-585. LAEA*, Vienna.

Banach, J. C., Lin, Z., & Lamsal, B. P., (2013). Enzymatic modification of milk protein concentrate and characterization of resulting functional properties. *LWT – Food Sci. Technol.*, *54*, 397–403.

Beilen, J. B., & Li, Z., (2002). Enzyme technology: An overview. *Curr. Opin. Biotech.*, *13*(4), 338–344.

Bhat, R., Ameran, S. B., Voon, H. C., Karim, A. A., & Tze, L. M., (2011). Quality attributes of starfruit (*Averrhoa carambola* L.) juice treated with ultraviolet radiation. *Food Chem.*, *127*, 641–644.

Bhat, R., & Sridhar, K. R., (2008). Nutritional quality evaluation of electron beam-irradiated lotus (Nelumbo nucifera) seeds. *Food Chem.*, *107*, 174–184.

Bhat, R., Sridhar, K. R., & Tomita-Yokotani, K., (2007). Effect of ionizing radiation on antinutritional features of velvet bean seeds (Mucuna pruriens). *Food Chem.*, *103*, 860–866.

Brault, D., D'Aprano, G., & Lacroix, M., (1997). Formation of freestanding sterilized edible films from irradiated caseinates. *J. Agri. Food Chem.*, *45*, 2964–2969.

Camillo, A., & Sabato, S. F., (2004). Effect of combined treatments on viscosity of whey dispersions. *Radiat. Phys. Chem.*, *71*, 103–106.

Chawla, S. P., Chander, R., & Sharma, A., (2009). Antioxidant properties of maillard reaction products obtained by gamma-irradiation of whey proteins. *Food Chem.*, *116*, 122–128.

Cho, S. Y., & Song, K. B., (1999). Effect of γ-irradiation on the physicochemical properties of soy protein isolate and whey protein concentrate. *Korean J. Food Sci. Technol.*, *31*(6), 1488–1494.

Cho, Y., & Song, K. B., (2000) Effect of γ-irradiation on the molecular properties of BSA and b-lactoglobulin. *J. Biochem. Mol. Biol.*, *33*, 133–137.

Cho, Y., Yang, J. S., & Song, K. B., (1999). Effect of ascorbic acid and protein concentration on the molecular weight profile of bovine serum albumin and β-lactoglobulin by γ-irradiation. *Food Res. Int.*, *32*, 515–519.

Choi, Y. S., Kim, Y. W., Hwang, K. E., Song, D. H., Jeong, T. J., Seo, K. W., Kim, Y. B., & Kim, C. J., (2015). Effects of gamma irradiation on physicochemical properties of heat-induced gel prepared with chicken salt-soluble proteins. *Radiat. Phys. Chem.*, *106*, 16–20.

Ciesla, K., Salmieri, S., & Lacroix, M., (2006). γ-Irradiation influence on the structure and properties of calcium caseinate-whey protein isolate based films. Part 2. Influence of polysaccharide addition and radiation treatment on the structure and functional properties of the films. *J. Agric. Food Chem.*, *54*, 8899–8908.

Clark, D. C., Kiss, I. F., Wilde, P. J., & Wilson, D. R., (1992). The effect of irradiation on the functional properties of spray-dried egg white protein. *Food Hydrocolloids*, *5*(6), 541–548.

Davies, K. J. A., & Delsignore, M. E., (1987). Protein damage and degradation by oxygen radicals III: modification of secondary structure and tertiary structure. *J. Bio. Chem.*, *262*(20), 9908–9913.

Diehl, J. F., (1995). *Chemical effects of ionizing radiation.* In *Safety of irradiated foods*, Diehl, J. F., Ed., Marcel Dekker: New York, pp. 43–88.

Diehl, J. F., (2002). Food irradiation: past, present and future. *Radiat. Phys. Chem.*, *63*, 211–215.

Dogbevi, M. K., Vachon, C., & Lacroix, M., (2000) Effect of gamma irradiation on the microbiological quality and on the functional properties of proteins in dry red kidney beans (Phaseolus vulgaris). *Radiat. Phys. Chem.*, *57*, 265–268.

Farkas, J., (2006). Irradiation for better foods. *Trends Food Sci. Techol.*, *17*, 148–152.

FDA., (1981). Irradiation in the production, processing, and handling of food; final rule. *21 CFR Part 179. Federal Register*, *51*, 13376–13399

Filali-Mouhim, A.;Audette, M., St-Louis, M., Thauvette, L., Denoroy, L., Penin, F., Chen, X., Rouleau, N., Le, C. J., Rossier, J., Potier, M., & Le, M. M., (1997). Lysozyme fragmentation induced by γ-radiolysis. *Int. J. Radiat. Biol.*, *72*, 63–70.

Gaber, M. H., (2005). Effect of γ-irradiation on the Molecular Properties of Bovine Serum Albumin. *J. Biosci. Bioeng.*, *100*, 203–206.

Garrison, W. M., (1987). Reaction mechanism in the radiolysis of peptides, polypeptides, and proteins. *Chem. Rev.*, *87*, 381–398.

Giroux, M., & Lacroix, M., (1998). Nutritional adequacy of irradiated meat – a review. *Food Res. Int.*, *31*(4), 257–264.

Gruener, L., & Ismond, M. A. H., (1997a). Effects of acetylation and succinylation on the physicochemical properties of the canola 12S globulin: Part I. *Food Chem.*, *6*(3), 357–363.

Gruener, L., & Ismond, M. A. H., (1997b). Effects of acetylation and succinylation on the functional properties of the canola 12S globulin. *Food Chem.*, *60*(4), 513–520.

Huang, X., Kanerva, P., Salovaara, H., Loponen, J., & Sontag-Strohm, T., (2013). Oxidative modification of a proline-rich gliadin peptide. *Food Chem.*, *141*, 2011–2016.

Kempner, E. S., (1993). Damage to proteins due to the direct action of ionizing radiation. *Q. Rev. Biophys.*, *26,* 27–48.

Ker, Y. C., & Toledo, R. T., (1992). Influence of shear treatments on consistency and gelling properties of whey protein isolate suspensions. *J. Food Sci.*, *57*, 82–90.

Kester, J. J., & Richardson, T., (1984). Modification of whey proteins to improve functionality. *J. Dairy Sci., 67,* 2757–2774.

Kinsella, J. E., (1982). Relationships between structure and functional properties of food proteins. In: *Food Proteins;* Fox, P. F., Condon, J. J., Ed., Applied Science Publishers Ltd: England, pp. 51–58.

Krochta, J. M., & Mulder-Johnston, D. C., (1997). Edible and biodegradable polymer films: challenge and opportunities. *Food Techol., 51,* 61–74.

Kuana, Y. H., Bhata, R., Patrasb, A., & Karim, A. A., (2013). Radiation processing of food proteins-A review on the recent developments. *Trends Food Sci. Techol., 30,* 105–120.

Lacroix, M., Le, T. C., Ouattara, B., Yu, H., Letendre, M., Sabato, S. F., Mateescu, M. A., & Patterson, G., (2002). Use of gamma irradiation to produce films from whey, casein and soya proteins: structure and functional characteristics. *Radiat. Phys. Chem., 63,* 827–832.

Lee, M., Lee, S., & Song, K. B., (2005b). Effect of γ-irradiation on the physicochemical properties of soy protein isolate films. *Radiat. Phys. Chem. 72,* 35–40.

Lee, S., Lee, S., & Song, K. B., (2003). Effect of gamma-irradiation on the physicochemical properties of porcine and bovine blood plasma proteins. *Food Chem., 82,* 521–526.

Lee, S. L., Lee, M. S., & Song, K. B., (2005a). Effect of gamma-irradiation on the physicochemical properties of gluten films. *Food Chem., 92,* 621–625.

Lee, Y., & Song, K. B., (2002). Effect of γ-irradiation on the Molecular Properties of Myoglobin. *J. Biochem. Mol. Biol., 35,* 590–594.

Le-Tien, C., Letendre, M., Ispas-Szabo, P., Mateescu, M. A., Delmas-Paterson, G., Yu, H. L., & Lacroix, M., (2000). Development of biodegradable films from whey proteins by cross-linking and entrapment in cellulose. *J. Agric. Food Chem., 48,* 5556–5575.

Li, C. P., Enomoto, H., Hayashi, Y., Zhao, H., & Aoki, T., (2010). Recent advances in phosphorylation of food proteins: A review. *LWT – Food Sci. Techol., 43,* 1295–1300.

Liu, J., Ru, Q., & Ding, Y., (2012). Glycation a promising method for food protein modification: Physicochemical properties and structure, a review. *Food Res. Int., 49,* 170–183.

Livney, Y. D., Corredig, M., & Dalgleish, D. G., (2003). Influence of thermal processing on the properties of dairy colloids. *Curr. Opin. Colloid Interface Sci., 8,* 359–364.

MacLean, W. C. J., Romana, G. L. D., Gatanaduy, A., & Graham, G. G., (1983). The effect of decortications and extrusion on the digestibility of sorghum by preschool children. *J. Nutr., 113*(10), 2071–2077.

MacLeod, G., Ames, J., & Betz, N. L., (1988). Soy flavor and its improvement. Crit. Rev. Food Sci. Nutr., *27,* 219–400.

McHugh, T. H., & Krochta, J. M., (1994). *Permeability properties of edible films.* In *Edible coating and films to improve food quality*; Baldwin, E., Krochta, J. M., Nisperos-Carriedo, M. O., Ed., Technomic Publishing: Lancaster, pp. 135–186.

Mezgheni, E., Vachon, C., & Lacroix, M., (2000). Bacterial use of biofilms cross-linked by gamma irradiation. *Radiat. Phy. Chem., 58,* 203–205.

Micard, V., Belamri, R., Morel, M. H., & Guilbert, S., (2000). Properties of chemically and physically treated wheat gluten films. *J. Agric. Food Chem., 48,* 2948–2953.

Mirmoghtadaie, L., Kadivar, M., & Shahedi, M., (2009). Effects of succinylation and deamidation on functional properties of oat protein isolate. *Food Chem., 114*(1), 127–131.

Molina, E., Papadopoulou, A., & Ledward, D. A., (2001). Emulsifying properties of high pressure treated soy protein isolate and 7S and 11S globulins. *Food Hydrocolloids, 15,* 263–269.

Moon, S., & Song, K. B., (2001). Effect of γ-irradiation on the molecular properties of ovalbumin and ovomucoid and protection by ascorbic acid. *Food Chem.*, *74*, 479–483.

Nicorescu, I., Loisel, C., Vial, C., Riaublanc, A., Djelveh, G., Cuvelier, G., & Legrand, J., (2008). Combined effect of dynamic heat treatment and ionic strength on denaturation and aggregation of whey proteins– Part I. *Food Res. Int.*, *41*, 707–713.

Ouattara, B., Canh, L. T., Vachon, C., Mateescu, M. A., & Lacroix, M., (2002). Use of γ-irradiation cross-linking to improve the water vapor permeability and the chemical stability of milk protein films. *Radiat. Phys. Chem.*, *63*, 821–825.

Periago, M. J., Vidal, M. L., Ros, G., Rincón, F., Martínez, C., Lopez, G., Rodrigo, J., & Martínez, I., (1998). Influence of enzymatic treatment on the nutritional and functional properties of pea flour. *Food Chem.*, *63*, 71–78.

Raikos, V., (2010). Effect of heat treatment on milk protein functionality at emulsion interfaces. A review. *Food Hydrocolloids.*, *24*, 259–265.

Ribotta, P. D., Colombo, A., & Rosell, C. M., (2012). Enzymatic modifications of pea protein and its application in proteinecassava and corn starch gels. *Food Hydrocolloids*, *27*, 185–190.

Roesch, R. R., & Corredig, M., (2003). Texture and microstructure of emulsions prepared with soy protein concentrate by high-pressure homogenization. *Food Sci. Techol. 36*, 113–124.

Sabato, S. F., Ouattara, B., Yu, H., Aprano, D. G., Le-Tien, C., Mateescu, M. A., & Lacroix, M., (2001). Mechanical and barrier properties of cross-linked soy and whey protein based films. *J. Agric. Food Chem.*, *49*, 1397–1403.

Schuessler, H., & Schilling, K., (1984). Oxygen effect in the radiolysis of proteins. *Int. J. Radiat. Biol.*, *45*, 267–281.

Shi, Y., Li, R., Tu, Z., Ma, D., Wang, H., Huang, X., & He, N., (2015). Effect of γ-irradiation on the physicochemical properties and structure of fish myofibrillar proteins. *Radiat. Phys. Chem.*, *109*, 70–72.

Shih, F. F., & Kalmar, A. D., (1987). SDS-catalyzed deamidation of oilseed proteins. *J. Agric. Food Chem.*, *35*, 672–675.

Simic, M. G., (1978). Radiation chemistry of amino acids and peptides in aqueous solutions. *J. Agric. Food Chem.*, *26*, 6–14.

Soliman, E. A., Eldin, M. M. S., & Futura, M., (2009). Biodegradable zein-based films: influence of γ-irradiation on structural and functional properties. *J. Agric. Food Chem.*, *57*, 2529–2535.

Song, H., Kim, B., Choe, J., Jung, S., Kim, K., Kim, D., & Jo, C., (2009). Improvement of foaming ability of egg white product by irradiation and its application. *Radiat. Phys. Chem.*, *78*, 217–221.

Soria, A. C., & Villamiel, M., (2010). Effect of ultrasound on the technological properties and bioactivity of food: a review. *Trends Food Sci. Techol.*, *21*(7), 323–331.

Speroni, F., Beaumal, V., De-Lamballerie, M., Anton, M., Añón, M., & Puppo, M., (2009). Gelation of soybean proteins induced by sequential high-pressure and thermal treatments. *Food Hydrocolloids 23*(5), 1433–1442.

Stewart, E. M., (2001). Food irradiation chemistry. In: *Food Irradiation: Principles and Applications*, Molins, R. A., Ed., John Wiley and Sons, Inc: New York, pp. 37–76.

Subirade, M., Loupil, F., Allain, A. F., & Paquin, P., (1998). Effect of dynamic high pressure on the secondary structure of β-lactoglobulin and on its conformational properties as determined by Fourier transform infrared spectroscopy. *Int. Dairy J.*, *8*, 135–140.

Surówka, K., & Zmudzinski, D., (2004). Functional properties modification of extruded soy protein using neutrase. *Czech J. Food Sci.*, *22*, 163–174.

Urbain, W. M., (1986). *Radiation Chemistry of Food Components of Foods*. In: *Food Irradiation*, Urbain, W. M., Ed., Academic Press Inc: London, pp. 37–81.

Vachon, C., Yu, H. L., Yefsah, R., Alain, R., St-Gelais, D., & Lacroix, M., (2000). Mechanical and structural properties of milk protein edible films cross-linked by heating and gamma irradiation. *J. Agric. Food Chem.*, *48*, 3202–3209.

Vu, K. D., Hollingsworth, R. G., Salmieri, S., Takala, P. N., & Lacroix, M., (2012). Development of bioactive coatings based on γ-irradiated proteins to preserve strawberries. *Radiat. Phys. Chem.*, *81*, 1211–1214.

World Health Organization., (1981). *Wholesomeness of irradiated food. Joint FAO/IAEA/ WHO Expert Committee*. WHO Techn Rep Ser 659.

World Health Organization., (1999). *High-dose irradiation: Wholesomeness of food irradiated with doses above 10 kGy. Report of a Joint FAO/IAEA/WHO Study Group*. WHO Tech Rep Ser 890: i–vi: 1–197.

Wolff, S. P., Garner, A., & Dean, R. T., (1986). Free radicals, lipids and protein degradation. *Trends Biochem. Sci.*, *11*, 27–31.

Yang, S. C., & Baldwin, R. E., (1995). Functional properties of eggs in foods. In: *Egg Science and Technology*; Stadelman, W. J., Cotterill, O. J., Ed.; Food Products Press: New York, pp. 405–463.

Zayas, J. F., (1997). *Introduction in Functionality of Proteins in Food*; Zayas, J. F., Ed., Springer-Verlag: Germany, pp. 1–5.

CHAPTER 6

EDIBLE COATINGS: POTENTIAL APPLICATIONS IN FOOD PACKAGING

LOVELEEN SHARMA, CHARANJIV SINGH SAINI, and
HARISH KUMAR SHARMA

*Department of Food Engineering and Technology (FET), Sant
Longowal Institute of Engineering and Technology, Longowal, Sangrur,
Punjab, 148106, India, E-mail: charanjiv_cjs@yahoo.co.in;
profh.sharma27@gmail.com*

CONTENTS

6.1   INTRODUCTION

Edible coating based on biodegradable materials is an area of research and has the potential use in food packaging (Pavlath and Orts, 2009). Edible coatings can render the same effect as modified atmosphere storage, and therefore may be considered as an alternative to enhance the postharvest life of fresh fruits and vegetables (Park, 1999).

Edible coatings are a thin layer of an edible material, which can be applied directly on the food surface and provides a barrier to moisture, oxygen, and solute movement for the food (Park, 1999). Edible coatings can be developed from proteins, polysaccharides, and lipids, alone or in combination and can be a substitute for synthetic plastic for food applications. Edible coatings can be consumed with a product, which is the main advantage over traditional synthetic films. The coatings are prepared from renewable, edible ingredients and hence are expected to degrade more rapidly than polymeric materials. Edible coatings can improve the organoleptic properties of packaged foods from the incorporation of different components such as flavorings, colorings, and sweetening agents. Edible coating is the covering of a consumable material applied and formed directly on the surface of a food product. These coatings improve and control surface conditions of the product. The presence and abundance of the material determine the barrier properties with regard to water vapor, oxygen, carbon dioxide, and lipid transfer in food systems (Guilbert et al., 1996). As edible coatings are generally consumed with the food product, the material used in edible coatings should be generally recognized as safe (GRAS) (Park et al., 1994; Krochta and Mulder-Johnston, 1997) and must conform to the regulations (Guilbert et al., 1996).

The major role of edible coatings is to restrict migration of oxygen, carbon dioxide, moisture, or any other solute materials. The coating should also act as a carrier for the different food additives such as antioxidants or antimicrobials and decrease the decay without any change in the quality of the food product. The modified atmosphere formed by edible coatings protects the food from the moment it is applied until it reaches the final consumer (Riberio et al., 2007). The edible coatings can provide a possibility to enhance the shelf-life of fresh-cut products by developing a semi-permeable barrier to gases and water vapor and decreasing respiration, water loss, and enzymatic browning (Baldwin et al., 1995).

The edible coating can be developed by proteins, polysaccharides, and lipids (Figure 6.1). Polysaccharides utilized for coatings contain starch derivatives, cellulose, pectin derivatives, seaweed extracts, exudate gums, microbial fermentation gums, and chitosan (Krochta and Mulder-Johnston, 1997). These coatings provide excellent barrier properties (oxygen, aroma, and oil) and structural integrity and strength, but are not worthwhile moisture

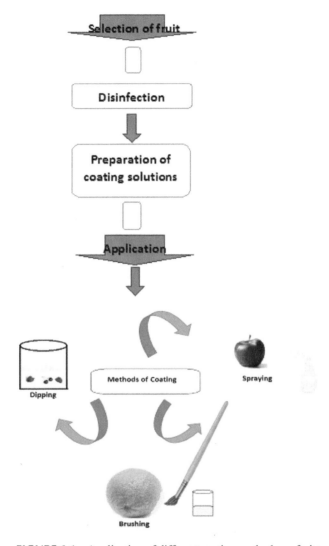

FIGURE 6.1 Application of different coating methods on fruits.

barriers due to their hydrophilic nature (Kester and Fennema, 1986; Krochta, 2002). Tightly packed, ordered hydrogen-bonded network structure and low solubility correspond to the oxygen barrier properties (Banker, 1966).

Protein-based coatings are found to be good oxygen barriers at low relative humidity (RH). Different types of proteins have been used to develop edible coatings. These include whey protein, gelatin, casein, corn zein, wheat gluten, soy protein, mung bean protein, and peanut protein (Gennadois et al., 1993; Bourtoom, 2008; Saglam et al., 2013). Water-soluble coatings were formed from whey protein isolate, whereas the protein is insoluble in the coating from heat-denatured solutions of whey protein isolate (WPI) (Perez-Gago et al., 2005). Whey protein concentrate (WPC) have a protein content generally in the range of 25% and 80%, while WPI has protein content above 90%. Transparent, colorless, flexible, and flavorless coatings with a poor moisture barrier were formed from whey proteins (McHugh and Krochta, 1994; Fairley et al., 1996).

Lipid coatings are good barriers to moisture loss mainly due to their hydrophobic properties. In addition to averting water loss, lipid coatings can decrease respiration and hence increase the shelf-life. These coatings can also bring shiny appearance on fruits and vegetables. A wide range of lipid substances, acetylated monoglycerides, surfactants, and natural waxes can be used for lipid-based coatings (Kester and Fennema, 1986).

Composite coatings can be developed as a stable emulsion or applied as a bi-layer. Lipid forms a second layer over the polysaccharide or protein in bi-layer composite coatings. Bi-layer coatings can be produced in one or two steps and provides the best moisture barriers. Lipid is dispersed and entrapped in the supporting matrix of protein or polysaccharide in case of emulsion composite coatings. Functional properties of coatings are improved by the addition of different additives to produce a biodegradable active packaging material. Substantial research has been performed to attempt natural active compounds as an alternative to synthetic ones, mainly due to safety concerns associated with synthetic active compounds (Gimenez et al., 2012).

6.2   HISTORY OF COATINGS

Edible coatings have been utilized since long to protect foods and prevent moisture loss. China was the first country to apply coatings on citrus in the 12th century (Hadenberg, 1967). It was later used in England to enhance the shelf-life of meat products (Contrereas-Medellin and Labuza, 1981). Coatings have been used as casings by using collagen for sausages (Becker, 1938, 1939), to add shine and prevent water loss from fruits and vegetables (Baldwin, 1994), and as a sugary coating on chocolates and confectionaries (Biquet and Labuza, 1988). Gelatin has been used as a coating material for meat products (Antoniewski et al., 2007). Cellulose, pectin derivatives, starch derivatives, seaweed extracts, exudate gums, chitosan, and microbial fermentation gums have been used as oxygen, aroma, and oil barriers for different food products (Kester and Fennema, 1986; Krochta and Mulder-Johnston, 1997; Krochta, 2002).

6.3 FUNCTIONS OF EDIBLE COATINGS

The recent usage of coatings is for fruits and vegetables to reduce moisture loss, prevent weight loss, improve appearance, and reduce gas transfer rates (Ribeiro et al., 2007; Talens et al., 2012; Aquino et al., 2015; Shah et al., 2016). Fruits and vegetables respire, where the cells consume oxygen and produce carbon dioxide. Fruits also release ethylene that is responsible for ripening and senescence (Wills et al., 1981). Shelf-life of fruits and vegetables can be enhanced by storage in controlled or modified atmospheres of relatively low oxygen and high carbon dioxide, which slows respiration rates resulting in decreased ethylene production and ripening process. Coatings can have the same effect as controlled and modified storages, depending upon their formulations. Coatings reduce gas permeation, developing an internal atmosphere of increased carbon dioxide and decreased oxygen concentration. Water loss can be prevented by hydrophobic coatings (Baldwin et al., 1997).

Coatings may act as carriers of many functional ingredients and enhance the shelf-life of processed foods by inhibiting microbial deterioration, flavors, rancidity, enzymatic browning, off flavors, etc. Such ingredients may include antimicrobial agents, antioxidants, spices, colorants, and neutraceuticals (Salmieri and Lacroix, 2006; Tapia et al., 2008) (Table 6.1).

6.4 COMPONENTS OF EDIBLE COATINGS

The main components of edible coatings are polymers, proteins, and polysaccharides (Park et al., 1994; Guilbert et al., 1996; Li and Barth, 1998). Sometimes, coating from waxes and resins with the inclusion of polymers to form either bilayer or composite coatings are used (Baldwin, 1994). Apart from polymers and waxes, other ingredients also have a prime role in the development of coatings. Plasticizers such as propylene glycol, glycerol,

TABLE 6.1 Application and Effect of Various Edible Coatings on Different Fruits

Edible coatings	Fruit/Vegetable	Effect	Reference
Carnauba Wax	Egg Plant	Shelf life enhanced up to 12 days	Singh et al. (2016)
Locus bean gum	Mandarin	Controls the growth of *Penicillium digitatum* and *Penicillium italicum*	Parafati et al. (2016)
Alginate and Pectin	Raspberry	Controlled microbial spoilage	Guerreiro et al. (2016)
Aloe gel and rosehip oil	Plum	2 fold increase in plum storability	Martinez-Romero et al. (2017)
Basil seed gum	Fresh cut apples	Improved microbial load	Hashemi et al. (2017)
Olive oil processing residues	Apple and strawberry	Improved microbial quality	Khalifa et al. (2016)
Carboxymethyl cellulose, ascorbic acid and calcium	Fresh cut apples	Maintained total anti-oxidant and vitamin C content, improved fruit quality	Saba et al. (2016)
Aloe vera and ascorbic acid	Strawberry	Maintained postharvest quality and reduced microbial load	Sogvar et al. (2016)
κ-Carageenan and tapioca starch with iron and ascorbic acid	Pumpkin	The Ascorbic acid degradation in the tissue was significantly reduced in the pumpkin	Genevois et al. (2016)
Gellan Gum	Mango bars	Improved post-harvest quality	Danalache et al. (2016)
Chitosan and aloe vera	Blueberry fruit	Delayed fungal decay and reduced weight loss	Vieira et al. (2016)
Opuntia *Ficus-indica* mucilage	Hayward kiwi-fruit slices	Maintained mechanical properties and flavour	Allegra et al. (2016)
Chitosan	Black radish	Reduced antibacterial activity	Jovanovic et al. (2016)
Chitosan and ascorbic acid coating	Pomegranate arils	Prolonged the shelf life up to 21 days	Özdemir and Gökmen (2016)
Chitosan and lemon essential oil	Strawberries	Modified the fruit volatile content	Perdones et al. (2016)

or sorbitol are added to control the flexibility and viscosity of coatings. The bulk of coatings are made up of solvent, ordinarily water or aqueous alcohol (Bai et al., 2003; Hagenmaier, 2004; Ribeiro, 2007).

6.5   PROTEIN-BASED COATINGS

The properties of proteins have made them an exceptional material for the development of coatings. Proteins are polymers made up of distinct amino acid sequences and molecular structure. The structures (secondary, tertiary, and quaternary) of proteins can easily be altered to ameliorate the protein configuration and protein interactions, which results in coating formation. Water vapor permeability and oxygen permeability are the properties that showed the potential of edible coatings. Different factors affecting the interactions during coating formations such as interaction between protein and small molecules like water, plasticizers, lipids, waxes, and other additives, physical treatments, and enzymatic treatments affect the mechanical and barrier properties of coatings (Park et al., 2002). Protein coating materials can be procured from various plant or animal sources. The different protein-based coatings such as corn-, wheat-, casein-, and whey-based protein coatings are described as under:

6.5.1   ZEIN-BASED EDIBLE COATINGS

Maize or corn (*Zea mays*) is one of the most important food and industrial crops. Zein is a protein from corn, and it is a biopolymer with the potential to produce biodegradable plastics (Biswas et al., 2009; Tihminlioglu et al., 2010). Zein consists of a group of alcohol-soluble proteins found in corn endosperm. High content of the nonpolar amino acids, i.e., leucine, alanine, and proline, present in these fractions make zein water insoluble. Zein coatings are generally prepared in alcohol or aqueous solutions of alcohol. Development of coatings involve hydrogen and hydrophobic bonds, while disulfide bonds participate little due to the low content of cysteine (Gennadios et al., 1994). Zein coatings are extremely brittle after drying on the surface of product, and therefore, they require a plasticizer to enhance the flexibility of the coating. Commonly used plasticizers for the preparation of zein coatings are glycerol, polyethylene glycol, polypropylene glycol, and different fatty acids. Bai et al. (2003) coated apples to improve shelf-life. Zein protein was used to coat apples as an alternative of other glossy coatings by dissolving zein in aqueous alcohol solution with propylene glycol. Gloss levels of apple surface were varied due to different ratios of zein and propylene glycol when compared to control and shellac-coated samples. Permeability to CO_2, O_2, and water vapors was significantly affected by the concentration of zein. The oxygen level decreased with the increasing concentration of zein in coating formulations. An optimum coating of 10% zein and 10% propylene glycol was applied to apples to maintain the overall fruit quality. Pena-Serna et al. (2016) determined the effect of zein-based blend coatings on the quality of cheese during ripening period of 56 days. Coatings were prepared by blending zein with glycerol, oleic acid, and xanthan gum. No significant effect of storage was observed for the biodegradable-coated cheese samples as compared to plastic-packaged and unpackaged cheese samples in terms of ash, protein, chlorides, and acidity. Further, coated cheese samples did not show any microbial contamination for 50 days, whereas noncoated samples showed spoilage after 21 days.

6.5.2   WHEAT PROTEIN-BASED EDIBLE COATINGS

Wheat gluten is the elastic and cohesive protein that remains after the removal of starch from wheat flour dough. The protein, gluten, consisting of small quantity of charged amino acids (lysine, histidine, and arginine) and a high quantity of nonpolar amino acids, aggregates easily due to hydrophobic interactions (Haard and Chism, 1996). According to solubility, wheat protein is classified into four primary fractions: globulins (soluble in dilute salt solutions), albumins (water soluble), gliadins (soluble in 70–90% ethanol), and glutenin (Haard and Chism, 1996). Upto 47% and 34% of glutenin and gliadin are present in wheat flour, respectively (Kinsella, 1982). Gliadin is a single chain peptide of approximately 40 kDa, consisting of intramolecular disulfide bonds (Kinsella, 1982), which attribute to strength, elasticity, and film formation. Molecular weight distribution of glutenin ranges from 100 and 1000 kDa, based on the number of intermolecular disulfide linkages. The strength of the protein matrix can be determined by the disulfide bonds present in gliadin and glutenin. Coating prepared from wheat gluten was used to increase the shelf life of refrigerated strawberries. Coatings were applied directly on the surface of fruits, and the fruit quality was scrutinized by weight loss, firmness, total soluble solids, titratable acidity, reducing sugars, and sensory evaluation. Coatings had a significant effect on the quality of fruit as compared to control (Tanada-Palmu and Grosso, 2005).

6.5.3   WHEY PROTEIN-BASED EDIBLE COATINGS

Whey protein is the leftover in milk serum after cheese or casein manufacturing. It is a globular protein that consists of four S-S bonds. Seven percent of the total whey protein consists of bovine serum albumin, which is a large globular protein, and the immunoglobulins and proteose-peptone fractions are 13% and 4%, respectively. Whey protein is divided into two categories, i.e., WPCs and WPIs, depending upon their industrial utilization. Heat denaturation of whey protein is necessary for the formation of films. Heating alters/modifies the three-dimensional structure of the protein and exposes the internal SH and hydrophobic groups (Shimada and Cheftel, 1988), which assist in intermolecular S-S and hydrophobic bonding upon drying (McHugh and Krochta, 1994). Perez-Gago et al. (2005) studied the effect of whey protein and hydroxyl methyl cellulose-based edible coatings on the shelf-life of fresh-cut apples. The apples were coated with composite coatings, and weight loss, color, and browning index were determined. Higher L^* value and lower a^* and b^* values indicated that whey protein coatings exerted an antibrowning effect. Weight loss did not reduce in coated samples, which might be due to higher relative humidity. Whey protein films exhibited antimicrobial properties, when incorporated with oregano, rosemary, and garlic essential oils. WPI containing 1.0–4.0% (wt/vol) ratios of oregano, rosemary, and garlic essential oils showed antimicrobial properties against *Staphylococcus aureus* (ATCC 43300), *Escherichia coli* O157:H7 (ATCC 35218), *Salmonella enteritidis* (ATCC 13076), *Lactobacillus plantarum* (DSM 20174), and *Listeria monocytogenes* (NCTC 2167).

6.6   POLYSACCHARIDE-BASED EDIBLE COATINGS

Polysaccharide-based coatings are composed of starch, alginate, cellulose ethers, chitosan, carrageenan, or pectin and have film-forming abilities with solvents. Coatings from polysaccharides impart excellent gas barrier properties due to the presence of the polymer chains (Baldwin et al., 1995). In the past few years, demand and utilization of polysaccharides has increased for the development of edible coatings, especially in the field of dairy, bakery, meat products, ready-to-eat meals, and fresh and fresh-cut fruits and vegetables. Starch and cellulose derivatives are widely used, and other sources such as gums and pectin are also gaining importance.

6.6.1 CELLULOSE AND DERIVATIVES

Cellulose materials are biodegradable and therefore have gained attention in the last decades as a prospective substitute for synthetic polymers. Regenerated cellulose films like cellophane have not been used for food applications as compared to their synthetic counterparts due to poor water vapor permeability (WVP). Transparent cellulose films showed good gas barrier properties in dry conditions and are comparable to low density polyethylene (LDPE) and high density polyethylene (HDPE) films (Su et al., 2010). The main cellulose and cellulose derivatives used for biodegradable/edible film development are microcrystalline cellulose, nanocrystalline cellulose, microfibrillated and nanofibrillated cellulose, methylcellulose (MC), hydroxypropyl cellulose (HPC), hydroxylpropyl methylcellulose (HPMC), and carboxymethylcellulose.

Microcrystalline or nanocrystalline cellulose is developed by acid hydrolysis of native cellulose to disrupt the reactive crystalline regions as fine crystals. Microfibril and nanofibril cellulose is prepared by passing the native cellulose under high pressure and shear force. Further, carboxy, methoxy, and hydroxypropyl substitutions affect the physical and chemical properties of cellulose. Sayanjali et al. (2011) evaluated the antimicrobial and physical properties of edible coatings based on carboxymethylcellulose containing potassium sorbate in fresh pistachios. Antimicrobial effects were studied against *Aspergillus* species by using the agar diffusion assay. Pistachios were coated with an antimicrobial film with three concentrations (1, 0.5, and 0.25/100 ml of film solution) of sorbate. All concentrations showed no mold growth. WVP values were increased with increased concentration of sorbate. Boumail et al. (2016) studied the effect of coatings on microbiological, sensorial, and physicochemical properties of pre-cut cauliflowers. Coatings were prepared from methylcellulose, maltodextrin, and starch with antimicrobial formulations of citral extract, lactic acid, and lemongrass essential oil. Six coatings containing different ratios of polysaccharides and antimicrobials were prepared and tested in vitro against *Listeria monocytogenes*. All selected coatings exhibited a total in vitro inhibition of bacteria. Choi et al. (2016) evaluated the effect of HPMC-based edible coatings enriched with oregano and bergamot essential oil on the quality attributes of "Formosa plums." The coating at the level of 2% was effective in decreasing the respiration rate, ethylene production, total weight loss, and total cell count, compared to the plums treated with other coatings. In addition, 2% coated plum was firmer and exhibited less surface color change than the control fruit at the end of storage.

6.6.2 STARCH AND DERIVATIVES

Starch is composed of amylose and amylopectin. Both are composed of glucose but differ in size and shape. Amylose is a linear molecule with (1-4)-α-D linkages of glycosidic units. Amylopectin is a branched macromolecule with glucose units, which are linked in a linear way with α (1→4) glycosidic bonds. Branching takes place with α (1→6) bonds, occurring every 24 to 30 glucose units, emerging as a soluble molecule that can be quickly degraded as it has many end points onto which enzymes can attach (Rodriguez et al., 2006). On the other hand, amylose contains very few α (1→6) bonds, or even none, which causes insolubility or slow hydrolysis of amylose.

Starch is a very common food hydrocolloid due to its low cost. The native and modified form of starch can render a vast range of functional properties in coatings. Tubers, legume, cereal grains, and certain fruits contain up to 30% to 85% starch (dry basis) (Zobel and Stephen, 2006). Starch exists as insoluble granules with some crystallinity and characteristic shape in its native state. Film formation ability and mechanical properties of aqueous native maize starches were evaluated by Palviainen et al. (2001). Saberi et al. (2016) recently optimized the physical and optical properties of coatings based on pea starch and guar gum. The effect of processing variables (pea starch, guar gum, and glycerol) on transparency, viscosity, solubility, moisture content, hunter parameters, and total color difference was determined. A linear effect of all three processing levels significantly influenced the response variables. The results exhibited that pea starch/guar gum films were produced successfully and can be used further for food packaging purposes.

6.6.3 PECTIN AND DERIVATIVES-BASED COATINGS

Pectins are the group of polysaccharides present as a main component of all cell walls. Pectin is composed of (1-4)-linked α-D-galacturonic acid units with single (1-2)-linked α-L-rhamnose residues (Ridley et al., 2001). Galactouronic acid units contain the carboxyl groups, which are partly esterified by methyl groups. Degree of esterification (DE) represents the dispersion of the methylated groups along the polysaccharide chains as well as the number of methyl esterified galacturonic acid residues versus the total number of galacturonic acid units (Solvia-Fortuny et al., 2012). To form gels, high-methoxyl pectins require a low pH (2.5 to 3.5) and a minimum amount of soluble solids (55% to 88%) (Lopes da Silva and Rao, 2006). Guerreiro et al. (2016) studied the effect of pectin and alginate-based coatings enriched with citral and eugenol essential oil on storage ability and nutritional and sensory quality of raspberry. Ascorbic acid was added as an antibrowning agent. Color, firmness, soluble solids content (SSC), weight loss, microbial growth, phenolic compounds (total phenolics, flavonoids, anthocyanins), sugars, organic acids, antioxidant activity (TEAC and ORAC), acetaldehyde, and CO_2 production were determined during 0, 5, 10, and 15 days of storage. All coatings were efficient in controlling microbial food spoilage. Acceptability of the coated raspberries was good up to 14 days, while the control fruit was acceptable only till the 7th day.

6.6.4   SEAWEED EXTRACTS

6.6.4.1   Alginates

Alginate is an alluring film-forming compound because of its non-toxicity, biodegradability, and low price (Vu and Won, 2013). Its functional properties, including suspending, thickening, stabilizing, film-forming, gel-producing, and emulsion-stabilizing, have been studied by different authors (Zactiti and Kieckbusch, 2006; Dhanapal et al., 2012). Seaweed is broken into two pieces and stirred in a hot solution of alkali to extract alginate, which gets dissolved in this solution and is extracted as sodium alginate. Lastly, alginate is separated by precipitation as alginic acid or calcium alginate (McHugh, 2003). Rojas-Grau et al. (2007) developed the apple puree-alginate coatings as a carrier of antimicrobial agents to enhance the shelf-life of fresh-cut apples. All antimicrobial coatings significantly inhibited the growth of psychrophilic aerobes, yeasts, and molds (Table 6.2). Alginate-based edible coatings with antibrowning agents were developed to maintain the sensory and antioxidant properties of fresh-cut pears (Oms-Oliu et al., 2008). N-acetylcysteine and glutathione were used as antibrowning agents. Both antibrowning agents inhibited the browning of fresh-cut pears for 2 weeks

TABLE 6.2 Effect of Edible Coatings with Different Antimicrobials Applied on Various Fruits or Vegetables

Functional Ingredients	Coating material	Fruit	Effect	Reference
Lemongrass oil	Chitosan and Tween	Grape berries	Improved microbiological safety against *Salmonella*	Oh et al. (2017)
Lemongrass oil	Sodium alginate	*Fuji* apples	Exhibited a faster and greater inactivation of *Escherichia coli* during storage	Salvia-Trujillo et al. (2015)
Citral and Eugenol	Sodium alginate and pectin	Raspberry	Edible coatings enriched with Citral and Eugenol were effective at reducing microbial spoilage	Guerreiro et al. (2015)
Thyme essential oil	Chitosan and Gelatin	Black reddish	Exhibited strong antimicrobial activity on the growth of *L. monocytogenes during storage*	Jovanović et al. (2016)
Alginate	Clove oil	Apples	Inhibited growth of *E.coli* in the first week of storage	Rojas-Grau et al. (2008)
Pectin	Transglutaminase-cross-linked whey protein	Carrots	Coating decreased food weight loss and prevented the microbial growth	Marquez et al. (2017)
Lemongrass oil	Chitosan and Tween	Grape berries	Improved microbiological safety against *Salmonella*	Oh et al. (2017)

and also reduced the microbial growth. Increased vitamin C content and polyphenols was observed in the case of coated fresh-cut pears as compared to noncoated samples. Azarakhsh et al. (2012) optimized the alginate-gellan-based edible coatings for fresh-cut pineapples. The influence of alginate-based (sodium alginate 0–2% (w/v), glycerol 0–2% (w/v), and sunflower oil 0.025% (w/v)) and gellan-based (gellan 0–1% (w/v), glycerol 0–1% (w/v), and sunflower oil 0.025% (w/v)) edible coatings on fresh-cut pineapple were studied by response surface methodology (RSM).

6.6.4.2 Carrageenan-Based Edible Coatings

Carrageenan exhibits great potentiality as a film-forming material because it consists of water-soluble polymers with a linear chain of partially sulfated galactans. The extraction of these sulfated polysaccharides can be carried out from the cell walls of various red seaweeds (*Rhodophyceae*) (Riberio et al., 2007). Different types of carrageenans were extracted from different seaweeds. Carrageenan as edible coatings have various applications on fresh and frozen meat, poultry and fish, sausage-casings, dry solid foods, granulation coated powders, and oily foods (Lee et al., 2003).

6.6.5 CHITOSAN

Chitosan is developed by deacetylation of chitin, which is the main component of the exoskeleton of crustaceans as well as a cell wall constituent of green algae and fungi. Chitin is formed of β-1, 4-linked linear polymer of 2-acetamido-2-deoxy-D-glucopyranosyl residues (BeMiller, 1965). Chitosan has been substantially used as coating applications due to its film-forming properties. Chitosan is insoluble in water; therefore, coatings from chitosan are also insoluble in water and exhibit good wet tensile strength properties. The ripening process of banana was delayed by applying chitosan-glycerol-based coatings. The concentration of chitosan and glycerol, 2.02% and 0.18%, respectively, are reported for the coating of Berangan banana (*Musa sapientum* cv. *Berangan)* (Jafarizadeh et al., 2011). Guerra et al. (2016) developed edible composite coatings from chitosan and mentha (*Piperita* L. or *x villosa* huds) essential oil and studied the effect on postharvest mold occurrence and quality of table grape cv. Isabella. The coatings were effective on delaying the mold growth on table grapes. Color and firmness were enhanced in the coated fruits. The coatings nonsignificantly affected the physicochemical properties of table grapes.

6.6.6 GUMS

Basil seed gum (BSG) is a typical gum, which can be isolated from *Ocimum basilicum L*, a member of genus *Ocimum*. The extracted BSG as a heteropolysaccharide made up of two major fractions: commonly known as glucomannan (43%) along with highly branched arabinogalactan, as the hydrophobic fraction, and xylan (24.29%), the accountable part for its hydrophilic properties. The hydrophilic portion can absorb water during soaking and swell into mucilage due to the presence of a polysaccharide substance (Hosseini-Parvar et al., 2010). Although BSG has various advantages such as low production cost, high availability, convenience of extraction and hydrophilic properties, its applications in the food industry particularly for food packaging is not well manifested (Hosseini-Parvar et al., 2010; Karimi and Kenari, 2016). Hashemi et al. (2017) developed BSG-based edible coatings containing *Origanum vulgare* subsp. *viride* essential oil for the preservation of fresh-cut apricots. Chemical attributes, microbial load, and sensory characteristics of coated fresh-cut apricot during cold storage at 4°C for 8 d were determined. Among all tested treatments, the BSG + 6% essential oil was analyzed as the most effective in reducing the microbial populations of apricot cuts.

Xanthan gum, produced as an exo-polysaccharide by *Xanthomonas campestris* under unfavorable conditions, is a GRAS compound for its utilization as an emulsifier or stabilizer. It develops a highly viscous solution in hot or cold water at low concentration with excellent stability over a varying range of temperature and pH. It is stable to enzymatic degradation. Further, in complex formulations, it accelerates the suspension of particulates, even for a long time (Sworn, 2009). Sharma and Rao (2015) investigated the effect of xanthan gum-based edible coatings containing cinnamic acid on the quality of fresh-cut pears. The coatings were effective in preventing the browning and extending the shelf-life. The enrichment of cinnamic acid as an antioxidant into xanthan gum-based edible coating exhibited significant ($p < 0.05$) decline in ascorbic acid, retardation of oxidative browning, reduction in antioxidant capacity, and degradation of total phenolics content as compared to uncoated fresh-cut pears and those coated with only xanthan gum. Zambrano-Zaragoza (2016) evaluated the effect of β-carotene release rate from nanocapsules incorporated with xanthan gum coatings on physical and physicochemical properties of fresh-cut cantaloupe. Coating formulations were prepared from xanthan gum alone (XG), xanthan gum combined with nanocapsules (Ncs/XG), and xanthan gum combined with nanospheres (Nsp/XG), nanocapsules (Ncs), and nanospheres (Nsp) and were compared to untreated fresh-cut melon in order to evaluate their preservation efficiency. Incorporation of β-carotene nanocapsules into a polysaccharide matrix improves the properties of the coatings, thereby increasing storage time to 21 days at 4 °C.

6.7 APPLICATIONS OF EDIBLE COATINGS

Total nonbiodegradability and increased utilization of synthetic packaging films has caused serious ecological problems. A shift in the use of various biodegradable materials, from the utilization of renewable agriculture feedstock to marine food processing industry wastes has gained focus as the requirement of a safe and eco-friendly atmosphere. This is the approach to increase conservation of natural resources, recyclability, and generation of new design and use. Protein and polysaccharide coatings are regarded as most multifaceted biomaterials due to their large number of applications. These macromolecules can be used to retain the quality of different food products. They act as potential carriers of antimicrobials and antioxidants, which help to enhance the shelf-life of the high-fat meat products such as sausages, fillets, or beef patties by preventing their oxidation and inhibiting microbial growth (Table 6.3). They are also used for supplementation of various food products by coating applications. Moreover, these macromolecules can be used to improve the moisture and oxygen barriers to enhance the shelf-life of fruits and vegetables.

6.8 CONCLUDING REMARKS

Utilization of biodegradable packaging materials will open up potential economic advantage to farmers and agricultural processors. Bi-layer or composite coatings are required to be developed that resemble synthetic packaging materials with excellent barrier and mechanical properties. For preserving food safety, structural integrity, and biodegradability, innovative techniques

TABLE 6.3 Effect of Edible Coatings with Antibrowning Agents and Antioxidants Applied on Various Fruits, Vegetables, or Nuts

Functional Ingredients	Coating material	Fruits/vegetable/nuts	Effect	Reference
Ascorbic acid and citric acid	Sodium alginate and pectin	Apples	Coatings and anti-browning agents maintained most quality attributes	Guerreiro et al. (2017)
N-acetylcysteine and gluta-thione	Alginate and gellan	Fresh cut pear	Prevented browning for 2 weeks without affecting firmness of fruit wedges	Oms-Oliu et al. (2008)
Potassium sorbate, sodium benzoate and nisin	Pectin	Persimmon slices	Microbial growth, colour, firmness, polyphenol oxidase (PPO) activity, visual quality and overall sensory flavour were maintained during storage.	Sanchis et al. (2016)
Calcium and ascorbic acid	Carboxymethyl cellulose	Fresh cut apples	Combination of CMC and AA controled surface browning and maintained total antioxidant, vitamin C and fruit quality attributes during storage.	Saba and Sogvar (2016)
Green tea extract	Chitosan	Fresh walnut kernel	Coating provided effective inhibition of lipid oxidation	Sabaghi et al. (2015)
Tea polyphenols	Alginate	Chinese winter jujube	The decline of total chlorophylls content, ascorbic acid and total phenol content was delayed	Zhang et al. (2016)

need to be adopted. Doubtlessly, biodegradation offers a striking route to environmental waste management. Therefore, a cost-effective biodegradable material that could replace the synthetic polymers is the need of the hour and requires concentrated and focused research.

KEYWORDS

- **antimicrobials**
- **edible coatings**
- **functional components**
- **polysaccharides**
- **protein**
- **shelf life**

REFERENCES

Allegra, A., et al., (2016). The influence of Opuntia *ficus-indica* mucilage edible coating on the quality of 'Hayward' kiwifruit slices. *Postharv. Biol. Technol.*, *120*, 45–51.

Antoniewski, M. N., et al., (2007). Effect of gelatin coating on the shelf life of fresh meat. *J. Food Sci.*, *72*, 382–387.

Aquino, A. B., de Blank, A. F., & Santana, L. C. L. de., (2015). Impact of edible chitosan–cassava starch coatings enriched with *Lippia gracilis* Schauer genotype mixtures on the shelf life of guavas (*Psidium guajava* L.) during storage at room temperature. *Food Chem.*, *171*, 108–116.

Azarakhsh, N. A., et al., (2012). Optimization of alginate and gellan-based edible coating formulations for fresh-cut pineapples. *Int. Food Res. J.*, *19*, 279–85.

Bai, J., et al., (2003). Formulation of zein coatings for apples (*Malus Domestica* Borkh). *Postharv. Bio. Technol.*, *28*, 259–68.

Baldwin, E. A., (1994). Edible coatings for fresh fruits and vegetables: past, present and future. In: *Edible Coatings and Films to Improve Food Quality*, Krochta, J. M., Baldwin, E. A., Nisperos-Carriedo, M., (eds.). Technomic, Lancaster.

Baldwin, E. A., Nisperos-Carriedo, M. O., & Baker, R. A., (1995). Use of edible coatings to preserve quality of lightly (and slightly) processed products. *Crit. Rev. Food Sci. Nutr.* *35*, 509–524.

Baldwin, E. A., et al., (1997). Using lipids in coatings for food products. *Food Tech. 51*, 56–61.

Banker, G. S., (1966). Film coating theory and practice. *J. Pharm. Sci.*, *55*, 81–89.

Boumail, A, (2016). Effect of antimicrobial coatings on microbiological, sensorial and physico-chemical properties of pre-cut cauliflowers. *Postharvest Bio Technol.*, *116*, 1–7.

BeMiller J. N., (1965). Chitosan. In: *Methods in Carbohydrate Chemistry, General Polysaccharides,* Gennadios, A. (ed.). Academic Press, New York.

Biquet B., & Labuza T. P., (1988). Evaluation of the moisture permeability characteristics of chocolate films as an edible moisture barrier. *J. Food Sci., 53,* 989–998.

Becker, O. W., (1939). Method of and apparatus for making artificial sausage casings. US Patent, *2*(161), 908.

Becker, O. W., (1938). Sausage casing. US patent. *2*(115), 607.

Biswas, A., et al., (2009). Surface modification of zein films. *Indus Crops Prod., 30,* 168–171.

Bourtoom, T., (2008). Factor affecting the properties of edible film prepared from mung bean proteins. *Int. Food Res. J. 15,* 167–180.

Choi, W. S., Singh S., & Lee Y. S., (2016). Characterization of edible film containing essential oils in hydroxypropyl methylcellulose and its effect on quality attributes of 'Formosa'plum (*Prunus salicina* L.). *LWT – Food Sci Technol., 70,* 213–222.

Danalache, F., et al., (2016). Optimisation of gellan gum edible coating for ready-to-eat mango (*Mangifera indica L.*) bars. *Int. J. Bio. Macro., 84,* 43–53.

Dhanapal, A, et al., (2016). Edible films from polysaccharides. *Food Sci. Qual. Manag., 3,* 9–18.

Fairley, P., et al. (1996). Mechanical properties and water vapor permeability of edible films from whey protein isolate and sodium dodecyl sulfate. *J. Agri. Food. Chem.,* 44, 438–443.

Gennadios, A., Weller C. L., & Testin R. F., (1993). Temperature effect on oxygen permeability of edible protein-based films. *J. Food Sci., 58,* 212–219.

Gennadios, A., et al., (1994). Edible coatings and films based on proteins. In: *Edible Coatings and Films to Improve Food Quality,* Krochta, J. M., Baldwin, E. A., Nisperos-Carriedo, M. (eds.). Technomic, Lancaster, PA.

Genevois, C. E., de Escalada Pla M. F., & Flores S. K., (2016). Application of edible coatings to improve global quality of fortified pumpkin. *Inn. Food Sci. Emerg. Technol. 33,* 506–514.

Gimenez, B., et al., (2012). Role of sepiolite in the release of active compounds from gelatin-egg white films. *Food Hydro., 27,* 475–486.

Guerra, I. C. D., et al., (2016). The effects of composite coatings containing chitosan and Mentha (*piperita L. or x villosa Huds*) essential oil on postharvest mold occurrence and quality of table grape cv. Isabella. *Innov. Food Sci. Emerg. Technol. 34,* 112–121.

Guerreiro, A. C., et al., (2017). The effect of edible coatings on the nutritional quality of 'Bravo de Esmolfe'fresh-cut apple through shelf-life. *LWT – Food Sci. Technol., 75,* 210–219.

Guerreiro, A. C., et al., (2016). The influence of edible coatings enriched with citral and eugenol on the raspberry storage ability, nutritional and sensory quality. *Food Pack Shelf Life, 9,* 20–28.

Guerreiro, A. C., et al., (2015). Raspberry fresh fruit quality as affected by pectin-and alginate-based edible coatings enriched with essential oils. *Sci. Horti., 194,* 138–146.

Guilbert, S., Gontard, N., & Gorris, L. G. M., (1996). Prolongation of the shelf life of perishable food products using biodegradable films and coatings *LWT: Food Sci. Technol., 29,* 10–17.

Hadenberg, R. E., (1967). Wax and related coatings for horticulture products. *Agr. Res. Bull.,* US Department of agriculture, Washington, D.C.

Haard, N. F., & Chism, G. W., (1996). Characteristics of edible plant tissue. In: *Food Chemistry.* Fennema, O. (ed.). Marcel Dekker, New York.

Hashemi, S. M. B., Khaneghah, A. M., & Ghahfarrokhi, M. G., Eş, I., (2017). Basil-seed gum containing *Origanum vulgare* subsp. viride essential oil as edible coating for fresh cut apricots. *Postharv. Bio. Technol.*, *125*, 26–34.

Hagenmaier, R. D., (2004). Fruit coatings containing ammonia instead of morpholine. *Proc. Fla. State Hort. Soc.*, *117*, 396–402.

Hosseini-Parvar, S. H., et al., (2010). Steady shear flow behavior of gum extracted from *Ocimum basilicum* L. seed: Effect of concentration and temperature. *J. Food Eng.*, *101*, 236–243.

Jafarizadeh, M., et al., (2011). Development of an edible coating based on chitosan-glycerol to delay 'Berangan' Banana (*Musa Sapientum* cv. Berangan) ripening process. *Int. Food Res. J.*, *18*, 989–997.

Jovanović, G. D., Klaus, A. S., & Nikšić, M. P., (2016). Antimicrobial activity of chitosan coatings and films against Listeria monocytogenes on black radish. *Rev Arg de Micro.* In Press, Corrected Proof.

Karimi, N., & Kenari, R. E., (2016). Functionality of coatings with salep and basil seed gum for deep fried potato strips. *J. Am. Oil Chem. Soc.*, *93*, 243–250.

Khalifa, I., et al., (2016). Improving the shelf-life stability of apple and strawberry fruits applying chitosan-incorporated olive oil processing residues coating. *Food Pack Shelf Life.*, *9*, 10–19.

Kinsella, J. E., (1982). Relationships between structure and functional properties of food proteins. In: *Food Proteins*, Fon and Codon (ed.). Springer.

Kester, J. J., & Fennema, O. R., (1986) Edible films and coatings: A review. *Food Tech.*, *40*, 47–59

Krochta, J. M., (2002). Proteins as raw materials for films and coatings, Definitions, current status and opportunities. In: *Protein Based Films and Coatings*, Gennadios, A. (ed.). CRC Press, Boca Raton.

Krochta, J. M., & Mulder-Johnston, C. D., (1997). Edible and biodegradable polymer films: Challenges and opportunities. *Food Tech.*, *51*, 61–74.

Labuza, T. P., & Contreras-Medellin R., (1997). Prediction of moisture protection requirements for foods. *Cereal Foods World. 26*, 335–343.

Lee, J. Y., et al., (2003). Extending shelf life of minimally processed apples with edible coatings and antibrowning agents. *LWT: Food Technol. 6*, 323–329.

Li P., & Barth, M. M., (1988). Impact of edible coating on nutritional and physiological changes in lightly processed carrots. *Postharv Bio Technol.*, *14*, 51–60.

Lopes da Silva, J., & Rao, M. A., (2006). Pectins: Structure, functionality and uses. In: *Food Polysaccharides and Their Applications*, Gennadios, A. (ed.). Boca Raton FL, CRC Press.

Martínez-Romero, D, et al., (2017). The addition of rosehip oil to aloe gels improves their properties as postharvest coatings for maintaining quality in plum. *Food Chem.*, *217*, 585–592.

Marquez, G. R., et al., (2017). Fresh-cut fruit and vegetable coatings by transglutaminase-crosslinked whey protein/pectin edible films. *LWT – Food Sci. Technol.*, *7*, 124–130.

McHugh, D. J., (2003). A guide to the seaweed industry. FAO fisheries technical paper, Rome Italy.

McHugh, T. H., & Krochta, M., (1994). Sorbitol vs glycerol-plasticize whey protein edible films: Integrated oxygen permeability and tensile property evaluation. *J. Agri. Food Chem.*, *42*, 841–845.

Morr, C. V., Ha, E. Y. W. (1993). Whey protein concentrate and isolates, Processing and functional properties. *Crit. Rev. Food Sci. Nutr. 33*, 431–476.

Oh, Y. A. et al., (2017). Comparison of effectiveness of edible coatings using emulsions containing lemongrass oil of different size droplets on grape berry safety and preservation. *LWT – Food Science Technol., 75*,742–750.

Oms-Oliu, Soliva-Fortuny, G. R., & Martín-Belloso O., (2008). Edible coatings with anti-browning agents to maintain sensory quality and antioxidant properties of fresh-cut pears. *Postharv Bio Technol., 50*, 87–94.

Özdemir, K. S., & Gökmen, V., (2016). Extending the shelf-life of pomegranate arils with chitosan-ascorbic acid coating. *LWT – Food Science and Technology*, In press, corrected proof.

Palviainen, P., et al., (2001). Corn starches as film formers in aqueous-based film coating. *Pharma Dev Technol., 6*, 351–359.

Parafati, L., et al., (2016). The effect of locust bean gum (LBG)-based edible coatings carrying biocontrol yeasts against *Penicillium digitatum* and *Penicillium italicum* causal agents of postharvest decay of mandarin fruit. *Food Micro., 58*, 87–94.

Pavlath A. E., & Orts W., (2009). Edible films and coatings: Why, What, and How? *In: Edible Films and Coatings for Food Applications*, Embuscado, M. E., Huber, K. C. (eds.). Springer, New York.

Park, H. J., Chinnan, M. S., & Shewfelt, R. L., (1994). Edible coating effects on storage life and quality of tomatoes. *J. Food Sci. 59*, 568–570.

Park, H. J., (1999). Development of advanced edible coatings for fruits. *Trends Food Sci Technol., 10*, 254–260.

Park, S. K., et al., (2002). Formation and properties of soy protein films and coatings, *Protein Based Films and Coatings*, Boca Raton, FL, CRC Press.

Perez-Gago, M. B, et al., (2005). Effect of whey protein- and hydroxypropyl methylcellulose-based edible composite coatings on color change of fresh-cut apples. *Posthav Bio Technol., 36*, 77–85.

Perdones, Á., et al., (2016). Effect of chitosan–lemon essential oil coatings on volatile profile of strawberries during storage. *Food Chem. 197*, 979–986.

Pena-Serna, C., Penna, A. L. B., & Lopes Filho, J. F., (2016). Zein-based blend coatings, Impact on the quality of a model cheese of short ripening period. *J. Food Eng., 171*, 208–213.

Riberio, C., et al., (2007). Optimization of edible coating composition to retard strawberry fruit senescence. *Postharv. Bio. Technol., 44*, 63–70.

Ridley, B. L., O'Neill M. A., & Mohnen D., (2001). Pectins: structure, biosynthesis, and oilgogalacturonide-related signaling. *Phytochem., 57*, 929–967.

Rodriguez, M., et al., (2006). Combined effect of plasticizers and surfactants on the physical properties of starch based edible films. *Food Res. Int., 39*, 840–846.

Rojas-Grau, M. A., et al., (2007). Apple puree-alginate edible coating as carrier of antimicrobial agents to prolong shelf-life of fresh-cut apples. *Postharv. Biol. Technol., 45*, 254–264.

Rojas-Grau, M. A., Tapia, M. S., & Martín-Belloso, O., (2008). Using polysaccharide-based edible coatings to maintain quality of fresh-cut Fuji apples. *LWT – Food Sci. Technol., 41*,139–147.

Saba, M. K., & Sogvar, O. B., (2016). Combination of carboxymethyl cellulose-based coatings with calcium and ascorbic acid impacts in browning and quality of fresh-cut apples. *LWT – Food Science Technol., 66*, 165–171.

Saberi, B., et al., (2016). Optimization of physical and optical properties of biodegradable edible films based on pea starch and guar gum. *Ind. Crops Prod.*, *86*, 342–352.

Sabaghi, M., et al., (2015). Active edible coating from chitosan incorporating green tea extract as an antioxidant and antifungal on fresh walnut kernel. *Postharv. Bio. Technol.*, *10*, 224–228.

Saglam, D., et al., (2013). Concentrated whey protein particle dispersions: Heat stability and rheological Properties. *Food Hydro.*, *30*, 100–109.

Salmieri, S., & Lacroix, M., (2006). Physiochemical properties of alginate /polycaprolactone-based films containing essential oils. *J. Agri. Food Chem.*, *54*, 0205–10214.

Salvia-Trujillo, L., et al., (2015). Use of antimicrobial nanoemulsions as edible coatings: Impact on safety and quality attributes of fresh-cut Fuji apples. *Postharv. Biol. Technol.*, *105*, 8–16.

Sanchís, E., et al., (2016). Browning inhibition and microbial control in fresh-cut persimmon (*Diospyros kaki Thunb. cv. Rojo Brillante*) by apple pectin-based edible coatings. *Postharv. Bio. Technol.*, *112*, 186–193.

Sayanjali, S., Ghanbarzadeh, B., & Ghiassifar, S., (2011). Evaluation of antimicrobial and physical properties of edible film based on carboxymethyl cellulose containing potassium sorbate on some mycotoxigenic *Aspergillus* species in fresh pistachios. *LWT – Food Sci Technol.*,44, 1133–1138.

Sharma, S., & Rao, T. R., (2015). Xanthan gum based edible coating enriched with cinnamic acid prevents browning and extends the shelf-life of fresh-cut pears. *LWT – Food Sci Technol.*, *62*, 791–800.

Shah, N. N., et al., (2016) *n*-Octenyl succinylation of pullulan: Effect on its physico-mechanical and thermal properties and application as an edible coating on fruits. *Food Hydro.* *55*, 178–188.

Shimada, K., & Cheftel, J. C., (1988). Sulfhydryl group disulphide bond interchange during heat induced gelation of whey protein isolate. *J. Agric. Food Chem.*, *37*,161–168.

Singh, S., (2016). Carnauba wax-based edible coating enhances shelf-life and retain quality of eggplant (*Solanum melongena*) fruits. *LWT – Food Science and Technology*, *74*, 420–426.

Sogvar, O. B., Saba, M. K., & Emamifar, A., (2016). Aloe vera and ascorbic acid coatings maintain postharvest quality and reduce microbial load of strawberry fruit. *Postharvest Biol. Technol.*, *114*, 29–35.

Solvia-Fortuny, Rojas-Grau, M. A., & Martin-Belloso, O., (2012). Polysaccharide coatings. In: *Edible Coatings and Films to Improve Food Quality,* Baldwin, E. A., Hagenmair, R. D., Bai, J. (eds.). Boca Raton, FL, Taylor and Francis.

Su, J. F., et al., (2010). Structure and Properties of Carboxymethyl Cellulose/soy Protein Isolate Blend Edible Films Crosslinked by Maillard Reactions. *Carbohy. Polym.*, *79*, 145–53.

Sworn, G., (2009). Xanthan gum. In: *Handbook of Hydrocolloids (2nd ed).* Phillips, G. O., Williams, P. A. (eds.). Boca Raton, CRC Press.

Talens, P., et al., (2012). Application of edible coatings to partially dehydrated pineapple for use in fruit–cereal products. *J. Food Eng.*, *112*, 86–93.

Tanada-Palmu, Patrícia, S., & Grosso, C., (2005). Effect of edible wheat gluten-based films and coatings on refrigerated strawberry (*Fragaria Ananassa*) quality. *Posthar Bio. Technol.*, *36*, 199–208.

Tihminlioglu, F., Atik, İ. D., Özen, B., (2010). Water vapor and oxygen-barrier performance of corn–zein coated polypropylene films. *J. Food Eng.*, *96*, 342–347.

Tapia, M. S., et al., (*2008*). Use of alginate and gellan-based coatings for improving barrier, texture and nutritional properties of fresh-cut papaya. *Food Hydro, 22*, 1493–1503.

Vieira, J. M., et al., (2016). Effect of chitosan–aloe vera coating on postharvest quality of blueberry (*Vaccinium corymbosum*) fruit. *Postharv Bio Technol, 116*, 88–97.

Vu, C. H. T., & Won, K., (2013). Novel water-resistant UV-activated oxygen indicator for intelligent food packaging. *Food Chem., 140*, 52–56.

Wills, R. H., et al., (1981). *Postharvest, an Introduction to the Physiology and Handling of Fruits and Vegetables*, AVI, Westport, CT.

Zactiti, E. M., & Kieckbusch, T. G., (2006). Potassium sorbate permeability in biodegradable alginate films: Effect of the antimicrobial agent concentration and crosslinking degree. *J. Food Eng, 77*, 462–467.

Zambrano-Zaragoza, M. L., et al., (2016). The release kinetics of β-carotene nanocapsules/ xanthan gum coating and quality changes in fresh-cut melon (cantaloupe). *Carbohydrate Poly*, In press.

Zhang, L., et al., (2016). Tea polyphenols incorporated into alginate-based edible coating for quality maintenance of Chinese winter jujube under ambient temperature. *LWT – Food Sci. Technol., 70*, 55–161.

Zobel, H. F., & Stephen, A. M., (2006). Starch: structure, analysis and application. In: *Food Polysaccharides and Their Applications,* Gennadios, A. (ed.). Boca Raton, FL, Taylor and Francis.

CHAPTER 7

NATURAL PIGMENTS: AN ALTERNATIVE TO SYNTHETIC FOOD COLORANTS

ROBINKA KHAJURIA

School of Bioengineering and Biosciences, Lovely Professional University, Phagwara, Punjab, 144411, India,
E-mail: robinkakhajuria@gmail.com

CONTENTS

7.1   COLOR ADDITIVES

Color is the first parameter to be noticed about a food product. A consumer's perception about an edible product is majorly influenced by the appearance of an edible product. The color of a food plays a profound role in flavor perception of a consumer (Joshi et al., 2003). The aim of the food-manu-facturing units behind adding color to food is to make it appealing so as to influence the consumer's decision to buy a product. The food color market is one of the fastest growing segments of the global food additives market. The demand for food color in global market increased from 3000 MT in 2005 to 15000 MT in 2015. The global natural food color market is estimated to be around US $1 billion, and it is continuously growing due to the growing demand for natural food colors (Lakshmi, 2014). According to the US Food Coloring Market report (2016), the food color market is estimated to reach $2.5 billion by 2020 (Food Colorants Market Report, 2016).

"A color additive is any dye, pigment or substance which when added or applied to a food, drug or cosmetic, or to the human body, is capable (alone or through reactions with other substances) of imparting color" (Amchova et al., 2015).

The major reasons behind the use of colors in foods include:

1. Compensate for the loss of color due to light, extreme temperatures, moisture, and storage conditions.
2. Mask natural variations in color or enhance natural colors.
3. Impart color to colorless foods so as to allow consumer to recognize a product by its color (Barrows et al., 2003).

According to the guidelines established by US Food and Drug Administration (FDA), permitted food colors are classified into two categories:

*   **Certified colors** include colors that are synthetically produced. These are used extensively as they impart uniform and intense colors to food. Besides this, they blend easily to create a wide array of hues and are less expensive. FDA has approved nine certified color addi-

TABLE 7.1 Certified Color Additives Approved for Use by FDA

Color	Year Approved	Uses
Citrus Red No. 2	1963	Skins of oranges
Orange B	1966	Casings or surfaces of frankfurters and sausages
FD&C Blue No. 1	1969	Food Generally
FD&C Red No. 3	1969	Food Generally
FD&C Yellow No. 5	1969	Food Generally
FD&C Red No. 40	1971	Food Generally
FD&C Green No. 3	1982	Food Generally
FD&C Yellow No. 6	1986	Food Generally
FD&C Blue No. 2	1987	Food Generally

Source: www.fda.gov.

tives for use in the US (Table 7.1). The certified food colors do not add any undesirable flavor to food.

- **Colors exempted from certification** comprise pigments derived from natural sources such as vegetables, minerals, or animals (Table 7.2). Nature-derived color additives are expensive and may impart unintended flavor to food.

Colors belonging to either of the category are subjected to rigorous safety standards before they are approved and listed for use in foods (Kanekar and Khale, 2014).

7.2 SYNTHETIC COLORING ADDITIVES: A GROWING CONCERN

Industrial food producers have been using synthetic food colors for at least a century. Synthetic dyes are preferred over natural colorants because of better brightness, cost effectiveness, and better stability (Tuli et al., 2014). However, over a period of time, several reports have been published raising concern over the effect of synthetic coloring agents on consumer's health. Feingold in 1975 proposed that synthetic food colors are capable of inducing adverse behavioral effects in children. He stated that increased sensitivity to

TABLE 7.2   Color Additives Approved for Use by FDA but Exempted from Certification

Color	Year Approved	Uses
Annatto extract	1963	Foods generally
Caramel	1963	Foods generally
β-Carotene	1964	Foods generally
Paprika	1966	Foods generally
Saffron	1966	Foods generally
Turmeric	1966	Foods generally
Dehydrated beets (beet powder)	1967	Foods generally
Riboflavin	1967	Foods generally
Grape color extract	1981	Non-beverage food
Ferrous Lactate	1996	Ripe olives
Sodium copper	2002	Citrus-based dry beverage mixes
Tomato lycopene extract	2003	Foods generally
Spirulina extract	2013	Candy and Chewing Gum

Source: www.fda.gov.

food additives is responsible for hyperactivity observed in certain children. US FDA, on the other hand, declared this preposition as inconclusive since there was no concrete evidence to prove Feingold's proposition (Weiss, 2012). However, Feingold did succeed in drawing the attention of scientific community toward the possible side effects of synthetic food colorants.

In 2008, the Centre for Science in the Public Interest (CSPI) Washington petitioned FDA to impose a ban on artificial food colorants as their consumption was being related to behavioral problems in children (CSPI Petition, 2010). In 2010, CSPI published a report entitled *Food Dyes: A Rainbow of Risks* that stated that the nine artificial dyes approved by FDA are likely to be carcinogenic in nature, and may lead to hypersensitivity reactions and behavioral problems especially in children. The average food dye consumption per person has increased tremendously in the US over the last decade. Among the FDA approved dyes, three dyes, viz., Red 40, Yellow 5, and Yellow 6, account for 90% of the food dyes.

According to the reports, approved dyes are inadequately tested. For instance, most of the chemical carcinogenicity studies are carried out on relatively small numbers of animals, do not include *in utero* exposures, and last for a period of 2 years (rodent equivalent of about 65 human years). Because cancers normally do not show up in a rodent until the third year (corresponding to the time when cancers are more likely to appear in humans), the 2-years time frame for standard bioassays may not be enough to identify carcinogenic chemicals (Lancaster and Lawrence, 1999).

Dyes such as Yellow 5, Yellow 6, and Red 40 contain benzene, which is considered as a human and animal carcinogen but is permitted in low and safe levels in dyes. FDA in 1985 published that ingestion of free benzidine raises the cancer risk to just under the concern threshold (1 cancer/1 million people). However, these dyes are also known to contain bound benzidene and in much greater amounts. But, the routine FDA tests measure only free benzidene and not the bound moieties. Intestinal enzymes have been reported to release bound benzidene, thus exposing the consumer to much higher amounts of carcinogens than those indicated by FDA's routine tests (Amachov et al., 2015). It has been a general policy of FDA not to comment on topics that are under review. In a statement released in response to the publication of *CSPI*, the International Association of Color Manufacturers asserted that they strictly adhered to current FDA protocols, highlighting that "the FDA has repeatedly stated that these colors are safe based on the available safety data" (Potera, 2010).

Among the food additives, synthetic food dyes have been found to be the most risky category of food additives. Orange 1 and Orange 2, now banned, have been reported to induce organ damage in animals. Orange B contains low levels of a cancer-causing contaminant. Although it was used only in sausage casings and is no longer used in the US, the ban on this dye was never finalized. Among the red dyes, Red 1 has been reported to induce liver cancer in animals, Red 2 has been reported as a possible carcinogen, and Red 4 at high levels was shown to damage the adrenal cortex of dogs. Red 32, now known as Citrus Red 2 which is used to color oranges at 2 parts per million, has been shown to damage internal organs in animals and may be a weak carcinogen. Yellow 1 and Yellow 2 at high dosages have been reported to cause intestinal lesions in animals, while Yellow 3 and Yellow 4 induced heart damage (Kobylewski and Jacobson, 2012).

Such reports about the negative impacts of synthetic dyes have led to a tremendous change in the attitude of the consumers toward synthetic dyes. Increasing public awareness about probable harmful effect of synthetic colorants has generated a worldwide demand for natural colorants. This surge in the demand for natural pigments and their success in the market in the last few years have led to an increased interest in plant and microbial pigments.

7.3 NATURAL PIGMENTS: AN ALTERNATIVE

Biological pigments also known as biochromes comprise pigments derived from natural sources such as plants or microorganisms. The different pigments are described in the following subsection.

7.3.1 PLANT PIGMENTS

Plant pigments include different kinds of molecules like carotenoids, anthocyanins, betalains, and porphyrins. Though biocolorants are extremely diverse in terms of their structure, tetrapyrrols, tetraterpenoids, and flavonoids form the three major groups of natural colorants (Rymbai et al., 2011). This section discusses the major plant pigments and their sources.

7.3.1.1 Carotenoids

Carotenoids are one of the most important pigments found in plants. They are soluble in lipids and are responsible for imparting yellow, orange, and red colors. Among carotenoids, carotenes such as α-carotene, β-carotene, β-cryptoxanthin, lycopene, and lutein and xanthophylls such as violaxanthin, neoxanthin, zeaxanthin, and canthxanthin are the most important pigments (Chattopadhyay et al., 2008).

Carotenes: ß-Carotene is an orange-yellow pigment used as a water dispersible emulsion. *Daucus carota* (carrot) is a rich source of ß-carotene; however, most of the commercially used ß-carotene is derived from algae (Barth et al., 1995). Oil palm, orange, apricot, mango, peach, and pepper are used extensively to increase ß-cryptoxanthin and ß-carotene concentrations of foods (Zeb and Mehmood, 2004; Moetensen, 2006). Besides being used as colorants, carotenes are also used as provitamin A in dietary supplements. Lycopene is a highly stable pigment that can withstand a wide range

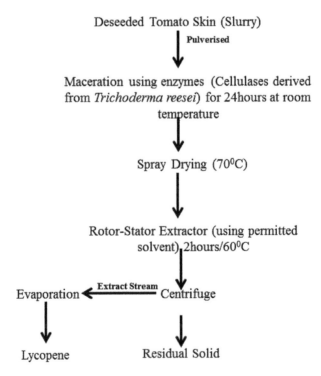

FIGURE 7.1 General scheme of lycopene production from tomato skin (adapted from Ishida et al., 2009).

of temperature and pH, and hence, it is used as common food colorant. It is commonly available in the form of powder that can be dispersed in cold water (Ryambi et al., 2011). The general method of lycopene has been explained in Figure 7.1. Although lycopene is normally associated with tomatoes, it is also found in about 70 plant species, including onion, red and *kapia* pepper, *Rosa rubiginosa* (rose hip), *Calendula officinalis* (marigold), *Taxus baccata* (yew), and *Citrullus lanatus* (watermelon). The major disadvantage of lycopene is that it is very prone to oxidative degradation and is expensive (Dweck, 2009).

Xanthophylls: are oxygenated carotenes that are normally orange to yellow in color. For example, rhodoxanthin is responsible for imparting yellow color to *T. baccata* (yew tree), while rubixanthin produces yellow color in *R. canina* (dog rose). Lutein is also a very common carotenoid found extensively in *Tagetes erecta* (marigold) flowers, making them the commercial source of this pigment (Siva, 2007).

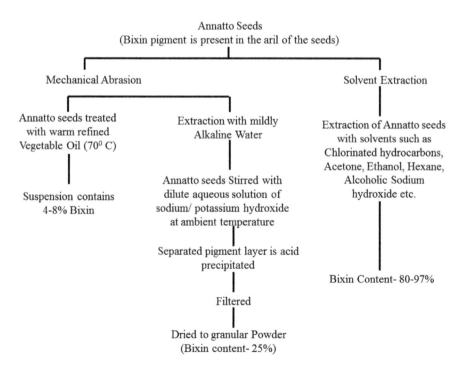

FIGURE 7.2 Pigment extraction from annatto seeds (adapted from Preston and Rickard, 1980).

Lutein: is a common carotenoid used in chicken feed. In addition to marigold, it is also found in Zucchini and green vegetables like cabbage, parsley, and spinach (Muntean, 2005).

Annatto: is a yellow-orange pigment that has been used for coloring dairy products for over two centuries. This pigment is derived from the aril of seeds of Achiote tree (*Bixa orellana*) (Figure 7.2). The chief coloring agents present in these pigments are the two carotenoids, bixin and norbixin. The fat-soluble pigment in the crude extracts is known as bixin. This bixin is then saponified to convert it to water-soluble norbixin. Higher concentrations of bixin impart orange color to foods, while norbixin is responsible for yellow color. Norbixin is used to color cheese as it binds to the proteins (Haila et al., 1996). A variety of food colorants are also obtained from *Capsicum annuum* (paprika). These include red carotenoids such as canthaxanthin and capsorubin and yellow pigments like ß-carotene, ß-cryptoxanthin, antheraxanthin, and zeaxanthin. The combinations of these pigments are responsible for bright orange to red orange colors in food products (Perez-Galvez et al., 2003). The saffron coloring pigment known as crocin is produced from the dried stigmas and styles of *Crocus sativa*. Crocin is a water-soluble pigment that is known to be the most expensive food coloring additive (Raina et al., 1996).

7.3.1.2 Flavonoids

Flavonoids are a group of widely distributed polyphenolic pigments that impart yellow color to plant. Till now, over four thousand flavonoids have been identified in plants. Based upon the structure of molecular backbone, flavonoids are divided into six major families, viz., isoflavones, flavones, flavonols, flavonols, flavanones, and anthocyanidins. The major flavone pigments include apigenin, quercetin, myricetin, kaempferol, luteolin, and tricin (Beecher, 2003).

Quercetin: is one of the most important flavonoids that is generally found in onions, apples and *Cruciferae* family. *Sambucus nigra* is the richest source of quercetin, but commercially, it is isolated from the inner bark of *Quercus tinctoria* (Filimon, 2010).

Anthocyanidins: are a class of flavonoids responsible for imparting orange-red, red-purple, and purple-blue color to flowers and fruits. Till now, around 540 different anthocyanin pigments have been identified. The most common anthocyanidins are peonidin (red), cyanidin (red-purple), malvidin (deep purple), delphinidin (blue-purple), pelargonidin (orange-red), and petunidin (purple). They are used in food products such as desserts, fruit fillings, gelatin, and confectionaries (Anderson and Francis, 2004). Petunidin is found abundantly in *Solanum scabrum* (Garden Huckleberry) while *Raphanus sativus* L is a rich source of pelargonidin. Other potential sources of red food colorants include *S. nigra, Oxalis triangularis, Brassica oleracea, Citrus sinensis, Hibiscus sabdariffa,* and *Aronia melanocarpa* (Giusti and Wrolstad, 1996).

7.3.1.3   Chlorophyll

Chlorophyll is a green pigment present in all higher plants and is responsible for photosynthesis. Plants are known to contain two forms of chlorophyll (a and b), which differ only in the substitution of the tetrapyrrole ring. Chlorophyll is not used as a colorant directly because of the liability of the coordinated magnesium and the associated color change. It is extracted from edible plants, alfalfa, grass, nettle and mulberry leaves (Mortensen, 2006).

7.3.1.4   Anthracenes

Anthracenes are generally found in the form of glycosides in young plants (Gilbert and Cooke, 2001). Different types of anthraquinones include mungistin, alizarin, and purpurin from Madder family; kermes, lac, and emodin from Persian berries; and napthoquinones. Anthraquinone dyes require a mordant, making the dying process a little complicated. Other plant yielding anthraquinone red dyes include *Galium tinctorium, G. mullugo* (wild madder), *R. cordifolia* Linn (Indian madder), *G. aperine* (goosegrass), and *G. verum* (yellow ladies bedstraw) (Deshmukh et al., 2011; Patel, 2011).

7.3.1.5   Betacyanins *(betalains)*

Betacyanins are extracted from *Beta vulgaris* (red beet) extract. These red dyes along with betaxanthins (yellows) were initially grouped with flavonoids but were later removed from that category due to structural difference (presence of nitrogen). Betanin forms the major component (95%) of the pigments present in these extracts. The beetroot extract contains a number of different pigments imparting yellow red, yellow, and bluish-red colors. These pigments have wide application in different food products such as beverages, candies, dairy, and cattle products (Im et al., 1990). Turmeric is another bright yellow pigment extracted from the rhizomes of *Curcuma longa* Linn. (Turmeric). It contains 2.5–6% curcuminoids, among which curcumin predominates. Indigo blue is extracted from dried leaves of *Indigofera* spp., which contains glucoside indican or Isatan B or Indigotin (Patel, 2011).

7.3.2 MICROBIAL PIGMENTS

Microbial pigments are promising alternative to plant-based color additives as they pose no seasonal production problems and have high productivity. Microorganisms produce a variety of pigments like carotenoids, melanins, flavins, indigo, monascins, and violacein (Dufosse, 2009). Microorganisms are considered as a preferred source for biopigments over plants in terms of

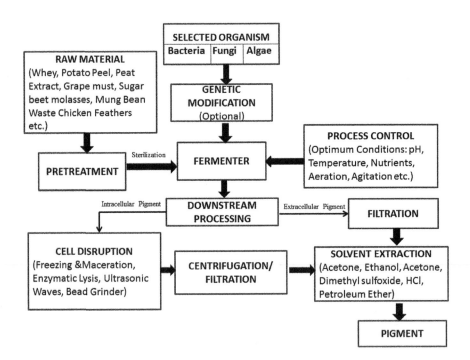

FIGURE 7.3   General scheme for pigment production.

their stability, cost efficiency, availability, yield, and easy downstream processing (Joshi et al., 2003). Microorganisms can be cultivated by solid state and submerged fermentation using agro-industrial residues as substrates. The general scheme for pigment production by fermentation is shown in Figure 7.3. Many of the microbial pigments not only act as coloring agents in various food processing and cosmetics industry but also possess anticancer, antioxidant, anti-inflammatory, and antimicrobial activities (Venil and Lakshmanaperumalsamy, 2009). Table 7.3 enlists some of the pigments produced by microbes and their current status. Microbial pigments can be divided into different categories depending on the source of the pigments. Suitable pigment-producing species should satisfy the following criteria in order to be used for commercial production of pigments:

1. Tolerance to pH, temperature, and mineral concentration
2. Ability to utilize a range of carbon and nitrogen sources
3. Possess moderate growth conditions and give reasonable color yield
4. Should be nontoxic and nonpathogenic.

7.3.2.1 Bacterial Pigments

Bacteria have been exploited commercially for pigment production. Over other pigment producers, bacteria offer certain distinctive advantages, owing to their short life cycle and compatibility to season and climate. A few examples of pigment-producing bacterial strains include *Serratia marcescens, Streptomyces coelicolor, Thislkalividrio versutus, and Chromobacterium violaceum* (Ahmad *et al., 2012*). *Pseudomonas aeruginosa is known to produce a wa*ter-soluble, blue-green pigment pyocyanin. *Pseudomonas* species are known to produce a yellow fluorescent pigment known as pyoyerdin under the conditions of iron limitation (Alemu, 2010). Phenazine produced by *Pseudomonas is used as a c*olorant in beverages, cakes, confectionaries, pudding, and decoration of food items (Saha et al., 2008). *Bradyrhizobium* sp. and *Halobacterium* sp. produce a dark red-colored carotenoid pigment known as canthaxanthin. *Rhodococcus maris* and *Corynebacterium* sp. are also being used for canthaxanthin production. The production of canthaxanthin is optimized in a medium containing hydrocarbons, ammonium phosphate, malt extract, minerals, vitamin B12, and a small amount of nonionic detergent. They are used for imparting color to farmed salmons (Rymbai et al., 2011). Similarly, *Flavobacterium* is known to produce zeaxanthin

TABLE 7.3 Microbial Pigments and Their Current Status

Pigment	Microorganism	Color	Status*	References
Canthaxanthin	*Bradyrhizobium* spp.	Orange	RP	Reyes et al., 1996
Astaxanthin	*Haematococcus pluvialis*, *Agrobacterium auran-tiacum*	Pink-red	RP	Duffose, 2006
Cycloprodigiosin	*Pseudoalteromonas denitrificans*	Red	DS	Kim et al., 1999
Granadaene	*Streptococcus agalactiae*	Orange–red	DS	George and Nizet, 2009
Indigoidine	*Corynebacterium insidiosum*	Blue	DS	Cude et al., 2012
Heptyl prodigiosin	*α-Proteobacteria*	Red	DS	Tuli et al., 2014
Prodigiosin	*Serratia marcescens*, *Pseudoalteromonas rubra*	Red	IP	Feher et al., 2008
Pyocyanin	*Pseudomonas* Spp.	Blue, green	IP	Tuli et al., 2014
Rubrolone	*Streptomyces echinoruber*		DS	Iacobucci and Sweeney, 1981
Staphyloxanthin	*Staphylococcus aureus*	Golden	-	Stevenson et al., 2002
Tryptanthrin	*Cytophaga/Flexibacteria AM13,1Strain*		-	Tuli et al., 2014
Undecylprodigiosin	*Streptomyces* spp.	Red	-	Stankovic et al., 2012
Violacein	*Janthinobacterium lividum*, *Pseudoalteromonas tunicate*, *Pseudoalteromonas* spp. *Chromobacterium violaceum*	Purple	-	Konzen et al., 2006
Xanthomonadin	*Xanthomonas oryzae*	Yellow	-	Rajagopal et al., 1997
Zeaxanthin	*Staphylococcus aureus*, *Flavobacterium* spp., *Paracoccus Zeaxanthinifaciens*, *Sphingobacterium Multivorum*	Yellow	DS	Tuli et al., 2014
Ankaflavin	*Monascus* spp.	Yellow	IP	Hsu et al., 2011

TABLE 7.3 (Continued)

Pigment	Microorganism	Color	Status*	References
Anthraquinone	Penicillium oxalicum	Red	IP	Venil and Lakshmanaperumalsamy, 2009
Canthaxanthin	Monascus roseus	Orange, Pink	-	Dufossé, 2009
Lycopene	Fusarium Sporotrichioides, Blakeslea trispora	Red	RS/DS	Giovannucci et al., 2002
Monascorubramin	Monascus spp.	Red	IP	Blanc et al., 1994
Naphtoquinone	Cordyceps unilateralis	Deep blood red	RP	Nematollahi et al., 2012
Riboflavin	Ashbya gossypi	Yellow	IP	Hong et al., 2008
Rubropunctatin	Monascus spp.	Orange	IP	Zheng et al., 2010
β-carotene	Blakeslea trispora, Fusarium sporotrichioides, Mucor, circinelloides, Neurospora crassa, Phycomyces, Blakesleeanus	Yellow-orange	IP	Dufossé, 2009
Astaxanthin	Haematococcus pluvialis	Red	-	Tuli et al., 2014
Astaxanthin	Phaffia rhodozyma, Xanthophyllomyces, Dendrorhous	Red, Pink-red	DS	Tuli et al., 2014
Melanin	Saccharomyces, Neoformans	Black	-	Vinarov et al., 2003
Torularhodin	Rhodotorula spp.	Orange-red	-	Ungureanu and Ferdes, 2012
Hemozoin	Plasmodium spp.	Brown– black	-	George and Nizet, 2009

*DS – Development stage; IP – Industrial production; RP – Research project.

(yellow pigment), which is used as an additive in poultry feed to increase yellow color of animal's skin and egg yolk. A mutant strain of *Flavobacterium* has been reported to produce zeaxanthin in a base medium containing glucose and corn steep liquor at a reduced temperature (Alcantara and Sanchez, 1999). *Streptomyces chrestomyceticus* subsp. *rubescens* has been used to produce lycopene. *Bacillus subtilis* has also been reported to produce a brown pigment (Joshi et al., 2003). *Agrobacterium auranticum, Paracoccus carotinifaciens*, and *Mycobacterium lacticola* are known to produce a pink-red pigment known as astaxanthin that is used as a food colorant (Tsubokura et al., 1999). Actinorhodin used as edible natural pigment and food colorant is synthesized by *S. coelicolor* (Zhang et al., 2006).

Corynebacterium insidiosum produces a blue pigment known as Indigoidine, while *Janthinobacterium lividum* is known to produce violacein- a purple pigment Prodigiosin. Red pigment is produced by a number of bacteria, viz., *Rugamonas rubra, Streptoverticillium rubrireticuli, Vibrio gaogenes, Alteromonas rubra, Serratia marcescens,* and *Serratia rubidaea* (Malik et al., 2012).

Bacteria such as *Cryptococcus neoformans, Azotobacter chroococcum, A. salinestris*, and Group B *Streptococcus* have been reported to synthesize melanin (Shivprasad and Page, 1989; Wang and Casadevall, 1994; Keith et al., 2007). The bacterium *Clavibacter michiganensis* subsp. *insidiosus* has been found to synthesize iodinine (Trutko et al., 2005). Carotenoids are reported in *Agromyces ramosus* and *Leifsonia sensulato* species (Collins and Bradbury, 1991; Reddy et al., 2003). Some *Alteromonas* species produce pigments ranging from lemon – yellow to violet in color. Most members of the *Microbacteriaceae* family produce yellow to orange or red-colored pigments of varying intensity (Trutko et al., 2005). Spores of *Bacillus megaterium* QM B1551 and *B. megaterium* KM produce red and yellow pigments (Venil and Lakshmanaperumalsamy, 2009). *Flexibacter* and *Sporocytophaga* spp. can produce a violet-colored pigment (Schindler and Metz, 1989), while *Vogesella indigofera* can produce a blue-colored pigment. Nianhong et al. (2009) reported the production of brown pigments in anoxygenic phototrophic bacteria such as *Chlorobium phaeovibroides* and *C. phaeobacteroides.*

7.3.2.2 Fungal Pigments

Filamentous fungi are known to produce a range of brightly colored pigments belonging to chemical classes of phenazines, carotenoids, flavins, quinones, melanins, and more specifically monascins, violacein, or indigo (Duffose et al., 2014). Anthraquinone pigments like chrysophanol, catenarin, helminthosporin, tritisporin, cynodontin, and erythroglaucin are produced by *Eurotium* spp.*, Fusarium* spp.*, Curvularia lunata,* and *Drechslera* spp. Yellow pigments epurpurins A to C are reported to be synthesized by *Emericella purpurea*. Azaphilone pigments such as falconensins A–H and falconensones A1 and B2 are produced by both *E. falconensis* and *E. fructiculosa (*Mapari et al., 2009).

In recent years, *Monascus* spp. has garnered a lot of scientific attention due to its ability to synthesize a number of polyketetides and azaphilones pigments. Major pigments produced by *Monascus* spp. have been categorized into three groups based on their color (Figure 7.4). These include the yellow pigment of monascin and ankaflavin, orange pigments of rubropunctain and monascoubrin and red pigments of rubropunctaminea and

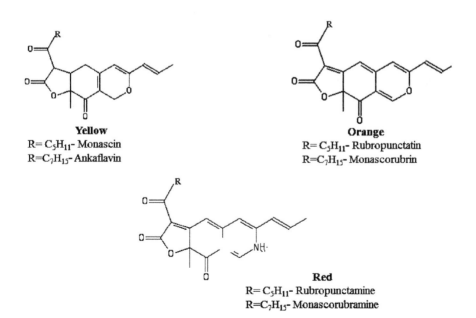

Yellow
R= C$_5$H$_{11}$- Monascin
R=C$_7$H$_{15}$-Ankaflavin

Orange
R= C$_5$H$_{11}$- Rubropunctatin
R=C$_7$H$_{15}$- Monascorubrin

Red
R= C$_5$H$_{11}$- Rubropunctamine
R=C$_7$H$_{15}$- Monascorubramine

FIGURE 7.4   Major pigments produced by *Monascus* spp.

monascorubramine (Patakova, 2012). The red and yellow pigments have been commercially used as food colorants in pigment extract forms and traditionally in the form of dried fermented red rice powder in southeast Asia for more than 1000 years (Feng, 2012).

Filamentous fungi belonging to the genera *Penicillium* sp. and *Aspergillus* sp. are known to produce a wide range of hydroxyanthraquinoid (HAQN) pigments with different color hues. For example, *Penicillium citrinum* and *P. islandicum* have been reported to produce pigment emodin. Arpink red™ known as Natural Red™ is produced commercially using the fungal strain *Penicillium oxalicum var. Armeniaca* CCM 8242 by a Czech Company. Mendez et al. (2011) reported the production of a red colored pigment by *P. purpurogenum*. Fungi from *Aspergillus* spp. are also known to produce a range of pigments. *A. glaucus, A. cristatus*, and *A. repens* produce yellow and red pigments, such as emodin (yellow), physcion (yellow), questin (yellow to orange-brown), erythroglaucin (red), catenarin (red), and rubrocristin (red) (Caro et al., 2012).

Although the abovementioned fungi produce a wide range of pigments, there has been a concern regarding the safety of these pigments. The fact that fungi belonging to *Monascus, Aspergillus*, and *Penicillium* sp. are known to co-produce toxic metabolites along with the target compound restricts their use as commercial natural colorants. Monascus pigments, for instance, are not approved as food colorants by European Union (EU) and FDA, due to the risk of the possible contamination by the mycotoxin citrinin. Similarly, *A. niger* is known to co-produce mycotoxins fumonisins and ochratoxins (Frisvad et al., 2011).

However, it needs to be considered here that these fungal producers are designated GRAS for the processes involved in food processing. This implies that the safe production of pigments from these microbes can be designed keeping the hazard analysis and critical control points in mind. In addition to this, traditional mutagenesis and metabolic engineering methods to eliminate the production of mycotoxins are being investigated as an alternative strategy.

It is also important to search for pigment-producing fungal strains that do not produce mycotoxins. Fungi belonging to *Talaromyces* spp. such as *Talaromyces purpurogenus, T. pinophilus, T. funiculosus*, and *T. aculeatus* produce Monascus-like polyketide azaphilone without the co-production of mycotoxins. *Epicoccum nigrum* is also known to yield a polyketide pigment known as orevactaene (Mapari et al., 2009). Similarly, a strain of *Dermocybe sanguinea* can produce a red HAQN glycoside dermocybin-1-β-D-glycopyranoside, along with other pigments such as emodin and physcion (Bechtold, 2009). Fungi such as *Blakeslea trispora* and *Phycomyces blakesleeanus* are known to produce β-carotene. The EU Committee considers that beta-carotene obtained by fermentation of *Blakeslea trispora* is equivalent to the chemically synthesized material used as food colorant and is therefore acceptable for use as a coloring agent in foodstuffs (Kim et al., 1997).

Pigment production has also been reported in yeast such as *Rhodotorula, Yarrowia lipolytica, Cryptococcus* sp., *Phaffia rhodozyma. P. rhodozyma* can produce a variety of carotenoids such as carotene, neurosporene, lycopene, echimenone, and astaxanthin (Andrews et al., 1976). *Xanthophyllomyces dendrorhous,* a red yeast, is reported to produce carotenoids such as astaxanthin and zeaxanthin (Roy et al., 2008). Commercial production of carotenoids using microorganisms has been achieved in case of astaxanthin by red yeast fermentation.

Marine fungi are another group of potential pigment producers because of their potential of producing bright colors, ranging from yellow to red, mainly belonging to polyketides. Physcion and macrosporin are yellow pigments extracted from the endophytic *Alternaria* sp., isolated from the fruit of the marine mangrove tree *Aegiceras corniculatum* (Huang et al., 2011). The orange questin, asperflavin (yellow color), and the brown 2-O-methyleurotinone have been reported to be produced by *Eurotium rubrum* found in marine mangrove plant *Hibiscus tiliaceus* (Li et al., 2009). Marine-derived *Penicillium bilaii* produces yellow pigments known *as* citromycetin and 2,3-dihydrocitromycetin. (Capon et al., 2007). Some of the other pigments include tetrahydroauroglaucin (yellow) and isodihydroauroglaucin (orange) from *Eurotium* sp., (Dnyaneshwar et al., 2002); flavoglaucin (yellow) by *Microsporum* sp. (Li et al., 2006), and blue pigment from *Periconia spp* (Cantrell et al., 2006). *Penicillium commune* G2M isolated from the mangrove plant *H. tiliaceus* has also been reported to synthesize a pale yellow oil characterized as 1-O-(2,4-dihydroxy-6 methylbenzoyl)-glycerol (Yan et al., 2010). Similarly, *Penicillium* sp. JP-1 isolated from the inner bark of an *Aegiceras corniculatum* tree is claimed to produce a red pigment named penicillenone (Lin, 2008).

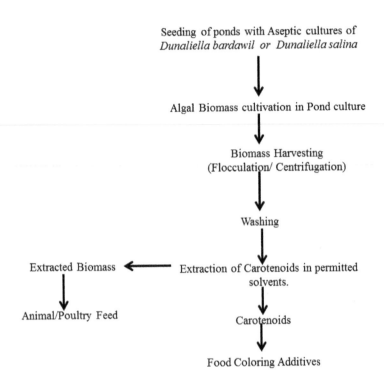

FIGURE 7.5   Carotenoid production using algae.

7.3.2.3 Algal Pigment

Algae is another potential source of natural pigments. For example, several species of marine microalgae such as *Dunaliella bardawil* and *D. salina* can synthesize β-carotene as their main carotenoid under high light intensity and high salinity conditions. A general flowsheet for the production of carotenoids using algae is shown in Figure 7.5. Microalgae species of *Rhodophyta, Cyanophyta*, and *Cryptophyta* are reported to produce red and blue pigments known as phycocyanins and phycoerythrins respectively. The alga *Haematococcus lacustris* is commercially used for the production of astaxanthin. Besides, *Haematococcus* cultures are also known to synthesize echineone and canthaxanthin (Yuan et al., 1997).

7.4 CONCLUDING REMARKS

For a long time, the use of synthetic coloring agents has dominated the food processing sector. However, toxicity issues associated with synthetic pigments have garnered an intense interest in natural colorants. The scrutiny and negative assessment of synthetic food dyes by the modern consumer have given rise to a strong demand of natural coloring additives. Both plants and microbes produce a wide range of pigments with different color hues. However, the high cost involved in implementation of modern technologies remains a major limiting factor for the commercial scale-up of natural pigments. It is therefore necessary to draw focus toward improving fermentation strategies, manipulation of external and cultural stimulants, and modifications of metabolic pathways to enhance pigment yield. As a matter of fact, agro-industrial byproducts produced throughout the world represent a low-cost source of substrates, which could be converted into a pool of chemical compounds with relevant applications in food industries. Further research is required to find solution to the problem of co-production of mycotoxins with fungal pigments, which prevents the regulatory bodies from approving a number of fungal pigments as food additives. Natural dyes also suffer from various drawbacks in terms of their stability to heat, light, or adverse pH. In-depth studies and research are needed to evaluate the real potential of natural colorants to develop economically viable processes for the production of these natural coloring additives.

KEYWORDS

- **bacterial pigments**
- **biochromes**
- **biocolorants**
- **carotenoids**
- **fungal pigments**
- **synthetic dyes**

REFERENCES

Ahmad, W. A., Ahmad, W. Y. W., Zakaria, Z. A., & Yusof, N. Z., (2012). Application of Bacterial Pigments as Colorant. *Springer Briefs in Molecular Science, 57*–74.

Alcantara, S., & Sanchez, S., (1999). Influence of carbon and nitrogen sources on *Flavobacterium* growth and zeaxanthin biosynthesis. *J. Ind. Mircobiol. Biot., 23*(1), 697–700.

Alemu, F., (2013). Isolation of *Pseudomonas fluorescens* species from rhizospheric soil of faba bean and assessment of their siderophores production. *Int. J. of Adv. Res., 1*(8), 203–210.

Amchova, P., Kotolova, H., & Ruda-Kucerova, J., (2015). Health safety issues of synthetic food colorants. *Regul. Toxicol. Pharm, 73*(3), 914–922.

Anderson, O. M., & Francis, G. W., (2004). Techniques of pigment identification, *Annual Plant Reviews: Plant Pigments and Their Manipulation, 14,* 293–341.

Andrews, A. G., Phaff, H. G., & Starr, M. P., (1976). Carotenoids of *Phaffia rhodozyma*, a redpigment fermenting yeast. *Phytochem., 15,* 1003–1007.

Barrows, J., Lipman, A., & Bailey, C., (2003). *Color Additives: FDA's Regulatory Process and Historical Perspectives.* Food Safety Magazine, U.S. Food and Drug Administration, Silver Spring.

Barth, M. M., Zhou, C., Kute, K. M., & Rosenthal, G. A., (1995). Determination of optimum conditions for supercritical fluid extraction of carotenoids from carrot (*Daucus carota* L.) tissue. *J. Agric. Food. Chem., 43,* 2876–2878.

Bechtold, T., (2009). Natural colorants – quinoid, naphthoquinoid and anthraquinoid dyes. *In Handbook of Natural Colorants.* Edited by Bechtold, T., & Mussak, R. John Wiley and Sons. Chapter 10, 151–182.

Beecher, G. R., (2003). Overview of Dietary Flavonoids: Nomenclature, Occurrence and Intake. *J. Nutr., 133,* 3248–3254.

Blanc, P. J., Loret, M. O., Santerre, A. L., Pareilleux, A., Prome, D., Prome, J. C., Laussac, J. P., & Goma, G., (1994). Pigments of *Monascus. J. Food. Sci., 59,* 862–865.

Cantrell, S. A., Casillas-Martinez, L., & Molina, M., (2006). Characterization of fungi from hyper saline environments of solar slatterns using morphological and molecular techniques. *Mycol. Res., 110,* 962–970.

Capon, R. J., Stewart, M., Ratnayake, R., Lacey, E., & Gill, J. H., (2007). Citromycetins and bilains A–C new aromatic polyketides and diketopiperazines from Australian marine-derived and terrestrial *Penicillium spp. J. Nat. Prod., 70,* 1746–1752

Caro, Y., Anamale, L., Fouillaud, M., Laurent, P., Petit, T., & Dufosse, L., (*2012*)**.** Natural hydroxyanthraquinoid pigments as potent food grade colorants: an overview. *Nat. Prod. Bioprospect., 2,* 174–193

Chattopadhyay, P., Chatterjee, S., & Sen, S. K., (2012). Biotechnological potential of natural food grade biocolorants. *Afri. J. Biotech., 7*(17), 2972–2985.

Collins, M. D., & Bradbury, J. F., (1991). *The Genera Agromyces, Aureobacterium, Clavibacter, Curtobacterium and Microbacterium.* In: The Prokaryotes, Balows, A., Trueper, H. G., Dworkin, M., Harder, W., Schleifer, K. H., Eds. Springer-Verlag Berlin Germany, 1355–1368.

CSPI Food Dyes, (2010): A Rainbow of Risks. Washington, DC: *Center for Science in the Public Interest Available*: http://tinyurl.com/2dsxlvd.

CSPI, petition to ban the use of yellow 5 and other food dyes, in the interim to require a warning on foods containing these dyes, to correct the information the food and drug administration gives to consumers on the impact of these dyes on the behavior of some children, and to require neurotoxicity testing of new food additives and food colors. Washington, DC: Center for Science in the Public Interest. Available: http://tinyurl. com/yk9ghx8.

Cude, W. N., Mooney, J., Tavanaei, A. A., Hadden, M. K., Frank, A. M., Gulvik, C. A., May, A. L., & Buchan, A., (2012). Production of the antimicrobial secondary metabolite indigoidine contributes to competitive surface colonization by the marine roseobacter *Phaeobacter sp.* strain Y4I. *Appl. Environ. Microbiol.*, *78*(14), 4771–4780.

Deshmukh, S. R., Wadegaonkar, V. P., Bhagat, R. P., & Wadegaonkar, P. A., (2011). Tissue specific expression of Anthraquinones, flavonoids and phenolics in leaf, fruit and root suspension cultures of Indian Mulberry (*Morinda citrifola* L.). *Plant Omics J.*, *4*(1), 6–13.

Dnyaneshwar, G., Devi, P., Supriya, T., Naik, C. G., & Parameswaran, P. S., (2002). Fungal metabolites: tetrahydroauroglaucin and isodihydroauroglaucin from the marine fungus, *Eurotium sp. In Proceedings of National Conference on Utilization of Bioresources — NATCUB-2002 October 24–25*. Edited by Sree, A., Rao, Y. R., Nanda, B., Misra, V. N., Bhubaneswar, Regional Research Laboratory, 453–457.

Duffose, L., (2006). Microbial production of food grade pigments, food grade pigments. *Food Technol. Biotechnol.*, *44*(3), 313–321.

Dufosse, L., (2009). *Pigments Encyclopedia of Microbiology*, *4*, 457–471.

Dufosse, L., Fouillaud, M., Caro, Y., Mapari, S. A. S., & Sutthiwong, N., (2014). Filamentous fungi are large-scale producers of pigments and colorants for the food industry. *Curr. Opin. Biotech.*, *26*, 56–61.

Dweck, A.C., FLS FRSC FRSPH – Technical Editor. Comprehensive focus on natural dyes. *Color Cosmetics. Pers Care, 2*(3), 57–69.

Feher, D., Barlow, R. S., Lorenzo, P. S., & Hemscheidt, T., (2008). A 2-substituted prodiginine, 2-(p-hydroxybenzyl) prodigiosin, from *Pseudoalteromonas rubra*. *J. Nat. Prod.*, *71*(11), 1970–1972.

Feng, Y., Shao, Y., & Chen, F., (2012). Monascus pigments. *Appl. Microbiol. Biotechnol.*, *96*, 1421–1440.

Filimon, R., (2010). Plants pigments with therapeutic potential from horticultural products, *Seria. Agro., 52*, 668–673.

Frisvad, J. C., Larsen, T. O., Thrane, U., Meijer, M., Varga, J., Samson, R. A., Kristian, F., & Nielsen, N., (2011). Fumonisin and Ochratoxin Production in Industrial *Aspergillus niger* Strains. *Plos. One*, *6*(8), e23496. doi:10.1371/journal.pone.0023496.

George, Y. L., & Nizet, V., (2009). Color me bad: microbial pigments as virulence factors Trends *Microbiol.*, *17*(9),406–413.

Gilbert, K. G., & Cooke, D. T., (2001). Dyes from plants: Past usage, present understanding and potential. *J. Plant Growth Regul.*, *34*, 57–69.

Giovannucci, E., Rimm, E. B., Liu, Y., Stampfer, M. J., & Willett, W. C., (2002). A prospective study of tomato products, lycopene, and prostate cancer risk. *J. Natl. Cancer Inst.*, *94*(5), 391–398.

Giusti, M. M., & Wrolstad, R. E., (1996). Characterization of red radish anthocyanins. *J. Food Sci.*, *61*(2), 322–326.

Haila, K. M., Lievonen, S. M., & Heinonen, M. I., (1996). Effects of lutein, lycopene, annatto and α-tocopherol on autoxidation of triglycerides. *J. Agric. & Food Chem.*, *44*, 2096–2100.

Hong, M. Y., Seeram, N. P., Zhang, Y., & Heber, D., (2008). Anticancer effects of Chinese red yeast rice versus monacolin K alone on colon cancer cells. *J. Nutr. Biochem., 19*(7), 448–458.

Hsu, L. C., Hsu, Y. W., Liang, Y. H., Kuo, Y. H., & Pan, T. M., (2011). Anti-tumor and anti-inflammatory properties of ankaflavin and monaphilone A from *Monascus purpureus* NTU 568. *J. Agri. Food Chem., 59*(4), 1124–1130.

http://www.businesswire.com/news/home/20160318005637/en/Food-Colorants-Market-Report-2016-2.5B.

http://www.fda.gov/ForIndustry/ColorAdditives/ColorAdditiveInventories/default.htm.

Huang, C. H., Pan, J. H., Chen, B., Yu, M., Huang, H. B., Zhu, X., Lu, Y. J., She, Z. G., & Lin, Y. C., (2011). Three bianthraquinone derivatives from the mangrove endophytic fungus *Alternaria sp* ZJ9–6B from the South China Sea. *Mar. Drugs, 9*, 832–843.

Iacobucci, G. A., & Sweeney, L. G., (1981). *Process for Enhancing the Sunlight Stability of Rubrolone*, US patent 4285985.

Im, J. S., Parkin, K. L., & Von Elve, J. H., (1990). Endogenous polyphenoloxidase activity associated with the Black ring defect in canned beet (*Beta vulgaris* L) root slices. *J. Food Sci., 55*(4), 1042–1059.

Ishida, B. K., Chapman, M. H., Randhava, S. S., & Randhava, S. S., (2009). Extraction of carotenoids from plant material. https://www.google.com/patents/US7572468.

Joshi,V. K., Attri, D., Bala, A., & Bhushan, S., (2003). Microbial Pigments. *Indian J. Biotech., 2*, 362–369.

Kanekar, H., & Khale, A., (2014). Coloring Agents: Current Regulatory Perspective for Coloring Agents Intended for Pharmaceutical and Cosmetic. *Int. J. Pharm. Phytopharmacol. Res., 3*(5), 365–373.

Keith, K. E., Killip, L. H., Moran, G. R., & Valvano, M. A., (2007). *Burkholderia cenocepacia* C5424 produces a pigment with antioxidant properties using a homogentisate intermediate. *J. Bacteriol., 189*, 9057–9065.

Kim, H. S., Hayashi, M., & Shibata, Y., (1999). Cycloprodigiosin hydrochloride obtained from *Pseudoalteromonas denitrificans* is a potent antimalarial agent. *Biol. Pharm Bull., 22*(5), 532–534.

Kim, S. W., Seo, W. T., & Park, Y. H., (1997). Enhanced production of β-carotene from *Blakeslea trispora* with span 20. *Biotechnol. Letters, 19*, 561–562.

Kobylewski, S., & Jacobson, M. F., (2012). Toxicology of food dyes. *Int. J. Occup. Environ. Med., 18*(3), 3220–3246.

Konzen, M., De Marco, D., Cordova, C. A., Vieira, T. O., Antônio, R. V., & Creczynski-Pasa, T. B., (2006). Antioxidant properties of violacein: possible relation on its biological function. *Bioorg. Med. Chem., 14*(24), 8307–8313.

Lakshmi, G. C., (2014). Food Coloring: The Natural Way. *Res. J. of Chem. Sci., 4*(2), 87–96.

Lancaster, F. E., & Lawrence, J. F., (1999). Food Additives Contaminants Part A, *16*(9), 381–390 doi: 10.1080/026520399283867.

Li, D. L., Li, X. M., & Wang, B. G., (2009). Natural anthraquinone derivatives from a marine mangrove plant-derived endophytic fungus *Eurotium rubrum*: structural elucidation and DPPH radical scavenging activity. *J. Microbiol. Biotechnol., 19*, 675–680.

Li, Y., Li, X., Lee, U., Kang, J. S., Choi, H. D., & Son, B. W., (2006). A new radical scavenging anthracene glycoside, asperflavin ribofuranoside, and polyketides from a marine isolate of the fungus *Microsporum*. *Chem. Pharm. Bull., 54*, 882–883.

Lin, Z., Zhu, T., Fang, Y., Gu, Q., & Zhu, W., (2008). Polyketides from *Penicillium sp.* JP-1, an endophytic fungus associated with the mangrove plant *Aegiceras corniculatum*. *Phytochem., 69*, 1273–1278.

Malik, K., Tokkas, J., & Goyal, S., (2012). Microbial Pigments: A review. *Int. J. Microbl. Res. Technol., 1*(4), 361–365.

Mapari, S. A. S., Meyer, A. S., Thrane, U., & Frisvad, J. C., (2009). Identification of potentially safe promising fungal cell factories for the production of polyketide natural food colorants using chemotaxonomic rationale. *Microb. Cell Fact, 8*, 24–30.

Mendez, A., Perez, C., Montanez, J. C., Martinez, G., & Aguilar, C. N., (2011). Red pigment production by *Penicillium purpurogenum* GH2 is influenced by pH and temperature. *J. Zhejiang. Univ-Sci. B. (Biomed Biotechnol.), 12*, 961–968.

Mortensen. A., (2006). Carotenoids and other pigments as natural colorants. *Pure Appl. Chem., 78*(8), 1477–1491.

Muntean, E., (2005). Production of a natural food coloring extract from the epicarp of *Cucurbita Pepo* L. Var. Giromontia Fruits. *Seria. F. Chemia., 8*(2), 65–68.

Nematollahi, A., Aminimoghadamfarouj, N., & Wiart, C., (2012). Reviews on 1, 4-naphthoquinones from *Diospyros L. J. Asian Nat. Prod. Res., 14*(1), 80–88.

Nianhong, C., Bianchi, T. S., McKee, B. A., & Bland, J. M., (2001). Historical trends of hypoxia on the Louisiana shelf: application of pigment as biomarkers. *Organic Geochemistry, 32*(4), 543–561.

Patakova, P., (2012). *Monascus* secondary metabolites: production and biological activity. *J. Ind. Mircobiol. Biot., 40*, 169–181.

Patel, N. K., (2011). Natural based sindoor. *Life Science Leaflets, 11*, 355–361.

Perez-Galvez, A., Martin, H. D., Sies, H., & Stahl, W., (2003). Incorporation of carotenoids from *paprika oleoresin* into human chylomicrons. *British J. Nutr., 8*, 787–793.

Potera, C., (2010). The artificial food dye blues. *Environmental Health Perspectives, 118*(10), A428.

Preston, H. D., & Rickard, M. D., (1980). Extraction and Chemistry of annatto. *Food Chem., 40*, 56–59.

Raina, B. L., Agrawal, S. G., Bhatia, A. K., & Gour, G. S., (1996). Changes in pigments and volatiles of saffron (*Crocus sativus* L.) during processing and storage. *J. Sci. Food Agri., 71*, 27–32.

Rajagopal, L., Sundari, C. S., Balasubramanian, D., & Sonti, R. V., (1997). The bacterial pigment Xanthomonadin offers protection against photodamage. *FEBS Lett., 415*, 125–128.

Reddy, N. S., Nimmagadda, A., & Rao, K. R. S., (2003). An overview of the microbial α amylase family. *Afri. J. Biotechnol., 2*, 645–648.

Reyes, F. G., Valim, M. F., & Vercesi, A. E., (1996). Effect of organic synthetic food colors on mitochondrial respiration. *Food Addit. Contam., 13*(1), 5–11.

Roy, S., Chatterjee, S., & Sen, S. K., (2008). Biotechnological potential of *Phaffia rhodozyma. J. Appl. Biosci., 5*, 115–122.

Rymbai, H., Sharma, R. R., & Srivastav, M., (2011). Biocolorants and its implications in Health and Food Industry – A Review. *Int. J. Pharm. Tech. Res., 3*(4), 2228–2245.

Saha, S., Thavasi, R., & Jayalakshmi, S., (2008). Phenazine pigments from *Pseudomonas aeruginosa* and their application as antibacterial agent and food colorants. *Res. J. Microbiol., 3*(3), 122–128.

Schindler, P. R. G., & Metz, H., (1989). Bacteria of the *Flexibacter Sporocytophaga* group and violet-colored bacteria as indicators of hygienic hazardous drinking water. *Zentralblatt fur Hygiene und Umweltmedizin, 189*(1), 29–36.

Shivprasad, S., & Page, W. J., (1989). Catechol formation and melanization by Na-dependent *Azotobacter chroococcum*: a protective mechanism for aeroadaptation. *Appl. Environ. Microbiol.*, *55*, 1811–1817.

Siva, R., (2007). Status of natural dyes and dye-yielding plants in India. *Curr. Sci.*, *92*(7), 916–925.

Stevenson, C. S., Capper, E. A., & Roshak, A. K., (2002). Scytonemin— a marine natural product inhibitor of kinases key in hyperproliferative inflammatory diseases. *Inflamm. Res.*, *51*(2), 112–114.

Stankovic, N., Radulovic, V., Petkovic, M., Vuckovic, I., Jadranin, M., Vasiljevic, B., & Nikodinovic-Runic, J., (2012). *Streptomyces sp.* JS520 produces exceptionally high quantities of undecylprodigiosin with antibacterial, antioxidative, and UV-protective properties. *Appl. Microbial Biotechnol.*, *96*(5), 1217–1231.

Trutko, S., Dorofeeva, L., Evtushenko, L., Ostrovskii, D., Hintz, M., Wiesner, J., Jomaa, H., Baskunov, B., & Akimenko, V., (2005). Isoprenoid pigments in representatives of the family *Microbacteriaceae*. *Microbiol.*, *74*, 284–289.

Tsubokura, A., Yoneda, H., & Mizuta, H., (1999). *Paracoccus carotinifaciens sp.* nov., a new aerobic Gram-negative astaxanthin-producing bacterium. *Int. J. Syst. Bacteriol.*, *49*(1), 277–282.

Tuli, H. S., Chaudhary, P., Beniwal, V., & Sharma, A. K., (2014). Microbial pigments as natural color sources: current trends and future perspectives. *J. Food Sci. Technol.*, DOI 10.1007/s13197-014-1601-6.

Unagul, P., Wongsa, P., Kittakoop, P., Intamas, S., Srikitikulchai, P., & Tanticharoen, M., (2005). Production of red pigments by the insect pathogenic fungus *Cordyceps unilateralis*. *J. Ind. Microbiol. Biotechnol.*, *32*, 135–140.

Ungureanu, C., & Ferdes, M., (2012). Evaluation of antioxidant and antimicrobial activities of torularhodin. *Adv. Sci. Lett.*, *18*(1), 50–53.

Vendruscolo, F., Bühler, R. M. M., de Carvalho, J. C., de Oliveira, D., Moritz, D. E., Schmidell, W., & Ninow, J. L., (2015). *Monascus:* a Reality on the Production and Application of Microbial Pigments. *Appl. Biochem. Biotechnol.*, DOI 10.1007/s12010-015-1880-z.

Venil, C. K., & Lakshmanaperumalsamy, P., (2009). An insightful overview on microbial pigment, prodigiosin. *Electron J. Biol.*, *5*(3), 49–61.

Wang, Y., & Casadevall, A., (1994). Susceptibility of melanized and nonmelanized *Cryptococcus neoformans* to nitrogen and oxygen derived oxidants. *Infect Immun.*, *62*, 3004–3007.

Weiss, B., (2012). Synthetic Food Colors and Neurobehavioral Hazards: The View from Environmental Health Research. *Environmental Health Perspectives.*, *120*(1), 1–5, doi:10.1289/ehp.1103827.

Yan, H. J., Gao, S. S., Li, C. S., Li, X. M., & Wang, B. G., (2010). Chemical constituents of a marine-derived endophytic fungus *Penicillium commune* G2M. *Molecules*, *15*, 3270–3275.

Yuan, J. P., Gong, X. D., & Chen, F., (1997). Separation and analysis of carotenoids and chlorophylls in *Haematococcus lacustris* by high performance liquid chromatography photodiode array detection. *J. Agric. Food Chem.*, *45*, 1952–1956.

Zeb, A., & Mehmood, S., (2004). Carotenoids contents from various sources and their potential health applications. *Pak. J. Nutr.*, *3*(3),199–204.

Zhang, H. C., Zhan, J. X., Su, K. M., & Zhang, Y. X., (2006). A kind of potential food additive produced by Streptomyces coelicolor: characteristics of blue pigment and identification of a novel compound, λ-actinorhodin. *Food Chem.*, *95*(2), 186–192.

CHAPTER 8

HURDLE TECHNOLOGY IN FOODS

PRAGATI KAUSHAL[1] and HARISH KUMAR SHARMA[2]

[1]*Department of Food Science and Technology, Punjab Agricultural University, Ludhiana, Punjab, India, E-mail: pragati_gndu@yahoo.co.in*

[2]*Department of Food Engineering and Technology, Sant Longowal Institute of Engineering and Technology, Longowal, Sangrur, Punjab, India, E-mail: profh.sharma27@gmail.com*

CONTENTS

8.1   NEED FOR TECHNOLOGIES

Different techniques are adopted by food processing industries to process the foods to meet the demand of the consumers. Traditionally thermal treatments have been used earlier in the food industry to provide the required food safety profiles and to enhance the shelf-life of the product. However, these thermal treatments result in damage of heat-labile compounds, resulting in loss of organoleptic and nutritional properties of food products. In recent years, the increased interest toward novel thermal and nonthermal technologies for food processing has gained industrial importance. The concern behind the thermal processing of food is loss of volatile compounds, nutrients, and flavor. Nonthermal processing is used to achieve better quality and acceptance and reduce the operational cost. Because these technologies allow the food sector to meet product safety and fulfilling the shelf-life requirements, these techniques have the ability to replace the existing traditional food preservation techniques. Therefore, various nonthermal technologies, e.g., high-power ultrasounds, pulsed electric fields, light technologies, and cold plasma, are widely employed for different food applications. The term "nonthermal" is used for technologies that are effective at temperatures below the lethal temperatures of microorganisms.

8.2   DIFFERENT EMERGING TECHNOLOGIES

Innovative food processing methods have better potential than other conventional food processing methods and are still an evolving challenging field for the food processors. The cost of equipment used in various nonthermal food processing methods is higher than that of equipment used in thermal processing of foods. After minimizing the investment costs and energy-saving potential of nonthermal processing methods, it can also be employed in small-scale industries. The various emerging technologies in food processing and preservation are hurdle technology, high pressure processing, ultrasonication, etc. High hydrostatic pressure, oscillating magnetic fields, high-intensity ultrasound, ultraviolet light, pulsed light, and ionizing radiation have the ability to inactivate microorganisms to varying degrees (Butz and Tauscher, 2002). Pulsed light is considered as an emerging non-thermal method capable of reducing the microbial population on contact materials. This nonthermal method uses short and intense pulses of light in the ultraviolet-near infrared (UV–NIR) range. Novel nonthermal technologies have the ability of preserving the nutritional and sensorial characteristics of fresh-like food products by inactivating the microorganisms at near-ambient temperatures. The lists of various thermal and nonthermal methods used in food processing and preservation are presented in Table 8.1.

8.3   HURDLE TECHNOLOGY

Hurdle technology is prominently used in the food industry for the gentle and effective preservation of foods. The shelf-life, nutritional, and sensory quality of foods depend on combined applications of various preservative

TABLE 8.1   Thermal and Nonthermal Methods of Food Processing and Preservation

S. No.	Type	Methods
1	Thermal methods	• Microwave heating
		• Radio frequency heating
		• Infrared heating
		• Ohmic heating
		• Sous vide processing
		• Induction heating
2	Nonthermal methods	• High hydrostatic pressure technology
		• Cold plasma
		• Pulsed electric field
		• High voltage arc discharge
		• Ultrasonication
		• Irradiation
		• Ultraviolet light treatment
		• Hurdle technology
		• Oscillating magnetic field
		• Ozonization

factors (called hurdles) (Leistner, 2000). In this technology, hurdles are deliberately combined for improving sensorial and nutritional quality of foods and enhancing the microbial stability of foods. Thus, the main aim of hurdle technology is to apply intelligent and appropriate mixing of hurdles for achieving the total quality of foods (Leistner, 2000). Examples of hurdles in a food system are low temperature during storage, high temperature during processing, increasing the acidity, lowering the water activity or redox potential, or the presence of preservatives. According to the intensity of the hurdles and the type of pathogens, the process can be controlled individually without affecting the safety of the food product. The most critical process in hurdle technology is homeostasis of microorganisms. Food preservation can be achieved by disturbing the homeostasis, i.e., balanced and stable internal environment of microorganisms.

Multitargeted approach is the best way for controlling microbial population in hurdle technology. More than 60 different hurdles have been recognized till date, which have proved beneficial in controlling microbial population. Chilling temperature is the major hurdle in case of refrigerated foods. However, if temperature variations occur during food distribution, then this hurdle breaks down and food spoilage or poisoning could happen. Therefore, additional hurdles should be incorporated as safeguards into chilled foods by using an approach called "invisible technology" (Leistner, 1999). The mechanism involved and different critical changes from the applications of hurdles are described as follows:

8.3.1 MECHANISM

The main mechanism involved in hurdle technology is the multitargeting approach of controlling microorganisms. Rather than using the single-targeting approach for microorganisms, the multi-targeted approach allows low intensity hurdles for inactivating and killing microorganisms, thereby improving the product quality. Moreover, to disturb simultaneously all the mechanisms involved, disturbing homeostasis is the best approach in this technology.

8.3.1.1 Homeostasis

Homeostasis, i.e., internal equilibrium, is the condition of uniformity and stability in the internal state of microorganisms. The concept behind homeostasis is well known in higher organisms, but this knowledge needs to be established in microorganisms responsible for spoilage of foods (Leistner, 2000). If the homeostasis of microorganisms is affected, then microbes do not multiply, till the re-establishment of homeostasis. Homeostasis can be affected by applying different hurdles. Therefore, food preservation temporarily or permanently can be achieved by disturbing the homeostasis or internal equilibrium of microorganisms in a food product.

8.3.1.2 Multitarget Preservation

Multitarget preservation can be achieved by hitting various targets like DNA and cell membrane within the microbial cell by using different hurdles simultaneously. This process disturbs the homeostasis. Therefore, the re-establishment of homeostasis and the activation of stress shock proteins become difficult. The switch on of genes for the synthesis of stress shock proteins will become difficult as dealing simultaneously with different stresses will demand more energy, which the microorganisms cannot deliver as they become metabolically exhausted (Yousef and Courtney, 2003). The optimum food preservation and microbial stability can be achieved by applying different hurdles in the food product at the same time. This synergistic approach can elucidate different targets within the microbial cells, resulting in grouping of hurdles in different classes according to the targets that will disturb the homeostasis in several ways. For this purpose, careful selection and intelligent mixing of hurdles will be a promising approach in food preservation using this technology.

8.3.1.3 Metabolic Exhaustion

The success of hurdle technology depends on ensuring metabolic exhaustion. The bacterial spores that survive even after the heat treatment can germinate under less favorable conditions as compared to vegetative bacteria. The bacterial spores will die more quickly if

- antimicrobial substances are present;
- the organisms are sublethally injured and damaged.

The microbes in the hurdle-treated stable products use their energy for homeostasis, thereby becoming metabolically exhausted. Therefore, during storage without refrigeration, some of the viable spores germinate, but the vegetative cells derived from these spores die. Thus, the spore count keeps on decreasing during storage, and the product becomes even safe during storage, especially at ambient temperature (Man and Jones, 2000). The microorganisms struggle for every possible repair mechanism to overcome this environment by using their energy. After complete use up of their energy, the microorganisms are metabolically exhausted and die automatically due to deficiency of energy. This process is referred as autosterilization.

8.3.1.4 Stress Reactions

Under stress, some of the bacteria become more resistant to heat and become more virulent due to generation of stress shock proteins. These stress shock proteins help the microorganisms to deal with various stress situations. This is the main obstruction in success of hurdle technology. Therefore, to avoid the synthesis of stress shock proteins, multitarget preservation needs to be followed. The synthesis of these proteins is influenced by heat, pH, water activity (a_w), and starvation. As a response to various hurdles and starvation, a stress shock protein is produced by some bacteria. These stress shock proteins could turn out to be problematic for the application of hurdle technology and may affect food preservation if only one hurdle has been applied. If the microorganism receives different stresses at the same time, then the activation of genes for the synthesis of stress shock proteins, which help organisms to cope with stress situation, would be difficult. Synthesis of many stress shock proteins due to continuous exposure to different stresses will be very energy consuming and would lead to metabolic exhaustion of the microorganism (Yousef and Courtney, 2003).

8.3.2   NEED OF HURDLE TECHNOLOGY

With the growing economy, there is an increased demand for fresh and minimally processed food products. Conventional food preservation methods create various changes in the sensory and nutritional quality of the foods. These conventional preservatives methods are based on the single parameter approach. The hurdle technology can bring out minimum nutritional changes in the product, which makes the product more valuable and acceptable than that obtained by any other conventional methods. Therefore, finding the combined effects of major factors affecting preservation of food products is necessary to formulate the correct application of preservation factors that can increase the microbial safety and quality of food products. However, hurdle technology needs to be applied without sacrificing the safety and microbial stability of foods, especially for foods that are stored without refrigeration.

8.3.3   DIFFERENT HURDLES AND THEIR ROLE

Each hurdle aims to inactivate and eliminate unwanted microorganisms in food. Common salt or organic acids can be used as hurdles to control microorganisms in foods. Many natural antimicrobials such as nisin, natamycin, and other bacteriocins, and essential oils derived from rosemary or thyme also work well (Sowjanya and Kumar, 2016). The major hurdles and their applications in food preservation are briefly presented in Table 8.2. There can be significant synergistic effects among different hurdles. For example,

TABLE 8.2   Major Hurdles in Food Preservation

S. No.	Hurdle	Application
1	Reduced redox potential	Removal of oxygen and addition of adsorbate
2	High temperature	Heating
3	Reduced water activity	Drying and curing
4	Increased acidity	Acid formation and addition
5	Low Temperature	Freezing and chilling
6	Bio-preservatives	Competitive flora such as microbial fermentation
7	Other preservatives	Nitrites, sulphites and sorbates

gram-positive bacteria include some of the spoilage bacteria such as *Clostridium*, *Bacillus*, and *Listeria*. A synergistic enhancement occurs if nisin is used against these bacteria in combination with antioxidants, organic acids, or other antimicrobials. Combining antimicrobial hurdles in an intelligent way means that other hurdles can be reduced, and yet, the resulting food can have superior sensory qualities (Malik and Sharma, 2014).

Depending on the intensity, every hurdle may have either positive or negative impact on foods. For example, the use of low temperature (chilling) below the critical limit of any food can lead to "chilling injury," whereas moderate chilling will prove to be beneficial for the extension of shelf-life as it retards microbial growth. Similarly, lowering the pH in fermented sausage inhibits the growth of pathogenic bacteria, but lowering beyond the required limit can also impair the taste. Therefore, a balanced intensity of any hurdle should be used for food preservation. The different hurdles and their role are presented in Table 8.3. The recommended substances for reducing water activity and pH are presented in Table 8.4.

8.3.4   DIFFERENT HURDLES IN FOOD

8.3.4.1   Physical Hurdle

8.3.4.1.1   Water Activity and Redox Potential

Microorganisms associated with foods grow best at relatively high water activity. Therefore, a reduction in water activity would limit the growth of majority of the microorganisms. The minimum water activity for microbial

TABLE 8.3 Different Hurdles and Their Role in Food Preservation

S. No.	Hurdle	Application	Effective Range	Application	Additional information
1	High temperature	Pasteurization	Mild heat treatment (63°C/30 min, 72°C/15 sec)	Destroy vegetative pathogens	Does not destroy spores
		Sterilization	Severe heat treatment moist heat treatment: 11–12°C /20 min/15 psi dry heat treatment: 160–180°C/2 h	Destroy spores	Some nutrient, quality destruction
		Blanching	Below 100°C	Inactivate natural enzymes	Commonly applied when food is to be frozen
2	Low temperature	Refrigeration	Ideally 0–4°C	Slows down microbial growth, respiration, enzymatic and chemical reactions	Some pathogens can grow e.g. *Clostridium botulinum* (Type E)
		Freezing	–18 to –30°C	Stops microbial growth and respiration	Quality depends on product, type and temperature
3	Reduced water activity (*aw*)		$aw<0.86$ (for pathogenic micro organisms) and $aw<0.62$ (for yeasts and moulds)	Lower is the water activity, longer will be the storage life	Free water can be removed by concentration, dehydration and freezing
4	Increased acidity (lowered pH)		pH<4.5	Slows down the growth of spoilage organisms and pathogens	Pathogens won't grow at pH<4.5

TABLE 8.4 Recommended Substances for Controlling Hurdles

S. No.	Hurdle	Recommended substances
1	Lowering water activity	Glucose, fructose, sucrose, sodium chloride, potassium chloride etc.
2	Lowering pH	Inorganic and organic acids
3	Preservatives	Sorbates, propionates, nitrates, nitrites, sodium benzoate, sulphur dioxide, sulphites etc.

growth is about 0.6, and below this water activity, the spoilage of foods is not microbiological. Therefore, food stability and safety could be enhanced by reducing the water activity. The commonly employed methods for decreasing water activity of foods are drying, addition of salt, humectants, and freezing. The water activity of foods strongly affects the metabolic rate, resistance, multiplication, and survival of microorganisms. The stability and safety of many foods are not based solely on water activity, but on the combined effects of several factors that could have an additive or synergistic effect.

The tendency of medium to accept or donate electrons, to oxidize or reduce, is termed as redox potential. The growth of microorganisms in a biological medium is dependent on Eh of the medium. The lower Eh value contributes to microbiological stability of food products. Not only aerobic but also many "facultative anaerobic bacteria" do not grow well at low Eh. This hurdle can be regulated by the type of package and packaging material.

8.3.4.1.2 *High Temperature (Sterilization, Pasteurization and Blanching)*

High temperature (F hurdle) is an energy intensive hurdle. The purpose of heat treatment in food preservation is to kill the microorganisms and inactivate the enzyme that become active in the subsequent storage of foods. The purpose of this hurdle is to inactivate/kill the enzymes and microorganisms. Pasteurization is combined with other hurdles like low temperature preservation. Blanching is always combined with preservation hurdles like freezing and drying.

Microbial growth can occur over a temperature range from about $-8°C$ up to $100°C$ at atmospheric pressure. Most bacteria, yeast, and molds grow best in the temperature range of about $16–38°C$. Some grow even at $0°C$ and others at a temperature as high as $100°C$. The optimum temperature for most of the bacterial, mold, and yeast growth is 35°C, 25–30°C, and 35–47°C, respectively. The lethal temperature for most of the bacteria is in the range of 82–93°C, but many bacterial spores still survive even at 100°C. When foods have high acid content (such as tomatoes, orange juice), there is no need to heat the food severely because acid increases the killing and destructive power of heat. A temperature of 93°C for 15 min is required to gain stability if sufficient acid is present.

8.3.4.1.3   Ultraviolet Radiation

The effective wavelength for destroying microorganisms is 260 nm. This hurdle cannot be used alone and hence is effectively combined with heat processing, chilling, packaging, etc.

8.3.4.1.4   Low Temperature (Chilling and Freezing)

Chilling (t hurdle) is also an energy intensive hurdle. The retardation of microbial and biochemical activities at low temperatures is the basis of preservation by cold. Cold temperature treatment only reduces the microbial activities and population. Some of the yeasts, molds, and bacteria are inhibited during freezing by lowering of a_w instead of temperature. Most spoilage organisms grow above 10°C. Cryophilic (psychrophilic) microorganisms even grow slowly from –4.4°C to –9.4°C. These organisms, however, do not produce toxins but cause deterioration.

Chilling temperature of foods vary between –1°C and +15°C (Leistner and Gorris, 1997). Chilling is generally used with other hurdles like curing and pasteurization. Freezing is also combined with hurdles like packaging and blanching.

8.3.4.1.5   Modified Atmospheric Packaging (MAP) and Active Packaging

Modified atmospheric packaging (MAP) means insertion of gas other than air in the package. MAP is generally used in combination with chilling. In MA storage, oxygen concentration is reduced, whereas the level of carbon dioxide is increased.

Aseptic packaging involves hermetic sealing of foodstuffs to prevent microbial recontamination of food stuffs. Packaging hurdles include clean room environment, active and intelligent packaging systems, barrier packaging materials, etc.

8.3.4.1.6 Ultrasonication

High intensity ultrasound waves (18 kHz–500 MHz) produce compression and expansion cycles results in creating cavitation that disrupt cellular structures of microorganisms.

8.3.4.1.7 Electromagnetic Energy

This results from a high voltage electrical field that creates friction between the molecules and reverses the polarity. Electric charge is built up in cell membranes of microorganisms, which destroy microbial cells. Examples include microwave energy, radio frequency energy, oscillating magnetic field pulses, and high electric field pulses.

8.3.4.1.8 Ionizing Radiation

Here, microorganisms are killed depending upon the dose of radiation. Because no heat is involved in the process, the product remains fresh after the treatment. Radiation dose of about 5 kGy is sufficient to kill microorganisms.

8.3.4.1.9 Photodynamic Inactivation

The inactivation of microorganisms, particularly bacteria, is achieved by the use of quenchers. Quenchers work by incorporating photosensitizers into the packaging material that act as a killing agent for microorganisms. This hurdle may not be used as a single hurdle.

8.3.4.1.10 Ultra-High Pressure

Pressures at 3000 bar and above destroys the cell membrane of microorganisms that results in cell leakage. This treatment results in enhancing the shelf-life of food products, particularly fruit-based products.

8.3.4.2 Microbial-Derived Hurdle

8.3.4.2.1 Competitive flora

This hurdle inhibits the growth of undesirable microorganisms in some foods. Example: competitive flora inhibits the growth of *Listeria monocytogenes* in the case of fresh meat (Leistner and Gorris, 1997).

8.3.4.2.2 Antibiotics

Very few antibiotics that are permitted can be used as a hurdle in foods, for example, natamycin can be used during ripening of cheese.

8.3.4.2.3 Protective or Starter Cultures

These cultures play an important role in food preservation. Fermentation can be initiated by the addition of microbial starter cultures. These cultures are added to both fermented and nonfermented foods. Example include fermented sausages, vegetable products, etc. This hurdle is always combined with other hurdles like chilling and curing.

8.3.4.2.4 Bacteriocins

These hurdles have antimicrobial activity. They are produced by lactic acid bacteria and are effective against bacterial spores. Nisin is a popular bacteriocin that provides protection against germinating bacterial spores

8.3.4.3 Chemical Hurdle

The increase in acidity enhances the microbiological stability of food products. This is achieved naturally by fermentation or artificially by the addition of acidulants. The acidity or alkalinity of an environment has a profound effect on the activity and stability of macromolecules such as enzymes. In general, bacterial growth is the fastest in the pH range of 6.0 to 8.0 (i.e., near the neutral pH). Yeast grows fastest in the pH range of 4.5 to 6.0. Filamentous fungi grow fastest in the pH range of 3.5 to 4.0. Lactobacilli and acetic acid bacteria grow optimally between pH 5 to 6.0. The ability of low pH to restrict microbial growth has been deliberately employed since the earliest times in the preservation of foods with lactic acid.

A preservative is defined as any substance that is capable of inhibiting, retarding, or arresting the growth of microorganisms. Chemical preservatives interfere with the cell membrane of microorganisms, their enzymes, or their genetic mechanisms. Chemical preservatives are added after the foods are processed. Several preservatives are frequently being used in the food industry within the permissible limits of legal standards. Apart from the above-mentioned chemicals, some spices and herbs are also used as preservatives. Antimycotic agents (like potassium sorbate, sodium benzoate) are added to intermediate moisture foods (IMF) to prevent mold growth. Antimicrobial agents control microorganisms by penetrating their cell walls and disrupting various activities, so that they cannot reproduce. Examples include: sodium benzoate, potassium sorbate, sodium nitrite, sodium lactate, nisin, sodium propionate, salt, natural spices, etc. These agents cannot be used alone. They are used in combination with other hurdles like packaging and chilling. Besides the above, antioxidants inhibit the oxidation of molecules by quenching the free radicals, thereby increasing the shelf-life of food products.

8.3.4.4 Miscellaneous Hurdles

- *Chitosan*: Chitosan is a polysaccharide that is effective against fungi, directly or indirectly.
- *Fatty acids*: Fatty acids like linoleic and arachidonic acids have inhibitory effect on gram-positive bacteria.
- *Monolaurin*: It is a food-grade glycerol monoester of lauric acid that has antimicrobial activity, particularly effective against gram-positive bacteria.
- *Chlorine*: Food products are generally dipped in a chlorine solution of 50–100 ppm concentration for a short time; this solution has disinfectant properties.

8.3.5 ADVANTAGES AND DISADVANTAGES

8.3.5.1 Advantages

1. It can avoid the acuteness of one hurdle for preservation.
2. It does not affect the integrity of food products.

3. It is applicable to both large- and small-scale industries.
4. Food remains stable and safe, with high sensory and nutritive value.

8.3.5.2 Disadvantages

1. This technique could provide varying results depending upon various bacterial stress reactions.
2. These foods are often not sterile products.
3. During storage of these foods, several microbiological, chemical, and biochemical reactions occur, which cause change in nutritional quality, flavor, texture, and color.

8.3.6 APPLICATIONS

Hurdle technology has wide applications in preservation of various food products like dairy products, fruits and vegetables, and meat and meat products. Hurdle technology was used in bakery product to prevent the fungal growth of common contaminants belonging to the genera *Aspergillus, Eurotium*, and *Penicillium* (Marin et al., 2002). Esteller et al. (2004) showed that different kinds of sugars alone or in combination could be used in bread making to improve products softness and shelf-life. The brief description of applications of hurdle technology is discussed as follows.

8.3.6.1 Shelf-Stable Products (SSPs)

Shelf-stable products (SSPs) are the heated moistened foods having a_w >0.90. These products are stored without refrigeration for weeks or months. SSPs are heated in sealed containers that avoid recontamination. The sensory and nutritional properties of food are improved by mild heat treatment (70–110°C). However, spores of bacilli and clostridia are still viable in the foods after mild heat treatment. These spores are inhibited by adjustment of pH, a_w, Eh, etc. Based on the main hurdle, SSPs are classified into four types, although additional hurdles promote the safety and stability of these products.

8.3.6.1.1 F-SSP

The main rationale behind F-SSP is the killing or inactivation of bacterial spores. The a_w of F-SSP must be lower than 0.96, whereas the pH should be between 6.0 and 6.5. A relatively mild heat treatment (F value 0.4) is given to the food products, which inactivates all vegetative microorganisms and damages spores (Leistner and Gorris, 1997). The vegetative cells/bacteria derived from such spores are inhibited by pH and a_w, which are not disadvantageous to the sensory parameters. Low Eh is preferred for food stability.

8.3.6.1.2 a_w-SSP

For safe and stable products of the a_w-SSP type, the followings need to be adopted:

i. The product should be heated to at least 75°C (internal temperature) in a sealed container.
ii. The product should have low Eh, because low Eh contributes to the inhibition of a_w-tolerant bacilli.
iii. Water activity must be adjusted to lesser than 0.95.

8.3.6.1.3 pH-SSP

In these products, vegetative microorganisms are inactivated by low pH and heat. In this process, the product having pH<5.0 and a_w around 0.98 is filled in containers and heated to 72–80°C (internal temperature). Heat resistance of bacteria and their spores is increased with decrease in water activity and diminished with decreasing pH (Lekogo et al., 2013). Pasteurized fruit and vegetable preserves and jelly sausages are the examples of pH-SSP.

8.3.6.1.4 COMBI-SSP

These products are stabilized by the combination of several hurdles. The application of hurdle technology with respect to specific segments of the foods is illustrated as follows:

8.3.6.2   Dairy Industry

Hurdle technology has been applied in dairy products widely. *Dudhchurpi*, a product of the Himalaya region that is made from milk of yak or cow, is self-stable for several months without refrigeration. Sensory analysis and microbial stability of *dudhchurpi* was optimized using combined hurdles like heating, acid coagulation, addition of sugar and sorbate, smoking, drying, and packaging in a closed container (Hossain, 1994).

Curd rice is a traditional dairy product popular in South India, whose shelf-life is 24 h at 30°C. Attempts were made to increase the shelf-life of curd rice by incorporating fresh ginger along with other spices. The culture pH and natural preservative like ginger have been identified as probable hurdles for improved shelf-life of curd rice (Balasubramanyam et al., 2004). Hurdle-treated brown peda could be best preserved up to 40 days at room temperature (30±1°C) without any quality loss (Panjagari et al., 2007). The quality and shelf-life of hurdle-treated paneer extended from 1 to 12 days at ambient temperature and 6 to 20 days at refrigeration temperature without affecting its physicochemical and sensory properties (Thippeswamy et al., 2011). Pina-Pérez et al. (2012) studied a possible synergistic effect of hurdle technology using a combination of pulsed electric field (PEF) technology and a natural ingredient, cinnamon, on the inactivation of *Salmonella typhimurium* in skim milk (SM). The maximum synergistic effect of hurdle technology was achieved by PEF treatment of 10 kV/cm for 3000 μs with the supplementation of 5% (w/v) cinnamon in SM.

Xu et al. (2014) extended the shelf life of soybean milk by the application of hurdle technology. Sodium dehydroacetate, nipagin complex esters, and nisin significantly inhibited the growth of the isolated strains. Khanipour et al. (2016) applied hurdle technology to inhibit the growth of *Clostridium sporogenes* in high moisture, low salt, ambient shelf-stable processed cheese analog (PCA). It was concluded that it is feasible to produce a high moisture and low salt PCA, if sufficient concentrations of potassium sorbate and nisin are added to the product.

8.3.6.3   Fruits and Vegetable Industry

Shelf-stable, ready-to-eat (RTE) intermediate moisture pineapple with increased shelf-life has been developed using hurdle technology. Osmotic dehydration, gamma radiation, and infrared drying reduced the microbial load successfully in pineapple slices, thereby increasing its shelf-life up to 40 days (Saxena et al., 2009).

Sankhla et al. (2012) preserved sugarcane juice by using hurdles like heat treatment, preservatives, irradiation, and various packaging materials. The applicability of these hurdles shows enhancement in the level of product safety and stability and therefore is recommended for the preservation of all kinds of food material. Maity and Raju (2015) developed ready-to-use tomato rasam paste using hurdle technology involving a_w, temperature, pH, and preservative. The paste was found to be acceptable on the sensory scale after 4 months of storage at ambient temperature. Arroyo et al. (2012) studied the combined effect of the simultaneous application of heat and ultrasonic waves under pressure (manothermosonication, MTS) on the survival of a strain of *Cronobacter sakazakii* in apple juice. Recovery on selective media with sodium chloride or bile salts revealed that a certain proportion of the survivors after MTS treatments were sublethally injured. Gupta et al. (2012) employed citric acid treatment in combination with gamma radiation and modified atmosphere packaging as hurdles for control of microorganisms and extending the shelf-life of minimally processed French beans.

Goyeneche et al. (2014) studied the effects of two hurdle technologies, citric acid application (CA) at 0.3%, 0.6%, and 0.9% and thermal treatments (IT) for 1, 2, and 3 min at 50°C, on the color of radish slices over 10 days of refrigerated storage. The single hurdle application of low citric acid concentration (0.3%) or intermediate immersion time (2 min) at 50°C minimized the color changes radish slices during storage. However, better results were obtained when two hurdles in series were applied. Luo et al. (2016) developed a hurdle approach that combined slightly acid electrolyzed water (SAcEW) and ultrasound (US) to improve the antimicrobial effect against *Bacillus cereus* as well as inhibition of the growth on potato. It was observed that SAcEW simultaneous with US treatment at 40°C for 3 min caused adequate synergic effects against *B. cereus* on potato as well as inhibition in the growth of *B. cereus* during storage at different from 5°C to 35°C. Tomadoni et al. (2016) developed optimum combination of ultrasound, vanillin, and pomegranate extract to improve the quality of strawberry juice by using temperatures response surface methodology.

8.3.6.4   Meat Industry

The combination of different hurdles is not universal for all meat products. Karthikeyan et al. (2000) used hurdle technology to extend the shelf life of the highly perishable indigenous traditional meat product *chevon* (capine) keema. The variable hurdles used were a_w, pH and vacuum packaging, whereas preservatives and heat treatment were the constant hurdles, which significantly improved the shelf life of keema. Chawla and Chander (2004) developed a number of ready-to-use shelf-stable meat products using a combination of hurdles (irradiation, reduced water activity, and vacuum packaging). Radiation treatment (2.5 kGy) resulted in complete elimination of *S. aureus* and *B. cereus* but not of *C. sporogenes*. The a_w of 0.85 and vacuum packaging of products prevented the growth of all three organisms inoculated into these samples during 3 months of storage at room temperature. Irradiation usefully inactivated yeast and molds which otherwise grow in the kababs after 2 months of storage.

Kanatt et al. (2006) developed shelf-stable, RTE shrimps using a combination of hurdles. The hurdles employed to cooked marinated shrimps included reduced a_w (0.85), packaging, and gamma irradiation (2.5 kGy). Microbiological analysis revealed a dose-dependent reduction in total viable count and *Staphylococcus* species. No significant changes in textural properties and sensory qualities of the product were observed on radiation treatment. These RTE shrimps were microbiologically safe and sensorially acceptable even after 2 months of storage at ambient temperature.

Thomas et al. (2008) evaluated the suitability of hot-boned pork and pork fat for processing shelf-stable pork sausages by using hurdle technology. The different hurdles incorporated were low pH, low water activity, dipping in potassium sorbate solution, vacuum packaging, and postpackage reheating. Hot-processing markedly increased the total plate counts of the sausages, but a significant difference was absent for anaerobic counts between treatments at any particular storage interval. Cold-processed sausages had higher lactobacillus counts throughout the storage period.

The effect of different hurdles such as pH, a_w, vacuum packaging, and postpackage treatment was observed in pork sausages at refrigerated temperature (Thomas et al., 2010). The combined effect of these hurdles on pork sausages inhibited the growth of yeast and molds up to 12 days. Shelf-stable, RTE pickle-type spiced buffalo meat products were also prepared and preserved by controlling different hurdles like pH and water activity (Malik and

Sharma, 2014). Several hurdles such as marination, cooking, and glycerol were applied in the production of shelf-stable chicken lollipop (Singh et al., 2014). Rodríguez-Calleja et al. (2012) developed high pressure-based hurdle strategy to extend the shelf-life of fresh chicken breast fillets. The combined effects of high hydrostatic pressure (HHP) and a commercial liquid antimicrobial edible coating consisting of lactic and acetic acid, sodium diacetate, pectin, and water followed by MAP extended the shelf-life of the sample.

8.3.6.5   Fermented Foods

In fermented foods, stability is achieved by using intelligent sequence of hurdles in different stages of fermentation and ripening. Hurdle technology has its wide applications in stability of fermented sausages, fermented vegetables, raw hams, and ripened cheese. Mini salami is produced either as fermented sausage ($a_w < 0.82$) or as dried Bologna-type sausage ($a_w < 0.85$) by using hurdle technology (Leistner and Gorris, 1997). On the other hand, stable salami-type fermented sausages can be produced at ambient temperature for extended periods of time. Important hurdles in the early stage of the ripening process of salami are the preservatives like salt and nitrite, which inhibit various bacteria present in the meat batter (Leistner, 1995). In fermented sausages, hurdles carefully and effectively inhibit various food-poisoning organisms (*Salmonella* spp., *L. monocytogenes, S. aureus, and Clostridium botulinum*) as well as other bacteria that might cause food spoilage (Man and Jones, 2000). In Japan, traditional fermented seafoods are typical examples of hurdle technology. The important hurdles in the early stages of fermentation of sushi are salt and vinegar. Fermentation of sushi employs hurdles that favor growth of desirable bacteria but inhibit the growth of pathogens (Sowjanya and Kumar, 2016). Salt and vinegar are the major hurdles in the early stages of fermentation. The hurdles can mainly control the growth of *C. botulinum* (Alasalvar, 2010), which can be achieved by controlling the pH.

8.4   CONCLUDING REMARKS

Hurdle technology can be applied in controlling and improving the overall quality of food products. The knowledge of various food safety management systems plays a major role in improving the overall quality of food products. Hurdle technology can effectively preserve foods without compromising the microbial stability and their safety. Therefore, the role of hurdle technology is very important in those foods that are stored without refrigeration. By proper selection and intelligent mixing of hurdles, excellent quality of food products having good shelf-life can be achieved. Hurdle technology with a thorough understanding of HACCP can prove to be more attractive technique in the food industry.

KEYWORDS

- homeostasis
- metabolic exhaustion
- multi target preservation
- photodynamic inactivation
- preservatives
- shelf stable products
- ultra violet radiation

REFERENCES

Alasalvar, C., (2010). Seafood Quality, Safety and Health Applications, John Wiley and Sons, ISBN 978-1-4051-8070-2, pp. 203.

Arroyo, C., Cebrián, G., Pagán, R., & Condón, S., (2012). Synergistic combination of heat and ultrasonic waves under pressure for *Cronobacter sakazakii* inactivation in apple juice. *Food Control.*, *25*, 342–348.

Balasubramanyam, B. V., Kulkarni, S., Ghosh, B. C., & Rao, K. J., (2004). *Application of Hurdle Technology for Large Scale Production of Curd Rice.* Annual Report, National Dairy Research Institute (Southern Campus), Bangalore.

Butz, P., & Tauscher, B., (2002). Emerging technologies: Chemical aspects. *Food Res. Int.*, *35*, 279–284.

Chawla, S. P., & Chander, R., (2004). Microbiological safety of shelf-stable meat products prepared by employing hurdle technology. *Food Control*, *15*, 559–563.

Esteller, M. S., Oliveira Yoshimoto, R. M., Amaral, R. L., & Silva Lannes, S. C., (2004). Sugar effect on bakery products. *Ciencia-e-Technologia-de-Alimentas.*, *24*, 602–607.

Goyeneche Rosario, Agüero María, V., Roura Sara, & Scal Karina Di., (2014). Application of citric acid and mild heat shock to minimally processed sliced radish: Color evaluation. *Postharvest Biol. Technol.*, *93*, 106–113.

Gupta, S., Chatterjee, S., Vaishnav, J., Kumar, V., Variyar, Prasad, S., & Sharma, A., (2012). Hurdle technology for shelf stable minimally processed French beans (*Phaseolus vul-*

garis): A response surface methodology approach. *LWT – Food Science and Technology*, *48*, 182–189.

Hossain, S. K. A., (1994). *Technological Innovation in Manufacturing Dudh Churpi*. PhD. Thesis, University of North Bengal, Siliguri, India, pp.122.

Kanatt, S. R., Chawla, S. P., Chander, R., & Sharma, A., (2006). Development of shelf-stable, ready-to-eat (RTE) shrimps (*Penaeus indicus*) using gamma-radiation as one of the hurdles. *LWT.*, *39*, 621–626.

Karthikeyan, J., Kumar, S., Anjaneyulu, A. S., & Rao, K. H., (2000). Application of hurdle technology for the development of Caprine keema and its stability at ambient temperature. *Meat Sci.*, *54*(1), 9–15.

Khanipour, E., Flint Steve, H., McCarthy Owen, J., Palmer, Jon, Golding Matt, Ratkowsky, David, A., Ross, T., & Tamplin, M., (2016). Modelling the combined effect of salt, sorbic acid and nisin on the probability of growth of *Clostridium sporogenes* in high moisture processed cheese analogue. *Int. Dairy J.*, *57*, 62–71.

Leistner, L., (1995). Stable and safe fermented sausages world-wide. In: *Fermented Meats*. Campbell-Platt G., Cook, P. E., (eds.). London, Blackie Academic and professional, pp. 160–175,.

Leistner, L., (1999). Combined methods for food preservation. In: *Handbook of Food Preservation*, Shafiur Rahman, M. (ed.), Marcel Dekker, New York, pp. 457–485.

Leistner, L., & Gorris, L. G. M., (1997). Food preservation by combined processes. *Final Report FLAIR Concerted Action No. 7,* Subgroup B, European Commission.

Leistner, L., (2000). Basic Aspects of Food Preservation by Hurdle Technology. *Int. J. Food Microbiol.*, *55*, 181–186.

Lekogo, B. M., Coroller, L., Mafart, P., & Leguerinel, I., (2013). Influence of long chain free fatty acids on the thermal resistance reduction of bacterial spores. *Food Nutr. Sci.*, *4*(9), 1–8.

Luo Ke, Young Kim, Seon, Wang Jun, & Oh Deog-Hwan, (2016). A combined hurdle approach of slightly acidic electrolyzed water simultaneous with ultrasound to inactivate *Bacillus cereus* on potato. *LWT – Food Science and Technology*, *73*, 615–621.

Maity, T., & Raju, P. S., (2015). Development of shelf stable tomato rasam paste using hurdle technology. *Int. Food Res. J.*, *22*(1), 171–177.

Malik, A. H., & Sharma, B. D., (2014). Shelf Life Study Of Hurdle Treated Ready-To- Eat Spiced Buffalo Meat Product Stored At 30±3 C For 7 Weeks Under Vacuum And Aerobic Packaging. *J. Food Sci. Technol.*, *51*, 832–844.

Man, C., & Jones, A., (2000). Shelf-Life Evaluation of Foods (2nd ed). In: *Scientific Principles of Shelf-Life Evaluation,* Singh, R.P. (ed). Aspen Publishing Inc, Gaithersburg, Maryland, pp. 3–22.

Marin, S., Guynot, M. E., Neira, P., Bernado, M., Sanchis, V., & Ramos, A. J., (2002). *Aspergillus flavus, Aspergillus niger* and *Penicillum corylophilum* spoilage prevention of bakery products by means of weak acid preservatives. *J. Food Sci.*, *67*, 2271–2277.

Panjagari, N. R., Londhe, G. K., & Pal, D., (2007). Effect of packaging techniques on self-life of brown peda, a milk based confection. *J. Food Sci. Tech.*, *47*, 117–125.

Pina-Pérez, M. C., Martínez-López, A., & Rodrigo, D., (2012). Cinnamon antimicrobial effect against *Salmonella typhimurium* cells treated by pulsed electric fields (PEF) in pasteurized skim milk beverage. *Food Res Int.*, *48*, 777–783.

Rodríguez-Calleja, J. M., Cruz-Romero, M. C., Sullivan, M. G. O., García-López, M. L., & Kerry, J. P., (2012). High-pressure-based hurdle strategy to extend the shelf-life of fresh chicken breast fillets. *Food Control*, *25*, 516–524.

Sankhla, S., Chaturvedi, A., Kuna, A., & Dhanlakshmi, K., (2012). Preservation of Sugarcane Juice Using Hurdle Technology. *Sugar Tech.*, *14*, 26–39.

Saxena, S., Mishra, B. B., Chander, R., & Sharma, A., (2009). Shelf Stable Intermediate Moisture Pineapple (*Ananas Comosus*) Slices Using Hurdle Technology. *LWT – Food Sci. Technol.*, *42*, 1681–1687.

Singh, V. P., Pathak, V., Nayak, N. K., & Goswami, M., (2014). Application of hurdle concept in development and shelf life enhancement of chicken lollipop. *Int. J. Curr. Microbiol. App. Sci.*, *3*(1), 355–361.

Sowjanya, K., & Kumar, K. V. P., (2016). Hurdle Technology in Food Preservation. *IOSR-JESTFT.*, *10*(5), 9–13.

Thippeswamy, L., Venkateshaiah, B. V., & Patil, S., (2011). Effect of modified atmospheric packaging on the shelf stability of paneer prepared by adopting hurdle technology. *J. Food Sci. Technol.*, *48*, 230–235.

Thomas, R., Anjaneyulu, A. S. R., & Kondaiah, N., (2008). Effect of hot-boned pork on the quality of hurdle treated pork sausages during ambient temperature (37 ± 1°C) storage. *Food Chem.*, *107*, 804–812.

Thomas, R., Anjaneyulu, A. S. R., & Kondaiah, N., (2010). Quality of hurdle treated pork sausages during refrigerated (4 ± 1°C) Storage. *J. Food Sci. Technol.*, *47*, 266–272.

Tomadoni, B., Cassani, L., Ponce, A., Moreira, M. R., & Agüero, M. V., (2016). Optimization of ultrasound, vanillin and pomegranate extract treatment for shelf-stable unpasteurized strawberry juice. *LWT – Food Science and Technology*, *72*, 475–484.

Xu, Xi-Lin, Guang-Li, Feng, Hong-Wei, Liu, & Xiaofeng, Li, (2014). Control of spoilage microorganisms in soybean milk by nipagin complex esters, nisin, sodium dehydroacetate and heat treatment. International Conference on Food Security and Nutrition. *IPCBEE* vol. 67, IACSIT Press, Singapore DOI: 10.7763/IPCBEE.

Yousef, & Courtney, P. D., (2003). Basics of stress adaptation and implications in new generation foods. In: *Microbial Stress Adaptation and Food Safety*, Yousef, A. E. & Juneja, V. K. (eds.), CRC Press LLC, Boca Raton, FL, USA.

CHAPTER 9

COMPOSITION, PROPERTIES, AND PROCESSING OF MUSHROOM

VIVEK KUMAR[1], HARISH KUMAR SHARMA[2], and ANJALI SRIVASTAVA[1]

[1]Harcourt Butler Technical University, Kanpur–208002, Uttar Pradesh, India

[2]Sant Longowal Institute of Engineering and Technology, Longowal–148106, Punjab, India

CONTENTS

9.1   INTRODUCTION

Mushrooms are macroscopic fungi that lack chlorophyll. They are a unique food because their production involves conversion of inedible plant wastes into palatable food with characteristic flavor and texture. Beside the flavor and texture, the nutritive and medicinal properties of mushrooms have enhanced their production and consumption across the world. World production of mushrooms is estimated about 12 million tones, and the annual growth rate is still above 8%. India too, though late starter, is fast catching up with mushroom production, and the current production has crossed one lakh ton mark with annual growth rate of above 15% (Singh, 2010). The venture is no more confined to the seasonal growing in the northern region; it has spread far and wide in the country. Besides the seasonal farmers, many big environmentally controlled units have also come up as export-oriented units. The country is proud to have the biggest mushroom unit of the world producing 200 tons of button mushroom per day, and its export accounts for about 25% of the US imports. Mushroom cultivation is a major food industry, which involves the bioconversion of cellulose waste into edible biomass. The cultivation of mushroom has a great potential for the production of protein-rich quality food and recycling of cellulose agro-residues and other wastes. Out of 38,000 known mushroom varieties, less than 25 are accepted as food, and few of them have assumed commercial significance (Singh et al., 1999). The most commercial varieties grown in India are *Agaricus bisporus* (white button mushroom), *Pleurotus* spp. (oyster mushroom), *Lentinus edodes* (shiitake or Japanese mushroom), *Volvarella volvacea* (paddy straw mushroom), *Flammulina velutipes* (winter mushroom), and *Auricularia polytricha* (Jew's ear mushroom) (Singh et al., 1999). Earlier, the cultivation of mushrooms was mostly carried out in hilly areas. From the adoption of scientific approaches in the cultivation techniques, researchers have succeeded in improving the quality and yields of mushroom along with a wider area of cultivation. The most widely cultivated edible mushroom is *A. bisporus* comprising 37.7% of the world's mushroom production, followed by *L. edodes* with 25.4% and *Pleurotus* spp. with 14.2% (Chang, 1999; Boa, 2004). Mushrooms have long been valued as the nutritional foods and are considered good source of carbohydrates (3–21%), digestible proteins (10–40%), and dietary fiber (3–35%) on dry weight basis. Mushroom proteins are rich in amino acids, glutamic acid (12.6–24.0%), aspartic acid (9.10–12.1%), and arginine (3.70–13.9%) but deficient in methionine and cysteine (Mattila et al., 2002; Cheung, 2008). Total digestible and nondigestible carbohydrate content of mushrooms varies between 35–70% depending on the species (Cheung, 2010). The fat content is generally low (2–10% db) and mainly consists of unsaturated fatty acids that accounts about 75% (w/w) of the total fatty acid present (Reis et al., 2012). Mushrooms are good source of vitamins, particularly riboflavin, thiamine, niacin, biotin, pantothenic acid, folic acid, and vitamin B12. They also contain minerals such as sodium, potassium, calcium, magnesium, and phosphorus along with microelements such as iron, copper, zinc, and manganese (Cheung, 2008). Processed mushrooms are nutritionally sound and good dietary component for vegetarians and are considered suitable for heart and persons with diabetes (Breene, 1990).

Mushrooms are useful in preventing diseases such as hypercholesterolemia, hypertension, and cancer and hence are reported as therapeutic foods (Bobek and Galbavy, 1999). Some recently isolated and identified compounds originating from mushrooms have shown medicinal properties such as immune modulator, cardiovascular, liver protective, antifibrotic, anti-inflammatory, antidiabetic, antiviral, and antimicrobial activities (Gunde-Cimerman, 1999; Ooi, 2000). Mushrooms are rapidly perishable commodities and deteriorate within a day after harvest. Therefore, fresh mushrooms have to be processed to extend their shelf-life.

Among the various methods employed for preservation, canning is considered as the most frequently adopted method on the commercial level. Mushroom can be processed in various ways such as drying and pickling to extend their shelf-life. Drying is a comparatively cost-effective method, and dried mushrooms packed in airtight containers can have a shelf-life of above 1 year (Bano et al., 1992). The dehydration of mushroom through a combination of hot air and microwave drying systems yielded a good quality product of satisfactory rehydration and flavor retention (Yang and Maguer, 1992). Pre-treatments through different ways, viz., washing with water or adding different concentrations of potassium metabisulfite (KMS), salt, and sugar, either by using one or in combination with each other may help in controlling enzymatic browning, preserving color, retaining flavor for a longer time, and stabilizing and/or modifying textural properties. Freezing is also a widely used technique for preservation, which also helps in retaining the nutritive value and other quality parameters like color, texture, and flavor without damaging the mushroom cells. The modified packaging of fresh mushrooms using high density polyethylene (HDPE), polypropylene (PP), laminated aluminum foil (LAF), high density polyethylene under vacuum (HDPEV), etc. and storing them at 0°C render good storage stability (Ajayi et al., 2015).

9.2   VARIETIES AND PRODUCTION

The Chinese were the first to grow edible mushrooms. As early as 600 AD, *Auricularia* was cultivated. The most common varieties grown in suitable ecological conditions (on a well-defined substrate and under full climatization) in Asian countries are *A. bisporus* (white button mushroom), *L. edodes* (shiitake or Japanese mushroom), *Pleurotus* spp. (oyster mushroom), *F. velutipes* (winter mushroom), *V. volvacea* (paddy straw mushroom), and *Auricularia polytricha* (Jew's ear mushroom) (Chang, 2008). The other species, including *A. blazei* (*A. brasiliensis*), *Grifola frondosa* (Maitake), and *Ganoderma lucidum* (Lingzhi) are also considered as important species as they have some special functional characteristics (Table 9.1).

These varieties also have significant difference in size, shape, flavor, texture, etc., which play an important role in their application in the food industry. *Tremella fuciformis, Pholiota nameko, Lepista nuda* (blewit), and *Coprinus comatus* (shaggy mane) also have significant medicinal and functional properties. Mushroom production is completely different from growing green plants. Mushrooms do not contain chlorophyll and therefore depend on other plant material (the "substrate") for their food. Each mushroom species generally prefers a particular growing medium, although some species can grow on a wide range of materials (Bano et al., 1992).

China has become the top-producing nation for all edible mushrooms, turning out over 46% of the world's supply, and India stands at the 16th position in the production of different varieties of mushroom (FAO, 2013). The share of India in the world production of mushrooms is 3%, as shown in Figure 9.1. The mushroom industry in developing countries (especially Asian countries) is often overwhelmingly focused on one mushroom species, *A. bisporus*. *A. bisporus* accounts highest for about 32% followed by *L. edodes* 25% and *Pleurotus* 14% among the edible fungi cultivated globally (Figure 9.2).

9.3   NUTRITIONAL COMPOSITION

Mushrooms are consumed for their nutritional value and/or palatability. The composition of edible mushrooms determines their nutritional value and sensory properties, which differs according to species. The variation in composition may also be due to substratum, atmospheric conditions, age,

TABLE 9.1 Characteristics and Production of Most Common Varieties of Mushroom

Scientific name	Common name	Characteristics	Global Production		Biological nature
			Share	Countries	
Agaricus bisporus	Champignon, Button mushroom	Creamy white to pale tan Mildest flavor Firm tiny to jumbo	32%	All over the world, particularly in Great Britain, the Netherlands and India	Leader in Production and Technology For growth of mycelium air temperature is 60°C for at least 4 hours. Then temperature is lowered to 50°C for 8 to 72 hours. CO_2 is maintained at 1.5 to 2% and the ammonia level drops below 10 ppm.
Lentinula edodes	Xiang-gu (Chinese), Shiitake (Japanese), Black forest/ Oak mushroom	Meaty tan to dark brown umbrella-like caps Distinctively smoky flavor	25%	Japan, China and Korea	A Mushrooming Mushroom Belongs to low temperature mushrooms Optimum temperature for mycelial growth is 23–25°C. Optimum temperature of fructification is 15°C C:N ratio of the substrate should be 25 to 40: 1
Pleurotus sajorcaju	Grey oyster mushroom, Phoenix-tail mushroom, Indian oyster	Delicate brown, grey, or reddish caps on grey-white stems Peppery flavor mild when cooked Velvety and trumpet shaped	14%	Britain and Ireland mainland Europe, Japan, North America.	A Mushroom of broad adaptability Mycelium growth (optimum) temp. is 23–28°C Fruiting body development (optimum) temp. is 18–24°C Optimum pH of the compost is 6.8–8.0 C:N ratio of the substrate should be 30–60: 1

TABLE 9.1 (Continued)

Scientific name	Common name	Characteristics	Global Production		Biological nature
			Share	Countries	
Volvariella volvacea	Paddy straw mushroom, Chinese mushroom	Distinctly delicate yet slightly musty flavor; tiny, bite-sized shape, long white stems with a bulbous base, and drooping yellow to brown cap with a partial veil	5%	Throughout East and Southeast Asia	A High-Temperature Cultivated Mushroom. The temperature, 28–36°C and relative humidity, 75–85% is required for mycelium growth. C:N ratio in the substrate-40–60: 1
Flammulina velutipes	Enokitake, gold mushroom Winter mushroom	Cap: 1–7 cm; convex, moist and sticky when fresh; bald. Color fairly dark orange brown to yellowish brown and Stem whitish to pale yellow and it becomes covered with a dark, rusty brown to blackish velvety coating as it matures.	3%	China, Japan, Vietnam and Korea	Low-temperature cultivated mushrooms

TABLE 9.1 (Continued)

Scientific name	Common name	Characteristics	Global Production		Biological nature
			Share	Countries	
Auricularia polytricha	Jew's ear, wood ear, jelly ear mushroom	Resembles the color of the tree bark it lives on, but it turns darker and becomes nearly black Texture is rubbery and jelly like and it smells very earthy, when cooked it absorbs flavor and take on a crunchy texture thick and smooth-skinned, though wrinkly	8%	Throughout temperate and subtropical zones worldwide, and can be found across Europe, North America, Asia, Australia, South America and Africa	Ideal temperature for cropping is 15–25°C which of course depends upon the strain used. Fruiting body development (optimum) temp. is 25–28°C
Agaricus blazei (*Agaricus brasiliensis*) **and** *Grifola frondosa*	God's mushroom, Royal Sun Agaricus, Himematsutake	Cap color white, grey to reddish-brown. Its surface has silk like fibres The shape of the cap begins as a hemisphere then it develops into a convex shape The mushroom's flesh tastes like green nuts and smells like almonds.	13%	Originally described from north-eastern United States and Canada, California, Hawaii, Great Britain, The Netherlands, Taiwan Philippines, Brazil.	Two Important Medicinal Mushrooms Belongs to middle temperature mushrooms. Optimum growth temp. for mycelium is 23°C to 27°C. Optimum developmental temp. of fruiting bodies is 18 to 25°C. Optimum pH of the compost – 6.5–6.8

TABLE 9.1 (Continued)

Scientific name	Common name	Characteristics	Global Production		Biological nature
			Share	Countries	
Ganoderma lucidum	Ling Zhi, Reishi	Red, black, blue/green, white, yellow, and purple		China, Japan, and other Asian countries	A Leader of Medicinal Mushrooms
		Soft red-varnished, kidney-shaped cap and white to dull brown pores underneath, depending on specimen age			Lack of availability was largely responsible for the mushroom being so highly cherished and expensive

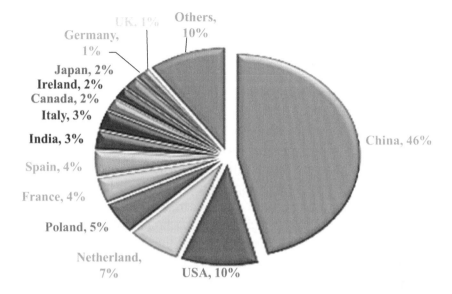

FIGURE 9.1   Country-wise production of mushroom (adapted from FAO, 2013).

and part of the fructification. Mushroom is rich in proteins and fibers but poor in fat. It contains all the essential amino acids but limited amount of the sulfur-containing amino acids cysteine and methionine (Chang, 1991).

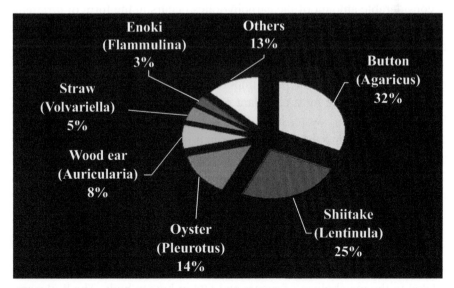

FIGURE 9.2   Production of different mushroom varieties in the world (adapted from FAO, 2013).

Therefore, it can supplement cereal-based formulations, which are generally deficient in lysine. Nutritional compositions of various species of mushroom are given in Table 9.2.

9.3.1 MOISTURE CONTENT

Moisture content of the fresh mushroom is generally in the range of 85–92%, and it varies with harvesting time, maturation period, and environmental conditions (Kamugisha and Sharan, 2005). High moisture content and water activity affect the texture and is responsible for short shelf-life of fruiting bodies. Therefore, moisture content can be decreased through various methods like cooking, drying and dehydration of mushroom for the extension of shelf-life. For instance, after boiling for 10 min, the weight of the cultivated species *A. bisporus* (portabella) decreases by 30%, but only 11–16% weight loss was recorded in the species *G. frondosa* (maitake) and *F. velutipes* (enokitake) (Dikeman et al., 2005).

9.3.2 CARBOHYDRATES

Carbohydrates are the major constituents of the dry matter of mushroom. A considerable amount of the carbohydrates occurs in the form of different sizes of polysaccharides. Glycogen and some indigestible forms of carbohydrate such as dietary fiber, cellulose, chitin, mannans, and glucans (Manzi and Pizzoferrato, 2000; Pizzoferrato et al., 2000; Manzi et al., 2001) are present in mushroom, which are important in the proper functioning of the alimentary tract. The most abundant polysaccharides are chitin and α- and β-glucans. Chitin is a water-insoluble structural polysaccharide (80–90% of dry matter in mushroom cell walls), which is indigestible for humans and apparently decreases the availability of other mushroom components (Kalac, 2013). In addition to chitin, it also contains other insoluble fiber compounds like hemicelluloses and pectic substances. The high proportion of insoluble fiber seems to be nutritionally desirable (Kalac, 2009). Mushroom glucans are also components of soluble and insoluble dietary fiber (Synytsya et al., 2009). β-glucans possess marked immunity-stimulating effects, resistive effect against allergies, and may also contribute to physiological processes related to the metabolism of sugars and fats in human body. The β-glucans content in oyster, shiitake, and split gill mushrooms are considered to be the

TABLE 9.2 Nutritional Composition of Various Species of Mushroom (Dry Weight Basis g/100 g)

Mushroom varieties	Carbohydrate (%)	Protein (%)	Fat (%)	Ash (%)	Energy (kcal/100g)	References
Agaricus bisporus	74.00	14.08	2.18	9.74	325	Reis et al., 2012
Pleurotus sajor-caju	55.30	37.40	1.00	6.30	-	Akyuz & Kirbag, 2010
Lentinula edodes	87.14	4.40	1.73	6.73	772	Reis et al., 2012
Pleurotus ostreatus	75.11	13.23	3.58	8.08	-	Akata et al., 2012
Flammulina velutipes	70.85	17.89	1.84	9.42	246	Pereira et al., 2012
Volvariella volvaceae	52.30	36.50	2.20	9.00	-	Hung & Nhi, 2012
Calocybe indica	48.50	21.40	4.95	13.10	-	Alam et al., 2008
Auricularia polytricha	88.60	7.20	1.70	2.50	-	Hung and Nhi, 2012
Auricularia auricula	76.86	19.27	0.82	3.05	391	Obodai et al., 2014

most effective (Rop et al., 2009). In edible mushroom, the dominant sugar is mannitol along with other sugars as glucose, galactose, trehalose, mannose, and fructose (Wannet et al., 2000).

9.3.3 PROTEIN

Protein is an important component of dry matter of mushrooms. On a dry weight basis, mushrooms normally contain 4.4% to 37.4% protein (Table 9.2), as compared to 7.3% in rice, 13.2% in wheat, 39.1% in soybean, and 25.2% in milk. Protein compounds constitute more than half of total nitrogen, and their content depends on the composition of the substratum, species of mushroom, size of pileus, and harvest time. The mushroom protein contains all the exogenous amino acids; however, the level of some of them is insufficient. Apart from essential amino acids, a considerable amount of alanine, glycine, arginine, histidine, glutamic acid, aspargic acid, proline, and serine can also be found in mushroom. The relative concentration of the 20 amino acids in protein is an index of quality of a protein. Twenty biologically active amino acids in humans were assayed in mushrooms and are represented in Table 9.3 (Beluhan and Ranogajec, 2011; Razak, 2013; Afiukwa et al., 2015; Woldegiorgis et al., 2015). Protein content virtually did not change during air-drying of mushrooms at 40°C or freezing up to −20°C, while a significance decrease was observed during boiling of fresh mushrooms (Barros et al., 2007).

9.3.4 FAT

Fat content in different species of mushrooms ranges from 0.82 to 4.95% (db) (Table 9.2). The fat content of mushroom is low; however, it contains unsaturated fatty acids, which constitute over 70% of the total fatty acids (Kalac, 2009). Hanus et al. (2008) examined 15 species of wild edible mushrooms belonging to the genus *Boletus* (*Phylum basidiomycota*) and found that linoleic acid (38–58%), oleic acid (15–42%), and palmitic acid (7–17%) were the major fatty acids present in various species. An appreciable high amount of linoleic acid in various species of mushrooms is a significant factor to consider mushrooms as a healthy food.

TABLE 9.3 Amino Acid Profiles of Some Common Mushrooms (g/100 g of Protein, db)

S.No	Amino acid	Abbreviation	A. bisporus	P. sajor-caju	L. edodes	P. ostreatus	F. velutipes	V. volvacea	A. polytricha	A. auricula
1	Aspartic acid	Asp	0.92	ND	0.79	1.76	2.59	5.21	ND	3.00
2	Glutamic acid	Glu	0.56	ND	0.23	0.46	2.99	2.10	ND	5.53
3	Serine	Ser	0.70	ND	0.20	0.40	0.21	4.47	ND	2.07
4	Asparagine	Asn	0.44	ND	0.36	0.75	ND	ND	ND	ND
5	Glutamine	Gln	0.49	ND	0.60	0.18	ND	ND	ND	ND
6	Histidine	His	ND	0.02	ND	ND	2.54	4.25	0.03	0.96
7	Glycine	Gly	3.00	ND	1.65	1.61	0.15	2.34	ND	1.32
8	Threonine	Thr	ND	0.05	ND	ND	5.21	5.57	0.02	1.74
9	Alanine	Ala	1.25	ND	0.65	0.18	1.95	5.77	ND	3.44
10	Arginine	Arg	0.95	ND	0.22	0.36	0.49	4.56	ND	3.37
11	Tyrosine	Tyr	0.55	ND	0.95	0.30	0.04	ND	ND	1.03
12	Cysteine	Cys-SS-Cys	0.64	ND	0.14	0.31	1.39	ND	ND	0.30
13	Valine	Val	ND	0.05	ND	ND	1.54	2.58	0.03	1.25
14	Methionine	Met	1.16	0.019	0.30	0.75	0.03	0.63	0.007	0.35
15	Tryptophan	Trp	ND	0.01	ND	ND	ND	ND	ND	ND
16	Phenylalanine	Phe	0.32	0.05	0.12	0.19	1.23	1.03	0.04	1.06
17	Isoleucine	Ile	0.32	0.04	0.19	0.22	0.44	1.64	0.01	1.03
18	Leucine	Leu	1.83	0.07	0.11	0.33	0.73	1.13	0.05	1.94
19	Lysine	Lys	0.29	0.06	0.09	0.14	5.68	ND	0.05	1.22
20	Proline	Pro	ND	ND	ND	ND	ND	ND	ND	0.20

ND: not detected.

9.3.5   VITAMINS

Mushrooms are rich in vitamin B, including riboflavin, niacin, and panto-thenic acid, which help to provide energy by breaking down carbohydrates fats and proteins. Riboflavin helps to maintain healthy red blood cells. Pantothenic acid helps with the production of hormones and also plays an important role in the functioning of the nervous system. Niacin promotes healthy skin and ensure proper functioning of the digestive and nervous sys-tems (Duyff, 2006).

9.3.6   MINERALS

Major minerals present in the mushrooms are potassium, sodium, phos-phorous, calcium, and magnesium. Total concentration of the major min-erals is approximately 56% to 70% of the total ash content. Potassium in particular is most abundant and accounts for around 45% of the total ash content. Sodium and calcium are present in approximately equal concen-trations in most varieties of the mushrooms, except in *L. edodes* in which calcium is present in an especially large amount. Mushrooms also contain other essential minerals like copper, zinc, iron, manganese, molybdenum, and cadmium in traces. Selenium functions as an antioxidant to protect body cells from damage that might lead to cancer, heart disease, and other diseases of aging (Duyff, 2006). It has also been found to be important for the immune system and fertility in men (*www.nlm.gov/medlineplus/ency/article/002414.htm*).

Foods from animal origin and grains are good sources of selenium, but mushrooms are among the richest sources of selenium and provide 0.57–19.46 mg/kg depending on the type, age, and place (Savic, 2009). Ergo-thioneine is a naturally occurring antioxidant that helps to protect the body cells. Mushrooms contains ergothioneine in the range of 0.4–2.1 mg/g of dry weight in the different varieties of mushroom, namely white button, crimini, portabellas, king oyster, maitake, oyster, and shitake (Dubost et al., 2006). Among these varieties, shitake has the highest and white button has the low-est concentration of ergothioneine. Copper helps in the formation of new red blood cells, which carry oxygen throughout the body and also keeps bones and nerves healthy. Potassium is an important mineral that helps in the main-tenance of normal fluid and mineral balance, which control blood pressure

(Duyff, 2006). Mushrooms contain 98–376 mg of potassium per 84 g (per serving), which is 3–11% of the daily intake (USDA, 2009).

9.3.7 FLAVOR-IMPARTING COMPONENTS

The flavor of mushroom plays an important role in the consumption of edible mushroom. The flavor-imparting compounds can vary markedly among species and culture conditions (Hadar and Dosoretz, 1991). The volatile fractions of chemical composition are largely responsible for mushroom flavor. Hundreds of odoros compounds have been identified in different variety of mushrooms. The derivatives of octane, 1-octene and 2-octene, alcohols and their esters with volatile fatty acids, and ketones form the very characteristic group of mushroom aroma (Fons et al., 2003). About 25 flavor compounds were identified by Venkateshwarlu et al. (1999) in three varieties of mushroom, namely *A. bisporus, Pleurotus florida*, and *Calocybe indica*. The most abundant compound was 1-octen-3-ol, and its concentration was highest in *P. florida*. The other important compounds were n-octanol, 3-octanol, 3-octanone, 2-octen-1-ol, n-pentanal, n-pentanol, 1-hexanol, benzyl alcohol, 2-octenal, and n-octanal. Certain nonvolatile components, namely L-glutamic acid, short-chain fatty acids, carbohydrates, proteins, and nonprotein nitrogenous substances like nucleotides contribute to the characteristic flavor of mushrooms. Other nonvolatile compounds such as free 5'-nucleotides of adenosine, guanosine, inosine, uridine, and xanthosine are reported in different varieties of mushrooms, *Boletus edulis, Cantharellus cibarius*, and *Craterellus cornucopioides* (Beluhan and Ranogajec, 2011).

9.4 MEDICINAL PROPERTIES

Mushrooms have become increasingly attractive as functional foods for their potential benefits on human health. Relevant nutritional and functional aspects of mushrooms include high fiber, low fat, and a low concentration of sodium as well as the presence of different compounds such as phenolic compounds, eritadenine, sterols (such as ergosterol), chitosan, and triterpenes. Commercial mushrooms have been found to be medically active in several therapies such as antitumor, antiviral, and immune-modulating treatments (Wasser and Weis, 1999). Functional compounds in mushrooms are able to alleviate cholesterolemia, modulate the immune system, and inhibit tumoral growth (Zhang et al., 2001). Other functional compounds in edible mushrooms are beta glucans, which are particularly effective in lowering blood cholesterol levels and glycemic response (Cheung, 1998; Bobek et al., 2001).

9.4.1 PREVENTION OF CARDIOVASCULAR DISEASES

Based on some metabolic markers (low-density lipoprotein (LDL), high-density lipoprotein (HDL), triacylglycerol, cholesterol, homocysteine, blood pressure, homeostatic function, and oxidative and inflammatory damage), it is proven that mushroom intake potentially reduces the risk of cardiovascular diseases (Yamada et al., 2002; Mori et al., 2008). According to Eva et al. (2010), mushroom intake clearly has a cholesterol-lowering effect or hypo-cholesterolemic effect by different mechanisms such as decreasing very low-density lipoproteins, improving lipid metabolism, inhibiting the activity of hydroxymethylglutaryl-coenzyme A (HMG-CoA) reductase, and consequently preventing the development of atherosclerosis. Antioxidant and anti-inflammatory compounds present in mushrooms may contribute to reduce the atherosclerosis risk. Mushroom varieties like *A. bisporus* (Neyrinck et al., 2009), *Auricularia auricula* (Cheung, 1996), *P. ostreatus* (Kajala et al., 2008), and *T. fuciformis* (Cheng et al., 2002) have the capabilities of reducing the level of LDL cholesterol (act as anti-atherosclerosis), while *P. citrinopiletus* (Hu et al., 2006) and *P. florida* (Bajaj et al., 1997) are highly effective in reducing the total lipid level and thus resulting in lowering the risk of cardiovascular diseases.

9.4.2 ANTIOXIDANT ACTIVITY

Phenolic substances are the major naturally occurring antioxidant compounds found in mushrooms, responsible for scavenging effects of free radicals (Yang et al., 2002). Levostatin, a cholesterol-lowering drug derived from *Pleurotus* species, and its analogs are reported to be the best therapeutic agents for correcting hypercholesterolemia (Endo, 1988). Jose and Janardhanan (2000) reported that the ethyl acetate and methanol extracts of *P. florida* exhibited potent hydroxyl radical scavenging and lipid peroxidation inhibition activities.

The concentrations of cysteine, methionine, and aspartic acid are reported to be higher in *P. ostreatus* than in other edible mushrooms such as *A. bisporus* (brown), *A. bisporus* (white), and *L. edodes* (Mattila et al., 2002). In addition, *P. ostreatus* has also been reported to possess excellent reducing power for ferric ions (Lin, 1999). Tree oyster mushrooms possess better antioxidant activity, reducing power, scavenging abilities, and higher total phenol content (Yang et al., 2002). The antioxidant activities of specialty commercial mushrooms such as shiitake, winter, and oyster are comparable (Mau et al., 2001; Mau et al., 2002).

9.4.3 IMMUNOREGULATORY ACTIVITY

Immunomodulatory/Immunoregulatory activity deals with the critical factors of protecting humans against numerous diseases. Mushrooms such as *L. edodes, Schizophyllum commune, G. frondosa,* and *Ganoderma lucidum* possess immunomodulatory activities due to the presence of bioactive compounds, including polysaccharides, proteoglycans, proteins, and triterpenoids (Guo et al., 2012). Numerous bioactive polysaccharides or polysaccharide-protein complexes present in mushrooms may enhance innate and cell-mediated immune responses. Stimulation of the host immune defense systems by bioactive polymers from medicinal mushrooms has significant effects on the maturation, differentiation, and proliferation of many kinds of immune cells in the host (Chihara, 1992). Other than polysaccharides, molecules like glucans are relatively resistant to stomach acid and act similar to the toll-like receptor 2 (a class of proteins that play a role in the immune system) and thus can be considered as a good immune modulator (Rop et al., 2009).

9.4.4 ANTICANCER AND ANTITUMOR ACTIVITY

The need for more effective and safe treatments for chemoprevention of human cancer has increased. Some natural products such as fruits and medicinal plants have shown antiproliferative activities (Li et al., 2013a; Li et al., 2013b; Zhou et al., 2016). Mushrooms show significant inhibitory activity against breast cancer, hepatocellular carcinoma (HCC), uterine cervix cancer, pancreatic cancer, gastric cancer and acute leukemia. In addition, antitumor compounds have been identified in various mushroom species (Zhang et al., 2006).

The mushrooms showing anticancer properties belong to the genus *Phellinus, Pleurotus, Agaricus, Ganoderma, Clitocybe, Antrodia, Trametes, Cordyceps, Xerocomus, Calvatia, Schizophyllum, Flammulinam suillus, Inonotus, Inocybe, Funlia, Lactarius, Albatrellus, Russula,* and *Fomes* (Patel and Goyel, 2012). Approximately 200 species of higher Basidiomycetes have been reported to exhibit antitumor activity (Yang and Jong, 1989; Mizuno, 1996). The black poplar mushroom *Agrocybe aegerita* is a popular edible mushroom with reported antitumor properties. In Russian medicine, an extract of Chaga (*Inonotus obliquus*) is used as an antitumor medicine (Wasser and Weiss, 1999).

The antitumor effects of shiitake feed have been reported by Kurashige et al. (1997). Hot water extract from the fruiting body of polyporaceae family showed a remarkable host-mediated antitumor activity (Ikekawa et al., 1968) due to presence of β-D-glucan (Mizuno et al., 1995; Wasser, 2002). The water extract of *Amauroderma rude* exerted anticancer activity (Pan et al., 2015). Ferreira et al. (2010) reported an unlimited source of compounds with potential antitumor and immune-stimulating properties, including low-molecular-weight (LMW) compounds, e.g., quinones, cerebrosides, isoflavones, catechols, amines, triacylglycerols, sesquiterpenes, steroids, organic germanium, and selenium, and high-molecular-weight (HMW) compounds, e.g., homo and heteropolysaccharides, glycoproteins, glycopeptides, proteins, RNA-protein complexes). Polysaccharides exert an anticancer effect by enhancing the host's immune system rather than by a direct cytotoxic effect (Wang et al., 1997).

Zou et al. (2015) reported that polysaccharides purified from the mushroom *Trametes robiniophila* (huaier) not only inhibited the proliferation but also suppressed the HCC tumor growth. The effect of fermented *Cordyceps sinensis* rich in selenium (Se-CS) on uterine cervical cancer in mice (Ji et al., 2014) showed that it was strongly effective in reducing the cancerous cell growth. The ceramide isolated from *A. aegerita* showed significant inhibition against COX-2 enzyme and the growth of stomach, breast and CNS cancer cell lines (Diyabalanage et al., 2008).

A new hemagglutinin was isolated from mushroom, *Boletus speciosus*, which showed anti-proliferative activity (Sun et al., 2014). An innovative strategy was suggested by Chen (2012) wherein β-glucans were used to deliver nanoparticles containing chemotherapeutic agents to the site of the colon cancer and, thus, improve the therapeutic efficacy.

9.5 PROCESSING OF MUSHROOMS

Freshly harvested mushrooms are highly perishable because of high moisture content. The shelf-life varies from 1–2 days at room temperature. Therefore, it should be marketed and consumed as soon as possible after harvest. However, the shelf-life can be enhanced for a long period by processing methods like drying, freezing, and canning.

9.5.1 DRYING

Drying is a classical method of food preservation, in which activity is reduced to such a level where biochemical and microbial activities of the product are checked. The dried mushrooms can be rehydrated by immersion in water before consumption.

9.5.1.1 Sun/Solar Drying

Sun drying is the most economical and suitable method in tropical and subtropical regions where plenty of sunshine is available. Farmers can use sun drying without any sophisticated equipment, provided the climate is hot and relatively dry. It is a low temperature drying method and therefore best suitable for heat-sensitive food materials. Suguna et al. (1995) conducted studies on dehydration of mushroom by sun drying, thin layer drying, fluidized bed drying, and solar bed drying. A natural convection solar cabinet dryer showed a drying time of 7 h at ambient temperature that varied between 29°C and 32°C. On the other hand, Kumar et al. (2013) reported that solar drying required 13 to 14 h drying time, whereas open sun drying required 19–20 h drying time. Performance of an efficient solar dryer for drying of oyster mushroom was investigated by Mustayen et al. (2015); they found that the mushroom dried at 50°C to 60°C with no effect on the color and the desired moisture content of 7.97% can be achieved without affecting the product color. A faster drying rate was observed in a solar tunnel dryer than in the natural open air drying (Basunia et al., 2011).

Drying combined with some pre-treatments appears to be a cost-effective method of preservation. Pre-treatments like blanching and irradiation also affect the quality of dried products. Blanched and chemically treated samples showed good sensory acceptability as compared to samples that were dried without giving any pretreatment. But the blanched sample was inferior to the chemical-treated sample probably because blanching increases the permeability of cell walls, which results higher shrinkage during subsequent drying. The pretreatment with 1% potassium metabisulfite renders best-quality dried mushroom in comparison to 0.5% citric acid and 0.5% potassium metabisulfite + 0.2% citric acid (Kumar et al., 2013). Irradiation, as a pre-treatment prior to drying, reduced the drying time of oyster mushroom slices. Pretreated slices dried faster than the control, with increasing dosage resulting in shorter drying time. Among the five different thin layer models, Page's model was the best to describe the solar drying behavior of slices exposed to lower doses of γ-radiation (Kortei et al., 2016).

9.5.1.2 Convective Hot Air Drying

Hot-air drying is considered as a comparatively simple, economical, and efficient method to extend the shelf-life of mushrooms. Hot-air drying of *A. bisporus* at 65°C was found to produce a product of desirable quality that was also acceptable to consumers (Lidhoo and Agrawal, 2008). However, Nour et al. (2011) reported that drying temperature of 50°C was better as it resulted in the dried products, having better rehydration characteristics, lesser shrinkage, and lighter color. The drying time was lesser in the case of oyster mushrooms than in button mushroom with cabinet drying. The time taken for drying to the safe moisture level was in the order of vacuum dryer > cabinet moisture dryer > fluidized bed dryer > microwave oven, as reported by Walde et al. (2006). The drying of *P. ostreatus* was best at 50°C in terms of highest sensory score (Hassan and Ghada, 2014). Mushroom slices treated with 1% sodium metabisulfite for 10 min prior to drying at 50°C prevented browning and resulted in satisfactory rehydration ratio. Drying of mushroom slices using hot air combined with an electro-hydrodynamic system can significantly reduce the drying time and results in a greater effective water diffusion coefficient, drying rate and reduced energy consumption. As such, this technique offers a promising solution to the drying industry for reduction of energy consumption (Dinani et al., 2014). The effect of hot air drying combined with microwave-vacuum drying was evaluated by Argy-ropoulos et al. (2011). The resulted dried product had superior quality with lower overall color variation, higher porosity, greater rehydration ratio, and softer texture in microwave-vacuum drying, when compared to the slices dried completely by conventional hot air.

9.5.1.3 Microwave Drying

Microwave drying is rapid, uniform, and energy efficient compared to conventional hot-air drying. Microwave energy combined with vacuum can be used for dehydration of heat-sensitive materials. If a suitable control system is used, microwaves provide high-quality products with lesser nutrient loss, more flavor retention, lesser color change with natural appearance, and subsequent good rehydration ability of dried products (Erle and Schubert, 2001). Microwave-vacuum drying resulted in 70–90% decrease in drying time, and the dried products had better rehydration characteristics than those with convective air drying. Microwave-vacuum-dried mushrooms also have more porous structure, which rehydrated more quickly and completely than the air-dried product (Giri and Prasad, 2007).

Microwave drying behavior of pretreated button mushroom was studied at different microwave power levels, and an optimum condition was established on the basis of rehydration ratio and sensory evaluation. Drying with pretreatment of blanching in boiling water for 3 min followed by steeping in solution of 0.1% KMS + 0.2% CA + 6% sugar + 3% NaCl at room temperature for 15 min at the microwave intensity of 400 W yielded an acceptable dehydrated product (Kar et al., 2004). Button mushroom slices of thickness 7.7 mm dried at microwave power of 202 W and 6.5 kPa pressure yielded the product with good quality attributes like color, texture, rehydration ratio, and sensory quality (Giri and Prasad, 2007). The mushroom slices dried with moderate microwave power level showed almost similar quality product as that obtained by freeze-drying (Rodriguez et al., 2007).

9.5.1.4 Freeze Drying

Freeze-dried samples of mushroom were more attractive in color and had good rehydration quality than the samples dried using hot air and vacuum drying. But the flavor of the freeze-dried mushroom was not significantly different from that of the hot air-dried mushrooms (Martinez-Soto et al., 2001). Hardness decreased with freeze drying and chewiness also varied quite significantly with freeze-drying, but cohesiveness did not exhibit any significant difference (Guiné and Barroca, 2011). Freeze-dried samples showed a porous structure that absorbed water more quickly and allowed rehydration to take place mainly at the extracellular level (Hernando et al., 2008). The effect of application of mid-infrared drying, before freeze drying and after freeze drying, showed that infrared drying after freeze drying had better color and aroma retention and good rehydration ratio (Wang et al., 2015). The appearance of the freeze-dried mushroom is similar to that of the fresh mushroom, except that the apparent density is 10 times lower.

9.5.2 FREEZING

Freezing is an increasingly popular method of mushroom preservation and offers a product with high nutritional value. Various methods of freezing like blast freezing, cryogenic freezing, plate freezing, and individual quick freezing have been applied to processing of many fresh commodities. Out of these methods, blast freezing is the most common method used in mushroom processing followed by cryogenic freezing (Jaworska and Bernaś, 2009). Blast freezers have advantages to rapidly bring down the temperature to −30°C and creating small ice crystals that gives lesser damage to mushroom cells.

The freezing of mushrooms is generally carried out by air blast freezing because it maintains the original shape in a better way but may have the problem of drip loss, when exposed to higher temperature. The different factors are involved in the quality loss of mushroom during frozen storage. The loss of firmness and decrease in the nutritive value can be the most prominent reasons during the frozen storage. In order to ensure good quality frozen mushrooms, preliminary processing like blanching or dipping can be beneficial.

9.5.3 CANNING

Canning is the most frequently adopted method on the commercial scale for the preservation of mushroom. About 38% of them are canned, holding a major share in world trade (Ravi and Siddiq, 2011). Through canning (sterilization), mushrooms can be stored for a period up to 2 years with storage costs being relatively low. *A. bisporus* is preferred more over other varieties of mushrooms for canning, but other species such as *P. ostreatus*, *L. edodes*, and *V. volvacea* and the wild *Cantharellus cibarius*, *Boletus edulis*, and *Lactarius deliciosus* are also canned or bottled (Bernaś et al., 2006; Ahlawat and Tewari, 2007). The canning process involves picking, sorting and grading, washing, blanching, filling of cans, brining, exhausting, seaming/can closing, processing and sterilization, cooling, labeling, and storage.

The optimization of the blanching time, soaking time, and concentration of ethylene diaminetetraacetic acid (EDTA), as an alternative to potassium metabisulfite and citric acid, was carried out for canning of *A. bisporus*. Arumuganathan et al. (2003) indicated that 5 min blanching time, 5 min soaking time, and 125 ppm EDTA was the best condition of pre-treatment before canning. These parameters significantly affected the whiteness of the mushroom and their keeping quality.

9.6 MUSHROOM-BASED PRODUCTS

Almost the entire domestic trade of mushrooms is in the form of fresh mushrooms, while all the export is in the preserved form (canned or freeze-dried). As mushrooms contain high moisture and are delicate in texture, it cannot be stored for more than 24 hours at the ambient conditions of the tropics. Weight loss, veil opening, browning, liquefaction, and microbial spoilage often make the product totally unsaleable. Effective processing techniques will not only diminish the postharvest losses but also result in greater remuneration to the growers as well as processors. Value can be added to the mushrooms at the various levels and to varied extent, right from grading to the readymade snacks or the main course item.

Improved and attractive packaging is another important but totally neglected area in mushrooms. It is still packed in unprinted plain polypropylene pouches in a number of countries, whereas attractive and labeled overwrapped trays are in vogue in the developed countries. A real value-added product in the Indian market is the mushroom soup powder. Technologies for the production of some other products like mushroom-based biscuits, nuggets, preserve, noodles, *papad*, candies, and readymade mushroom curry in retort pouches have been developed but are yet to be popularized. Attractive packaging of the value-added products is yet another area that may be called as the secondary value addition. While small growers may add value by grading and packaging, the industry may require the processed products for better returns as well as improvement in the demand, which will have a cascading positive effect on the production.

9.6.1 MUSHROOM SOUP POWDER

A soup is a viscous liquid food that is made by boiling vegetables or meat in water until the flavor is extracted, forming a broth. Soups are commonly used as appetizers to stimulate the appetite and aid in the flow of digestive juices in the stomach (Singh and Chaudhary, 2015). Experiments were conducted to prepare good quality ready-to-make mushroom soup powder using quality mushroom powder produced from button mushroom and oyster mushroom dried in the dehumidifying air cabinet drier. Dried button mushroom slices or whole oyster mushrooms were finely ground in pulverizers and passed through a 0.5-mm sieve. As shown in Figure 9.3 (Rai and Arumganathan, 2008), the mushroom soup powder is prepared by mixing this powder with milk powder, corn flour, and other ingredients. This means that the soup powder has to be mixed with equal quantity of water for the preparation of good quality mushroom soup with characteristic aroma and taste.

Singh (1996) developed a ready-to-reconstitute mushroom soup powder utilizing the vacuum concentrated whey, a by-product of dairy industry. Mushroom whey soup powder was also prepared by spray drying of

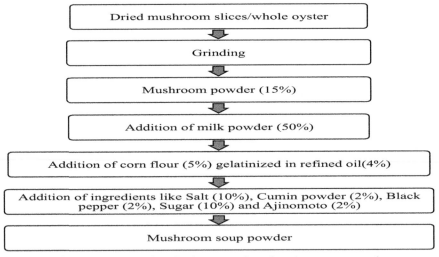

FIGURE 9.3   Flow chart for the preparation of mushroom soup powder.

cooked mushroom-whey blend containing seasonings (onion, garlic, and ginger) (Singh et al., 2003). The quality of the prepared powder showed that the dispersibility decreased during storage. The soup powder reconstituted well, when boiled in water for 2 min, and the reconstituted soup was considered acceptable with overall acceptability score of 7.1 on a 9-point hedonic scale.

White button mushroom powder was prepared using microwave-dried mushrooms, which were pre-treated with 0.5% KMS + 0.2% citric acid for 30 min (Kumar, 2015). Wheat flour, salt, fat, sugar, onion, garlic powder, dried pieces of mushroom, and skimmed milk powder were mixed with mushroom powder in different proportions to formulate different recipes for soup powder and packed in polypropylene pouches and stored at room temperature. Good overall acceptability was found during 12 months of storage period.

9.6.2 MUSHROOM NUGGETS

Nuggets, popularly known as *warrian* in Indian household, is prepared commonly from black gram, soybean powder, *urad dal* (white lentil) powder, etc. and are a rich source of protein, but lack in few essential amino acids, fibers, and antioxidant properties. Therefore, efforts were made to improve the nutritional value of the nuggets with the supplementation of mushroom. White button mushrooms are widely grown throughout world and are rich in lysine and other essential amino acids.

Sharma et al. (2015) conducted a study to prepare lysine-enriched nuggets using white button mushrooms. The flow diagram for the preparation of mushroom nuggets is shown in Figure 9.4. White button mushrooms were supplemented in different proportions from 10% to 50% in a black gram paste. The results indicated that the nutritional value, antioxidant activity, and lysine content of the nuggets increased with an increase in mushroom proportions. On the basis of physico-chemical, cooking, and textural properties, the black gram to mushroom ratio of 80:20 was the best.

9.6.3 MUSHROOM KETCHUP

Ketchup is a common and popular product relished for its typical taste and texture. It is made by concentrating the juice or pulp of the fruits or veg-

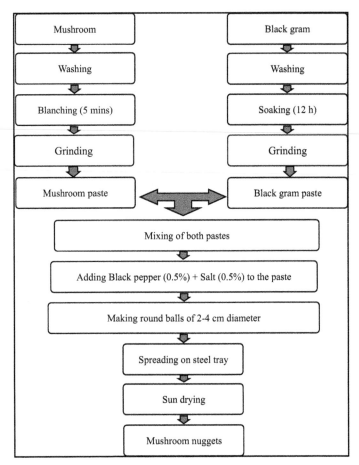

FIGURE 9.4   Flow chart for preparation of mushroom nuggets.

etables without seeds and pieces of skin. It does not flow freely and is highly viscous in nature. It also has more sugar and less acid. Freshly harvested button mushrooms are washed, sliced, and Cooked in water of half of its amount for 20 min. A mushroom paste is prepared using a mixer-grinder. The paste is placed in a stainless steel vessel, and a spice bag is immersed in it. About one-third quantity of sugar and salt are added to the paste and mixed thoroughly. After some heating, the thickening agent starch is added and boiled till the paste reduces to one-third of its original volume. The spice bag is removed and squeezed to extract its essence, and the remaining quantity of sugar and salt is then added. At the end, a small quantity of vinegar is added and mixed thoroughly. The final total soluble solids (TSS) is maintained at 35°Brix, and the product is filled in sterilized bottles.

The process for manufacturing of mushroom ketchup was optimized by Bhuiyan and Rana (2012). Optimal combination of sugar and vinegar were 76 g and 50 ml, respectively, for per kg raw mushroom paste to make the ketchup. Onion 12 g, garlic 1 g, cardamom, black pepper, cinnamon, and red chilli 0.5 g each, clove 0.5 g, salt 5 g, starch 1 g, and sodium benzoate (0.7 g/kg finished product) were the common ingredients. Tomato-mushroom ketchup with 50% tomato pulp and 50% mushroom pulp showed highest organoleptic scores. There was a significant increase in protein, crude fiber, and ash content, while TSS, acidity, total sugars, and vitamin C decreased significantly with an increase in the level of mushroom pulp in the ketchup (Kumar and Ray, 2016).

9.6.4 MUSHROOM CANDY

Mushroom candy is prepared in a very similar manner to fruit candy. Wakchaure (2008) suggested the whole procedure for manufacturing mushroom candy; according to this procedure, after harvesting, the mushroom is washed and halved into two pieces. The pieces are blanched for 5 min in 0.05% of potassium metabisulfite solution. After draining the excess water, the pieces are treated with sugar at the rate of 1.5 kg sugar per kg of blanched mushroom. Initially, the sugar potion is divided into three equal parts. On the first day, blanched mushrooms are covered with one part of sugar and kept for 24 h.

Next day, the same mushrooms are covered with the second part of sugar and are kept overnight, and on the third day, the mushrooms are removed from the sugar syrup. The sugar syrup is boiled with the third part of sugar and 0.1% citric acid to bring its concentration up to 70°Brix. Mushrooms are mixed with this syrup, and the contents are again boiled for 5 min to bring its concentration upto 72°Brix.

After cooling, the mushrooms are removed from the syrup and drained for 30 min. The drained mushrooms are placed on the sorting tables to separate and reject defective and unwanted pieces. Finally, the mushroom pieces are subjected to drying in a cabinet dryer at 60°C for about 10 h. As soon as these pieces become crispy, they are taken out, packed in polypropylene bags, and sealed. The mushroom candy can be stored up to 8 months with excellent acceptability and good taste.

9.6.5 MUSHROOM-BASED PASTA/NOODLES

Pasta/noodles are a cereal-based food that is popular worldwide because of its low cost, easy preparation, sensory properties, and long shelf-life (Bergman et al., 1994). Substitution of semolina with mushroom flour can improve the nutritive value and functional properties of pasta. Wet noodles prepared from wheat flour containing 3%, 5%, and 7% oyster mushroom and oak mushroom showed improved protein content, fiber content, and overall acceptability (Kim, 1998). Nutritious and acceptable quality of spaghetti was obtained from wheat flour supplemented with mushroom powder (Amin, 2016). Devina et al. (2008) developed a process for the preparation of noodles containing white button mushrooms. Noodles with good sensory and functional properties could be prepared using 20 g mushroom meal, 38 g wheat flour, 20 g potato meal, 0.2 g baking powder, and 2 ml edible oil. Salama (2007) prepared pasta products with oyster mushroom mycelia powder and showed that replacement of 5%, 10%, or 15% of semolina with mushroom powder resulted in pasta with acceptable cooking properties and best sensory scores for color, flavor, mouth feel, elasticity, and overall acceptability.

Durum wheat semolina was partially substituted with three different varieties of mushrooms (white button, shitake, and porcini) in pasta (Lu et al., 2016). Cooking loss and firmness of pasta were increased with the increase in mushroom powder. Porcini mushroom incorporation significantly decreased the swelling index, water absorption index, and moisture content of the cooked pasta, while no noticeable effect was observed to control in white button and shiitake mushrooms.

The effect of β-glucan-rich fractions (BGRFs) from *P. eryngii* mushroom powder on the quality of common wheat pasta was evaluated (Kim et al., 2016). The results indicated that the pasting properties, except the breakdown viscosity, were reduced with increasing percentages of BGRFs. The prepared pasta was reddish brown in color with low L^* value and high a^* value with an increase in the proportion of BGRFs. Pasta prepared from common wheat flour and 4% BGRFs was very similar to semolina pasta with regard to cooking loss. Conclusively, wheat flour containing 4% BGRFs could be used to produce pasta with an improved quality and texture similar to that of semolina pasta.

9.6.6   MUSHROOM BISCUIT/COOKIE/CAKE/BREAD

Biscuits are very popular, ready to eat, convenient, inexpensive, and an important product in human diet. Very delicious and crunchy mushroom biscuits were prepared using mushroom powder and other essential ingredients like *maida*, sugar, oil, baking powder, ammonium bicarbonate, salt, vanilla, milk powder, and glucose. Fat is used in biscuits to improve the softness of the biscuits. Ammonium bicarbonate is an aerating agent, which increased volume and improved textural quality of the biscuits. The general flow chart for mushroom cookies as suggested by Rai and Arumganathan (2008) is represented in Figure 9.5.

Sharma et al. (1991) successfully prepared biscuits from mushrooms. All the ingredients are mixed in a mixer for 3 to 5 min. The dough is kept at 30°C in an oven for 90 min, spread to a thickness of 2 to 4 mm over a cleaned platform, cut into circular or rectangular shapes (required shape), and then baked for 10–12 min at 210°C in a laboratory baking oven.

The primary ingredient used for the production of biscuits is wheat flour, which is deficient in several nutrients including essential amino acids, vitamins, minerals, and dietary fiber. Mushroom being a good source of protein, essential amino acids especially lysine and tryptophan, vitamins, minerals

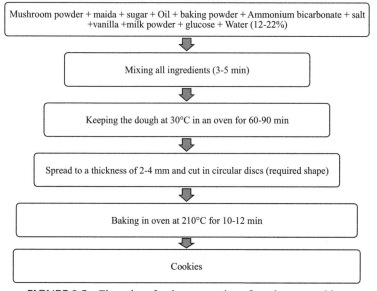

FIGURE 9.5 Flow chart for the preparation of mushroom cookies.

and dietary fiber, can be supplemented with wheat flour for the preparation of bakery goods. Prodhan et al. (2015) developed biscuits enriched with mushroom (*P. sajor-caju,* PSC) powder. Biscuits with 15% mushroom powder had higher protein content, dietary fiber, and acceptable sensory attributes in comparison to control.

A good quality shiitake mushroom biscuits were developed by replacing 10% wheat flour with treated and untreated dehydrated shiitake mushroom powder (Singh et al., 2016). Crude protein, iron, zinc, phosphorus, and calcium content were significantly higher in the mushroom biscuits than in the control sample. Replacement of wheat flour in biscuits with different levels (10%, 20%, and 30% of the mixture) of mushroom powder (MP) and sweet potato flour (SPF) in equal proportion (1:1) was carried out (Ibrahium and Hegazy, 2014). The levels of protein, fiber, ash, Fe, Ca, K, P, and indispensable amino acids (IAAs) were increased with an increase in the MP/SPF mixture. Chemical score of the most limited amino acid lysine in wheat biscuit (37%) was increased to 81% after supplementation with 30% MP/SPF mixture. The incorporation of 20% MP/SPF considerably improved the nutritional and sensory quality of biscuits. Nutrient composition and sensory parameters of butter biscuits incorporated with various levels of gray oyster mushroom (PSC) powder have also been studied (Rosli et al., 2012). The incorporation of PSC powder up to 4% to replace wheat flour improved crispiness and flavor, protein, dietary fiber, and β-glucan but did not affect the fat content of the butter biscuits.

Chomdao et al. (2005) developed a mushroom cookie formulation with four dried mushrooms (Jew's black ear, Hungarian, straw, and Puethan mushroom powders). Cookies prepared with Jew's black ear mushroom powder had the best properties. Eissa et al. (2007) used oyster mushroom (PSC) flour as a partial replacement for wheat flour in the production of biscuits to improve their nutritional quality.

Breads were prepared from wheat flour supplemented with oyster mushroom (*P. ostreatus*) powder, and the effects of the supplementation of this powder on dough rheology and bread quality were examined (Hyung et al., 2005). The loaf volume and peak and final viscosities decreased with the increased amount of the oyster mushroom powder. Dark color and coarse crumb texture were observed in the breads with oyster mushroom powder. Sensory evaluation revealed that the addition of 1% oyster mushroom powder could be supplemented to make an acceptable quality of bread. Wheat-mushroom bread was developed (Ndung et al., 2015) with supplementation of *P. ostreatus* mushroom powder in the ratio of 0%, 5%, 10%, and 15%. Supplementing mushrooms in wheat bread increased protein, minerals, B-vitamins, and amino acids compared to wheat bread alone. High nutritional quality bread can be prepared with the addition of 10% mushroom flour without affecting the sensory quality. Oyster mushroom powder could be added to wheat flour up to 10% without any detrimental effect on the sensory properties of bread (Okafor et al., 2012).

Chang and Ki (2004) prepared a sponge cake containing mushroom (*P. eryngii*) powder. Batters prepared with the added mushroom powder showed higher specific gravity and viscosity. The addition of mushroom powder decreased the cake height and volume. The cake with good sensory quality was prepared from 3% or 5% mushroom powder.

9.6.7 MUSHROOM MATHRI AND PAPAD

Mathri (traditional salty snack) is popular deep fat-fried Indian snack traditionally prepared from refined wheat flour. Kumar et al. (2006) developed *mathri* prepared from white mushroom powder and refined wheat flour to improve the quality of the product and add variety to traditional Indian snacks. A mixture containing 9.0 g mushroom powder, 91.0 g refined wheat flour, 0.5 g baking powder, and 20.0 g hydrogenated vegetable oil gave a product of good organoleptic quality.

Mathri prepared with the fortification of oyster (PSC) mushrooms powder at a level of 10% showed similar organoleptic quality as refined wheat flour *mathri* (Verma and Singh, 2016). Tyagi and Nath (2005) prepared *papads* with dried mushroom (*P. florida*) powder up to 40% level. Incorporation of mushroom powder up to 25% in black gram *papads* or up to 10% in green gram (*mung* beans) or black gram-green gram *papads* gave highly acceptable products. The fat absorption index and expansion of *papads* during frying was reduced with an increase in mushroom powder in the mixture.

9.6.8 MUSHROOM CHIPS

Potato chips are the most preferred and accounts nearly 33% of the global snacks market. Mushroom chips are the rich sources of nutritive values such as proteins, fibers, vitamins, and minerals. For the preparation of mushroom chips, freshly harvested button mushrooms are washed, sliced (2 mm), and blanched in 2% brine solution. The mushroom slices are dipped overnight in a solution of 0.1% of citric acid + 1.5% of NaCl + 0.3% of red chilli powder. After draining off the solution, the mushroom slices are subjected to drying in a cabinet dryer at 60°C for 8 h. Then, the slices are fried in refined oil at 160°C. Spices powder can be spread over the fried slices to enhance the flavor of the chips (Rai and Arumganathan, 2008). Fresh mushroom chips were prepared by soaking the slices in different solutions of 1% salt, 1% KMS, 0.5% citric acid, and in combinations of all for 5 min, partially dehydrated for 2 minutes, and then fried in oil (Desayi, 2012). Amongst the different pretreatments, soaking in combined solution of 1% salt+ 1% KMS + 0.5% citric acid for 5 min along with partial dehydration for 2 min had higher organoleptic scores.

Vacuum frying uses lower pressure and temperature rather than atmospheric deep fat frying to improve the quality attributes of fried products (Diamante et al., 2015). The vacuum frying time in general affects the color and oil uptake of chips significantly. Moisture content declines significantly after vacuum frying at 90°C and 100°C for 3 min. The increase in frying time and temperature generally causes a decrease in the moisture content of mushroom chips, while the oil content and crispness increased (Tarzi et al., 2011). Vacuum frying at 90°C with a pressure of 4.25 kPa for 12.5 min produced mushroom chips with acceptable quality. In another study, Charoen et al. (2015) suggested the optimal vacuum frying condition, viz., 120°C-15 min and spinning speed at 1200 rpm under vacuum of 700 mmHg. The fried chips had high fiber (25.62%) and protein (17.47%) content and low moisture content (2.15%). Process optimization of vacuum fried rice-straw mushroom (*V. volvacea*) stem chips was also carried out by Suryatman and Ahza (2016). The optimum conditions for a 2-mm slice thickness chips were reported as vacuum frying at 100 °C for 3 min based on the color and low oil uptake.

9.7   PACKAGING AND STORAGE OF MUSHROOMS

Packaging plays a very important role in handling, marketing, and consumption of the produce/product and protects the quality of the product during storage and transport. Packaging of mushrooms from the production site to the consumer, including packaging for export market, is an important aspect of postharvest handling.

Oyster mushrooms are harvested and stem-cut, and adhering straw, if any, is removed. The cleaned mushrooms are packed in polypropylene bags of about

100-gage thickness with perforations having a vent area of above 5%. Although the perforation causes slight reduction in weight during the storage, it helps to maintain the freshness and firmness of the produce. Storage of *Dhingri* at very low temperatures especially in nonperforated polypacks results in condensation of water, sliminess of the surface, and softening of the texture. Paddy straw mushrooms are packed in polythene bags as well as tray packs.

As the very low temperature storage causes frost injury and deterioration in quality, the best way of storage is at 10–15°C in polythene bags with perforations (Pathak et al., 1998). For transportation of mushrooms in Taiwan, they are packed in bamboo baskets with an aeration channel at the center, and dry ice wrapped in a paper is placed above the mushrooms. Packing in wooden cases for transport by rail or boat is practiced in China (Saxena and Rai, 1990). Button mushrooms are packed in many ways according to the retail, wholesale, and transport requirements. Saxena and Rai (1988) stored button mushrooms in polypropylene bags of less than 100-gage thickness with perforations having a vent area of about 5% and observed browning and reduction in weight during storage at 15°C.

Mushrooms were best preserved in nonperforated bags kept at 5°C. It is suggested that button mushroom should be stored in polystyrene or pulp board for transporting to the long distances, instead of using polyethylene bags.

Langerak (1972) reported that chipboard box, lined inside with an absorbent paper and outside with a waterproof paper, was one of the best packing materials for mushrooms. It was also observed that mushrooms discolored faster in the PVC boxes with the perforated cover than in closed boxes of chipboard, and desiccation was much less in the closed cardboard box than in the PVC box. However, Dhar (1992) found that fruiting bodies of summer white button mushroom (*A. bisporus*) could be stored without a significant loss of quality for 6 days at 15°C in nonperforated packs without any chemical treatment or washing in water. De la Plaza et al. (1995) found that the use of an oriented polypropylene (OPP) film can double the storage period compared to a PE film and maintains the mushroom quality for at least 2 days at 18°C.

Modified atmosphere packaging (MAP) of mushrooms has been shown to delay senescence and maintain the quality of mushrooms during postharvest storage (Saray et al., 1994; Roy et al., 1995; Tano et al., 1999). Mushrooms covered with a PVC film had a shelf-life of 5–7 days at 15–21°C, as compared to those left uncovered which had a shelf-life of 2–4 days under similar conditions. Application of nitric oxide in combination with MAP can extend the storage life of button mushroom up to 12 days (Jiang et al., 2011).

Ramanathan et al. (1992) studied the quality of oyster mushrooms stored in different thickness of polyethylene bags under controlled atmosphere storage and found that 300-gage and 150-gage polyethylene bags maintained the keeping quality of mushroom up to 20 days at 15% CO_2 and 1% O_2 gas composition.

Zheng et al. (1994) conducted studies on controlled atmosphere storage of fresh button mushrooms and found that 8% O_2 + 10% CO_2 was the best atmosphere for maintaining the quality of fresh mushrooms. Popa et al. (1998) studied the use of a humidity absorber (silica gel) in the packages during the modified atmosphere storage and found washing in chlorinated water and incorporation of dehumidifiers decreased the microbial contamination and extended the shelf-life of *A. bisporus*.

Arumuganathan and Rai (2004) conducted studies to identify the suitable packaging materials for the mushroom products. The different packaging materials used were polythene, polypropylene, lug bottles, laminated pouches, PVC-wrapped trays, plastic jars, and tin cans. The suitability and adoptability of these packaging materials were studied in terms of keeping quality during the storage period and are given in Table 9.4. The addition of chitosan, as a supplementary packaging material, can extend the shelf life of

TABLE 9.4 Suitable Packaging Materials for Mushroom Products (Arumuganathan and Rai, 2004)

S. No.	Name of the packaging material	Mushroom Product	Storage Period
1	Polythene bag	Dried Mushroom	2 Months
		Mushroom Powder	2 Months
2	Polypropylene	Mushroom Candy	6 Months
		Mushroom Soup Powder	6 Months
		Mushroom Powder	6 Months
		Dried Mushroom	3 Months
		Mushroom Chips	3 Months
3	Lug bottles	Mushroom Pickles	1 Year
4	PET Jar	Mushroom Biscuits	3 Months
		Mushroom Candy	6 Months
5	Butter paper	Mushroom Candy	3 Months
		Mushroom Biscuits	2 Months
6	PVC wrapped trays	Mushroom Nuggets	1 Month
7	Laminated pouches	Mushroom Curry	1 Year
8	Tin cans	Canned Mushroom	1 Year

mushroom at ambient temperature as compared to control (Sungsan et al., 1998). Mushrooms stored in an anti-fogging film at the ambient temperature with chitosan prolonged the shelf-life to 10 days, which was 4 days longer than the control treatment.

9.8 CONCLUDING REMARKS

There is an increasing demand for good quality mushroom and the value-added products thereof at competitive rates in both domestic and global export market. To be successful in both domestic and global market, it is required to produce high quality fresh mushrooms and processed products. Mushrooms or compounds extracted from mushrooms may aid in the treatment of certain types of cancer, boost the immune system, and reduce the risk of coronary heart disease. Besides being rich in protein, mushrooms are also rich in fiber and vitamins such as thiamine, riboflavin, niacin, biotin, and ascorbic acid. Information about physicochemical and nutritional quality of several mushroom species is still somewhat limited. Research is also needed to validate the medicinal properties of mushroom and to isolate new compounds. The present processing and preservation techniques like drying, freezing, and canning should be improved in order to get more value-added products of greater shelf-life. Due to high moisture content, mushroom is considered among the highly perishable vegetables; therefore, its storage and transportation are a big challenge. Hence, packaging condition and the use of different packaging materials taking their dimensions into consideration for increasing the shelf life of packaged fresh mushrooms may also be explained and disseminated.

KEYWORDS

- chemical composition
- medicinal properties
- mushroom
- packaging
- processing
- value added products

REFERENCES

Afiukwa, C. A., Ebem, E. C., & Igwe, D. O., (2015). Characterization of the proximate and amino acid composition of edible wild mushroom species in abakaliki, Nigeria. *A. A. S. C. I. T. J. Biosci.*, *1*(2), 20–25.

Ahlawat, O. P., & Tewari, R. P., (2007). *Cultivation Technology of Paddy Straw Mushroom* (*Volvariella volvacea*), Technical Bulletin: National Research Center for Mushroom (Indian Council of Agriculture), Solan, India, pp. 1–44.

Ajayi, O., et al., (2015). Effect of packaging materials on the chemical composition and microbiological quality of edible mushroom (*Pleurotus ostreatus*) grown on cassava peels. *Food Sci. Nutr.*, *3*, 284–291.

Akata, I., Ergonul, B., & Kalyoncu, F., (2012). Chemical compositions and antioxidant activities of 16 wild edible mushroom species grown in Anatolia. *Int. J. Pharmacol.*, *8*, 134–138.

Akyuz, M., & Kirbag, S., (2010). Nutritive value of wild edible and cultured mushrooms. *Turk. J. Biol.*, *34*, 97–102.

Alam, N., et al., (2008). Nutritional analysis of cultivated mushrooms in Bangladesh – *Pleurotus ostreatus, Pleurotus sajor-caju, Pleurotus florida* and *Calocybe indica*. *Mycobiol.*, *36*(4), 228–232.

Amin, A., (2016). Development and evaluation of oyster mushroom powder supplemented spaghetti. *J. Nutr. Food. Sci.*, DOI: 10.4172/2155–9600.C1.024. *6*(4), 50.

Argyropoulos, D., Heindl, A., & Müller, J., (2011). Assessment of convection, hot-air combined with microwave vacuum and freeze-drying methods for mushrooms with regard to product quality. *Int. J. Food Sci. Technol.*, *46*, 333–342.

Arumuganathan, T., & Rai, R. D., (2004). Suitable packaging materials for mushroom products. *Beverage and Food World*, *31*(10), 46.

Arumuganathan, T., Rai, R. D., Chandrasckar, V., & Hcmakar, A. K., (2003). Studies on canning of button mushroom Agaricus bisporus for improved quality. *Mushroom Res.*, *12*(2), 117–120.

Bajaj, M., Vadhera, S., Brar, A. P., & Soni, G. L., (1997). Role of oyster mushroom *(Pleurotus florida)* as hypocholesterolemic/antiatherogenic agent. *Indian J. Exp. Biol.*, *35*, 1070–1075.

Bano, Z., Rajarathnam, S., & ShashiRekha, M. N., (1992). Mushroom as the unconventional single cell protein for a conventional consumption. *Indian Food Packer*, *46*(5), 20–31.

Barros, L., et al., (2007). Effects of conservation treatment and cooking on the chemical composition and antioxidant activity of Portuguese wild edible mushrooms. *J. Agric. Food Chem.*, *55*, 4781–4788.

Basunia, M. A., et al., (2011). Drying of fish sardines in Oman using solar tunnel dryers. *J. Agricultural Sci. Technol.*, *B1*, 108–114.

Beluhan, S., & Ranogajec, A., (2011). Chemical composition and non-volatile components of Croatian wild edible mushrooms. *Food Chem.*, *124*, 1076–1082.

Bergman, C. J., Gualberto, D. G., & Weber, C. W., (1994). Development of a high-temperature-dried soft wheat pasta supplemented with cowpea (*Vigna unguiculata* L. Walp). Cooking quality, color, and sensory evaluation. *Cereal Chem.*, *71*, 523–527.

Bernaś, E., Jaworska, G., & Kmiecik, W., (2006). Storage and processing of edible mushrooms. *Acta. Sci. Pol. Technol. Aliment.*, *5*(2), 55–23.

Bhuiyan, M. H. R., & Rana, M. S., (2012). Ketchup development from fresh mushroom. *Bangladesh Res. Pub. J.*, *7*(2), 182–188.

Boa, E., (2004). Wild Edible Fungi—A global overview of their use and importance to people. FAO, Rome., 21–148.

Bobek, P., & Galbavy, S., (1999). Hypercholesterolemia and anti-atherogenic effect of oyster mushroom (*Pleurotus ostreatus*) in rabbit. *Nahrung.*, *45*, 339–342.

Bobek, P., Nosalova, V., & Cerna, S., (2001). Effect of pleuran (beta glucan from *Pleurotus ostreatus*) in diet or drinking fluid on colits rats. *Mol. Nutr. Food Rev.*, *45*, 360–363.

Breene, W. M., (1990). Nutritional and medicinal value of specialty mushrooms. *Journal of Food Production*, *53*(10), 883–894.

Chang, H. J., & Ki, H. S., (2004). Quality characteristics of sponge cakes with addition of *Pleurotus eryngii* mushroom powders. *J. Korean Soc. Food Sci. Nutr.*, *33*, 716–722.

Chang, S. T., (1991). Cultivated mushroom In *Handbook of Applied Mycology*. Marcel Dekker, New York, *3*, pp. 221–240.

Chang, S. T., (2008). *Training Manual on Mushroom Cultivation*. Asian and pacific centre for agricultural engineering and machinery (APCAEM), United Nations. *http://www.unapcaem.org/publication/TM-Mushroom.pdf.* Last accessed on 18 Jan, 2017.

Chang, S. T., (1999). World production of cultivated edible and medicinal mushrooms in 1997 with emphasis on *Lentinus edodes* (Berk.) in China. *Int. J. Med. Mushrooms.*, *1*, 291–300.

Charoen, R., Lakerd, S., & Kornpetch, C., (2015). Development of seasoned gray oyster mushroom chips using vacuum frying process. *Food Applied Biosci. J.*, *3*(2) 100–108.

Chen, J., (2012). Recent advance in the studies of β-glucans for cancer therapy. *Anti-Cancer Agents in Med. Chem.*, *13*, 679–680.

Cheng, H. H., Hou, W. C., & Lu, M. L., (2002). Interactions of lipid metabolism and intestinal physiology with *Tremella fuciformis* Berk edible mushroom in rats fed a high-cholesterol diet with or without Nebacitin. *J. Agric. Food Chem.*, *50*, 7438- 7443.

Cheung, P. C. K., (2008). Nutritional Value and Health Benefits of Mushrooms, In: *Mushrooms as Functional Foods*. Cheung, P. C. K. (ed.). John Wiley & Sons Inc., Hoboken, NJ, USA.

Cheung, P. C. K., (1998). Plasma and hepatic cholesterol levels and fecal neutral sterol excretion are altered in hamsters fed Straw mushroom diets. *J. Nutr.*, *128*, 1512–1516.

Cheung, P. C. K., (1996). The hypocholesterolemic effect of two edible mushrooms: *Auricularia auricula* (tree-ear) and *Tremella fuciformis* (white jelly-leaf) in hypercholesterolemic rats. *Nutr. Res.*, *16*, 1721–1725.

Cheung, P. C. K., (2010). The nutritional and health benefits of mushrooms. *Nutrition Bulletin.*, *35*, pp. 292–299.

Chihara, G., (1992). Immunopharmacology of Lentinan, a polysaccharide isolated from *Lentinus edodes* its application as a host defence potentiator. *Int. J. Oriental Med.*, *17*, 57–77.

Chomdao, S., et al., (2005). Development of mushroom cookies, processing and acceptability testing. *Food.*, *35*, 293–301.

De la Plaza, J. L., et al., (1995). Effect of the high permeability to O_2 on the quality changes and shelf-life of fresh mushrooms stored under modified atmosphere packaging. *Mushroom Sci.*, *14*(2), 709–716.

Desayi, D., (2012). Development, recovery and sensory evaluation of fresh mushroom chips. *Int. J. Food Agr. Vet. Sci.*, *2*(2), 190–193.

Devina, V., et al., (2008). Development and quality evaluation of white button mushroom noodles. *J. Food Sci. Technol.*, *45*(6), 513–515.

Dhar, B. L., (1992). Postharvest storage of white button mushroom *Agaricus bitorquis*. *Mushroom Res., 1*, 127–130.

Diamante, L. M., et al., (2015). Vacuum frying foods: products, process and optimization. *Int. Food Res. J., 22*(1), 15–22.

Dikeman, C. L., et al., (2005). Effects of stage of maturity and cooking on the chemical composition of select mushroom varieties. *J. Agric. Food Chem., 53*, 1130–1138.

Dinani, S. T., et al., (2014). Drying of mushroom slices using hot air combined with an electrohydrodynamic (EHD) drying system. *Drying Technol., 32*, 597–605.

Diyabalanage, T., et al., (2008). Health-beneficial qualities of the edible mushroom, *Agrocybe aegerita*. *Food Chem., 108*, 97–102.

Dubost, N. J., et al., (2006). Identification and quantification of ergothioneine in cultivated mushrooms by liquid chromatography-mass spectroscopy. *Int. J. Med. Mushrooms, 8*, 215–22.

Duyff, R., (2006). *American Dietetic Association's Complete Food and Nutrition Guide.* Third Addition. Wiley & Sons. N.J.

Eissa, H. A., Hussein, A. S., & Mostafa, B. E., (2007). Rheological properties and quality evaluation of Egyptian Balady bread and biscuits supplemented with flours of ungerminated and germinated legume seeds or mushroom. *Pol. J. Food Nutr. Sci., 57*, 487–496.

Endo, A., (1988). Chemistry biochemistry and pharmacology of HMGCoA reductase inhibitors. *Klin Wochenschr., 66*(10), 421–27.

Erle, U., & Schubert, H., (2001). Combined osmotic and microwave-vacuum dehydration of apples and strawberries. *J. Food Eng., 49*, 193–199.

Eva, G., (2010). Edible mushrooms: Role in the prevention of cardiovascular diseases. *Fitoterapia., 81*, 715–723.

FAO (2013) Food and Agricultural Organization of the United Nations. World mushroom & truffles: production. http://faostat.fao.org/beta/en/#data. Last accessed on 15.12.2016.

Ferreira, I. C. F. R., (2010). Compounds from wild mushrooms with antitumor potential. *Anti-Cancer Agents in Medicinal Chemistry., 10*, 424–436.

Fons, F., et al., (2003). Volatile compounds of genera Cantharellus, Craterellus and Hydnum. *Cryptogamie, Mycologie., 24*, 367–376.

Giri, S. K., & Prasad, S., (2007). Optimization of Microwave-Vacuum Drying of Button Mushrooms Using Response-Surface Methodology. *Drying technol., 25*(5), 901–911.

Guiné, R. P. F., & Barroca, M. J., (2011). Influence of freeze-drying treatment on the texture of mushrooms and onions Croat. *J. Food Sci. Technol., 3*(2), 26–31.

Gunde-Cimerman, N., (1999). Medicinal value of the genus *Pleurotus* (Fr.) P. Karst. (Agaricales S.R., Basidiomycetes). *Int. J. Medicinal Mushrooms., 1*, 69–80.

Guo, C. X., Choi, M. W., & Cheung, P. C. K., (2012). Mushroom and immunity. *Curr. Top. Nutraceutical Res., 1*, 31–41.

Hadar, Y., & Dosoretz, C. G., (1991). Mushroom mycelium as a potential source of food flavor. *Trends in Food Sci. Technol., 2*, 214–218.

Hanus, L. O., Shkrob, I., & Dembitsky, V. M., (2008). Lipids and fatty acids of wild edible mushrooms the genus Boletus. *J. Food Lipids, 15*, 370–383.

Hassan, F. R. H., & Ghada, M. M., (2014). Effect of pretreatments and drying temperatures on the quality of dried *pleurotus* mushroom spp. *Egypt J. Agric. Res., 92*(3), 1009–1022.

Hernando, I., et al., (2008). Rehydration of freeze-dried and convective dried boletus edulis mushrooms: effect on some quality parameters. *J. Food Sci., 73*(8), 356–62.

Hu, S. H., et al., (2006). Antihyperlipidemic and antioxidant effects of extracts from *Pleurotus citrinopileatus*. *J. Agric. Food Chem.*, *54*, 2103–2110.

Hung, P. V., & Nhi, N. N. Y., (2012). Nutritional composition and antioxidant capacity of several edible mushrooms grown in the Southern Vietnam. *Int. Food Res. J.*, *19*(2), 611–615.

Hyung, H. G., Soo, K. Y., & Seoup, S. G., (2005). Effect of oyster mushroom powder on bread quality. *J. Food Sci. Nutr.*, *10*, 214–218.

Ibrahium, M. I., & Hegazy, A. I., (2014). Effect of replacement of wheat flour with mushroom powder and sweet potato flour on nutritional composition and sensory characteristics of biscuits. *Current Sci. Int.*, *3*(1), 26–33.

Ikekawa, T., et al., (1968). Antitumor action of some Basidiomycetes, especially *Phellinus linteus*. *J. Cancer Res.*, *59*, 155–157.

Jaworska, G., & Bernaś, E., (2009). The effect of preliminary processing and period of storage on the quality of frozen *Boletus edulis* (Bull: Fr.) mushrooms. *Food Chem.*, *113*, 936–943.

Jiang, T., et al., (2011). Integrated application of nitric oxide and modified atmosphere packaging to improve quality retention of button mushroom (*Agaricus bisporus*). *Food Chem.*, *126*(4), 1693–1699.

Jose, N., & Janardhanan, K. K., (2000). Antioxidant and antitumour activity of *Pleurotus florida*. *Curr. Sci.*, *79*(7), 941–943.

Kajala, I., Simmoncic, R., & Belay, G., (2008). *Bratisl. Lek. Listy*, *109*, 267–72.

Kalac, P., (2013). A review of chemical composition and nutritional value of wild-growing and cultivated mushrooms. *J. Sci. Food Agric, 93*, 209–218.

Kalac, P., (2009). Chemical composition and nutritional value of European species of wild growing mushrooms: A review. *Food Chem.*, *113*, 9–16.

Kamugisha, D., & Sharan S., (2005). Nutritional composition of milky mushroom (*Calocybe indica*) cultivated on paddy straw amended with *Ragi* Flour. Karnataka, *J. Agric. Sci.*, *18*(4), 1048–1051.

Kar, A., et al., (2004). Microwave drying characteristics of button mushroom (*Agaricus bisporus*). *J. Food Sci. Technol.*, *41*(6), 636–641.

Kim, S., et al., (2016). Effect of *Pleurotus eryngii* mushroom β-glucan on quality characteristics of common wheat pasta. *J. Food Sci.*, *81*, C835–C840.

Kim, Y. S., (1998). Quality of wet noodles prepared with wheat flour and mushroom powder. *Korean J. Food Sci. Technol.*, *30,* 1373–1380.

Kortei, N. K., et al., (2016). Modeling the solar drying kinetics of gamma irradiation-pretreated oyster mushrooms (*Pleurotus ostreatus*). *Int. Food Res. J.*, *23*(1), 34–39.

Kumar, A., Singh, M., & Singh, G., (2013). Effect of different pretreatments on the quality of mushrooms during solar drying. *J. Food Sci. Technol.*, *50*(1), 165–170.

Kumar, K., & Ray, A. B., (2016). Development and shelf-life evaluation of tomato-mushroom mixed ketchup. *J. Food Sci. Technol.*, *53*(5), 2236–2243.

Kumar, K., (2015). Studies on development and shelf life evaluation of soup powder prepared by incorporation of white button mushroom (*Agaricus bisporus* L.). South Asian, *J. Food Technol. Env.*, *3–4*, 219–224.

Kumar, S., Nirankar, N., & Tyagi, R. K., (2006). Development and evaluation of button mushroom (*Agaricus bisporus*) *mathri* using response surface methodology. *J. Food Sci. Technol.*, *43*, 186–189.

Kurashige, S., Akuzawa, Y., & Endo, F., (1997). Effects of *Lentinus edodes, Grifola frondosa* and *Pleurotus ostreatusa* administration on cancer outbreak, and activities of macrophages and lymphocytes in mice treated with a carcinogen, N-butyl-N-butanolnitroso-amine. *Immunopharmacology and Immunotoxicology*, *19*, 175–183.

Langerak, D. I., (1972). The influence of irradiation and packaging upon the keeping quality of fresh mushrooms. *Mushroom Sci.*, *8*, 221–230.

Li, F., et al., (2013a). Antiproliferative activities of tea and herbal infusions. *Food Funct.*, *4*, 530–538.

Li, F., et al., (2013b). Antiproliferative activity of peels, pulps and seeds of 61 fruits. *J. Funct. Foods*, *5*, 1298–1309.

Lidhoo, C. K., & Agrawal, Y. C., (2008). Optimizing temperature in mushroom drying. *J. Food Process. Preserv.*, *32*(6), 881–897.

Lin, H. C., (1999). Evaluation of taste quality and antioxidant properties of edible mushrooms. *Master's Thesis*, National Chung-Hsing University, Taichung, Taiwan.

Lu, X., et al., (2016). How the inclusion of mushroom powder can affect the physicochemical characteristics of pasta. *Int. J. Food Sci. Technol.*, *51*, 2433–2439.

Manzi, P., Aguzzi, A., & Pizzoferrato, L., (2001). Nutritional value of mushrooms widely consumed in Italy. *Food Chem.*, *73*, 321–325.

Manzi, P., & Pizzoferrato, L., (2000). Beta-glucans in edible mushrooms. *Food Chem.*, *68*(3), 315–318.

Martinez-Soto, G., Ocana-Camacho, R., & Peredes-Lopez, O., (2001). Effect of pretreatment and drying on the quality of oyster mushrooms (*Pleurotus ostreatus*). *Drying Technol.*, *19*(3–4), 661–672.

Mattila, P., et al., (2002). Basic composition and amino acid contents of mushrooms cultivated in Finland. *J. Agric. Food Chem.*, *50*, 6419–6422.

Mau, J. L., Chao, G. R., & Wu, K. T., (2001). Antioxidant properties of methanolic extracts from several ear mushrooms. *J. Agric. Food Chem.*, *49*(11), 5461–7.

Mau, J. L., Lin, H. C., & Song, S. F., (2002). Antioxidant properties of several specialty mushrooms. *Food Res. Int.*, *35*, 519–526.

Mizuno, T., (1996). Development of antitumor polysaccharides from mushroom fungi. *Food Ingred. J. Jpn.*, *167*, 69–85.

Mizuno, T., Sakai, T., & Chihara, G., (1995). Health foods and medicinal usages of mushrooms. *Food Rev. Int.*, *11*(1), 69–81.

Mori, K., et al., (2008). Antiatherosclerotic effect of the edible mushrooms *Pleurotus eryngii* (Eringi), *Grifola frondosa* (Maitake), and *Hypsizygus marmoreus* (Bunashimeji) in apolipoprotein E-deficient mice. *Nutr. Res.*, *28*, 335–42.

Mustayen, A. G. M. B., et al., (2015). Performance evaluation of a solar powered air dryer for white oyster mushroom drying. *Int. J. Green Ener.*, *12*(11), 1113–1121.

National Institutes of Health. Medline Plus. *www.nlm.gov/medlineplus/ency/article/002414. htm.*

Ndung'u, S. W., et al., (2015). Nutritional composition, physical qualities and sensory evaluation of wheat bread supplemented with oyster mushroom. *Americ. J. Food Technol.*, *10*(6), 279–288.

Neyrinck, A. M., et al., (2009). Dietary supplementation with chitosan derived from mushrooms changes adipocytokine profile in diet-induced obese mice, a phenomenon linked to its lipid-lowering action. *Int. Immunopharmacol.*, *9*, 767–773.

Nour, V., Trandafir, I., & Ionica, M. E., (2011). Effects of pre-treatments and drying temperatures on the quality of dried button mushrooms. *South Western Journal of Horticulture Biology and Environment*, *2*(1), 15–24.

Obodai, M., et al., (2014). Evaluation of the chemical and antioxidant properties of wild and cultivated mushrooms of ghana. *Mol.*, *19*, 19532–19548.

Okafor, J. N. C., et al., (2012). Quality characteristics of bread made from wheat and nigerian oyster mushroom (*Pleurotus plumonarius)* powder. *Pakistan J. Nutr.*, *11*, 5–10.

Ooi, V. E. C., (2000). *Medicinally Important Fungi.* In: Science and Cultivation of Edible Fungi. Van Griensven (ed.). Balkema, Rotterdam, pp. 41–51.

Pan, H. H., et al., (2015). Purification and identification of a polysaccharide from medicinal mushroom Amauroderma rude with immunomodulatory activity and inhibitory effect on tumor growth. *Oncotarget.*, *6*, 17777–17791.

Patel, S., & Goyal, A., (2012). Recent developments in mushrooms as anti-cancer therapeutics a review. *Biotech.*, *2*, 1–15.

Pathak, V. N., Yadav, N., & Gaur, M., (1998). Mushroom production and processing technology. *Agrobot.*, 1–179.

Pereira, E., et al., (2012). Towards chemical and nutritional inventory of Portuguese wild edible mushrooms in different habitats. *Food Chem.*, *130*, 394–403.

Pizzoferrato, L., et al., (2000). Solid-state 13 C CP MAS NMR spectroscopy of mushrooms gives directly the ratio between proteins and polysachharides. *J. Agric. Food Chem.*, *48*(11), 5484–5488.

Popa, M., et al., (1998). *Agri-Food Quality-II: Quality Management of Fruits and Vegetables from Field to Table.* Hagg, M., & Evers, A. M. (eds.). Finland, pp. 177–181.

Prodhan, U. K., et al., (2015). Development and quality evaluation of mushroom (*Pleurotus sajor-caju)* enriched biscuits. Emirates, *J. Food Agric.*, *27*(7), 542–547.

Rai, R. D., & Arumganathan, T., (2008). *Post-Harvest Technology of Mushrooms*: Technical Bulletin. Nation research centre for mushroom (ICAR), Solan, HP.

Ramanathan, M., Sreenarayanan, V. V., & Swaminathan, K. R., (1992). Controlled atmosphere storage for mushroom. *Traction.*, *2*(2), 63–67.

Ravi, R., & Siddiq, M., (2011). Edible mushrooms: Production, processing and quality. *Handbook of Vegetables and Vegetable Processing.* Ames, IA: Wiley-Blackwell, pp. 643–661.

Razak, D. L. A. B., (2013). *Cultivation of Auricularia Polytricha Mont.* Sacc (black jelly mushroom) using oil palm wastes. Thesis: Faculty of Science, University of Malaya, Kuala Lumpur.

Reis, F. S., Barros, L., Martins, A., & Ferreira, I. C. F. R., (2012). Chemical composition and nutritional value of the most widely appreciated cultivated mushrooms: an inter-species comparative study. *Food and Chem. Toxicol.*, *50*, 191–197.

Rodriguez, R., et al., (2005). Kinetic and quality study of mushroom drying under microwave and vacuum. *Drying Technol.*, *23*, 2197–2213.

Rop, O., Mlcek, J., & Jurikova, T., (2009). Beta-glucans in higher fungi and their health effects. *Nutr. Rev.*, *67*, 624–631.

Rosli, W. I. W., Nurhanan, A. R., & Aishah, M. S., (2012). Effect of partial replacement of wheat flour with oyster mushroom (*Pleurotus sajor-caju*) powder on nutritional composition and sensory properties of butter biscuit. *Sains Malaysiana*, *41*(12), 1565–1570.

Roy, S., Anantheswaran, R. C., & Beelman, R. B., (1995). Fresh mushroom quality as affected by modified atmosphere packaging. *J. Food Sci.*, *60*(2), 334–340.

Salama, M. F., (2007). Preparation and evaluation of pasta prepared from semolina flour and oyster mushroom myclia powder. *Egyptian J. Food Sci.*, *35*, 59–70.

Saray, T., et al., (1994). The importance of packaging and modified atmosphere in maintaining the quality of cultivated mushrooms (*Agaricus bisporus* L.) stored in chill chain. *Acta. Horticulturae, 368*, 322–326.

Savic, M., (2009). Matica Srpska Proceedings for Natural Sciences, *116*, 7–14.

Saxena, S., & Rai, R. D., (1990). *Postharvest Technology of Mushrooms*. Technical Bulletin No.2, NRCM, Solan.

Saxena, S., & Rai, R. D., (1988). Storage of button mushrooms (*Agaricus bisporus*). The effect of temperature, perforation of packs and pre-treatment with potassium metabisulphite. *Mushroom J. Tropics, 8*, 15–22.

Sharma, A., et al., (2015). Functional and textural properties of Indian nuggets assorted with mushroom for lysine enrichment. *J. Food Sci. Technol., 52*(6), 3837–3842.

Sharma; R. C., Jandaik, C. L., & Bhardwaj, S. R., (1991). Enrichment of cookies/biscuits with mushroom prepared from unmarketable mushroom portions. *Advances in Mushroom Sci., 102*.

Singh, A. K., et al., (1999). Physicochemical changes in white button mushroom (*Agaricus bisporus*) at different drying temperatures. *Mushroom Res., 8*, 27–30.

Singh, J., et al., (2016). Development and evaluation of value added products. *Int. J. Current Res., 8*(3), 27155–27159.

Singh, M., (2010). *Technologies for Mushroom Production*. ICAR, Solan, Himachal Pradesh.

Singh, S., & Chaudhary, G., (2015). Quality evaluation of dried vegetables for preparation of soups. *Indian Res. J. Genetics Biotechnol., 7*(2), 241–242.

Singh, S., (1996). Effect of whey concentration on the quality of whey based mushroom soup powder. *Mushroom Res., 5*, 33–38.

Singh, S., Ghosh, S., & Patil, G. R., (2003). Development of a mushroom whey soup powder. *Int. J. Food Sci. Technol., 38*(2), 217–224.

Suguna, S., et al., (1995). Dehydration of mushroom by sun drying, thin layer drying, fluidized bed drying and solar cabinet drying. *J. Fd. Sci. Technol., 34*, 284–288.

Sun, J., (2014). A novel hemagglutinin with antiproliferative activity against tumor cells from the hallucinogenic mushroom *Boletus* speciosus. *BioMed. Res. Int.*, 340–467.

Sungsan, C., (1998). The effect of supplementary package materials for keeping freshness of fresh mushroom at ambient temperature. *J. Industrial Crop Sci., 2*, 52–57.

Suryatman, S., & Ahza, A. B., (2016). Process optimization of vacuum fried rice-straw mushroom (*Volvariella Volvacea*) stem chip making. *J. Food Sci. Eng., 6*, 109–120.

Synytsya, A., et al., (2009). Glucans from fruit bodies of cultivated mushrooms *Pleurotus ostreatus* and *Pleurotus eryngii*: structure and potential prebiotic activity. *Carbohydrate Poly., 76*, 548–556.

Tano, K., et al., (1999). Atmospheric composition and quality of fresh mushrooms in modified atmosphere packages as affected by storage temperature abuse. *J. Food Sci., 64*(6), 1073–1077.

Tarzi, B. G., et al., (2011). Process of optimization in vacuum frying of mushroom using response surface methodology. *World Applied Sci. J., 14*, 960–966.

Tyagi, R. K., & Nath, N., (2005). Effect of addition of mushroom (*Pleurotus florida*) powder on quality of papad. *J. Food Sci. Technol., 42*, 404–407.

U.S. Department of Agriculture, (2009). Agricultural research service, USDA nutrient data laboratory. USDA national nutrient database for standard reference, release *22*,. *www. ars.usda.gov/nutrientdata*.

Venkateshwarlu, G., Chandravadana, M. V., & Tewari, R. P., (1999). Volatile flavour components of some edible mushrooms (Basidiomycetes). *Flavour Fragr. J., 14*, 191–194.

Verma, A., & Singh, V., (2016). Organoleptic evaluation of mushroom powder fortification in *Ravaldli* and *Mathari*. *Food Sci. Res. J.*, *7*(1), 40–44.

Wakchaure, G. C., (2008). Mushrooms – Value added products, In: *Mushrooms-cultivation, marketing and consumption*. National research centre for mushroom (ICAR), Solan, HP, pp. 233–238.

Walde, S. G., et al., (2006). Effects of pretreatments and drying methods on dehydration of mushroom. *J. Food Eng.*, *74*, 108–15.

Wang, H., Zhang, M., & Adhikari, B., (2015). Drying of shiitake mushroom by combining freeze-drying and mid-infrared radiation. *Food and Bioproducts Process.*, *94*, 507–517.

Wang, S. Y., et al., (1997). The anti-tumor effect of *Ganoderma lucidum* is mediated by cytokines released from activated macrophages and T lymphocytes. *Int. J. Cancer*, *70*(6), 699–705.

Wannet, W. J. B., et al., (2000). HPLC detection of soluble carbohydrates involved in mannitol and trehalose metabolism in the edible mushroom *Agaricus bisporus*. *J. Agric. Food Chem.*, *48*(2), 287–291.

Wasser, S. P., & Weis, A. L., (1999). Medicinal properties of substances occurring in higher Basidiomycetes mushrooms current perspectives (review). *Int. J. Med. Mushrooms*, *1*, 31–62.

Wasser, S. P., (2002). Medicinal mushrooms as a source of antitumor and immunomodulating polysaccharides (mini review). *Appl. Microbiol. Biotechnol.*, *60*, 258–274.

Woldegiorgis, A. Z., et al., (2015). Proximate and amino acid composition of wild and cultivated edible mushrooms collected from Ethiopia. *J. Food Nutr. Sci.*, *3*(2), 48–55.

Yamada, T., Oinuma, T., & Niihashi, M., (2002). Effects of Lentinus edodes mycelia on dietary-induced atherosclerotic involvement in rabbit aorta. *J. Atheroscler. Thromb.*, *9*, 149–56.

Yang, D. C., & Maguer, M. L., (1992). Mass transfer kinetics of osmotic dehydration of mushrooms. *J. Food Process. Preserv.*, *16*(3), 215–231.

Yang, J. H., Lin, H. C., & Mau, J. L., (2002). Antioxidant properties of several commercial mushrooms. *Food Chem.*, *7*, 229–235.

Yang, Q. Y., & Jong, S. C., (1989). Medicinal mushrooms in China. *Mushr. Sci.*, *9*(1), 631–643.

Zhang, M., Cheung, P. C. K., & Zangh, L., (2001). Evaluation of mushroom dietary fibre (non-starch polysaccharides) from sclerotia of *Pleurotus tuber-regium* (Fries) Singer as a potental antitumor agent. *J. Agric. Food Chem.*, *49*, 5059–5062.

Zhang, M., et al., (2006). Polysaccharides from mushrooms A review on their isolation process, structural characteristics and antitumor activity. *Trends Food Sci. Technol.*, *18*, 4–19.

Zheng, Y. H., et al., (1994). Preliminary study on color fixation and controlled atmosphere storage of fresh mushrooms. *J. Zhejiang Agric. University*, *20*(2), 165–168.

Zhou, Y., et al., (2016). Dietary natural products for prevention and treatment of liver cancer. *Nutrients*, *8*, 156.

Zou, Y. M., et al., (2015). A polysaccharide from mushroom Huaier retards human hepatocellular carcinoma growth, angiogenesis, and metastasis in nude mice. *Tumor Biol.*, *36*, 2929–2936.

CHAPTER 10

CHANGES IN PHYSICAL, FUNCTIONAL, AND NUTRITIONAL CHARACTERISTICS OF EXTRUDATES DURING EXTRUSION

S. A. WANI, M. A. PARRAY, and P. KUMAR

Department of Food Engineering and Technology, Sant Longowal Institute of Engineering and Technology, Longowal 148106, Punjab, India

CONTENTS

10.1   INTRODUCTION

Extrusion cooking is a high-temperature short-time (HTST) process in which raw materials, starchy and/or proteinaceous, are plasticized and cooked in an extruder barrel by the effect of mechanical shear, moisture, and temperature, resulting in molecular transformation and chemical alterations (Havck and Huber, 1989; Castells et al., 2005). As compared to other heating processes, this technique possesses some unique features, as the material is subjected to intense mechanical shear and generation of temperature. The covalent bond in biopolymers is broken, and intense structural disruption causes modification of functional properties of feed ingredients (Carvalho and Mitchelle, 2000). Additionally, the extrusion process denatures enzymes, destroys antinutritional factors such as hemagglutinins, phytates, tannins, and trypsin inhibitors in the finished product, and retains natural flavors and colors of foods (Fellows, 2000; Bhandari et al., 2001).

Extruded products are popular because they are ready-to-eat products with crispy texture. They are, however, often considered as junk food because of their composition that is mainly based on carbohydrates and fat. Addition of healthy ingredients into breakfast cereals and snacks can improve their nutritional value and overall acceptability (Hirth et al., 2014). The influence of extrusion cooking on the nutritional value and chemical composition of balanced extruded products is very complex and requires further examination (Brennan et al., 2013).

10.2 EFFECTS OF EXTRUSION COOKING ON PHYSICAL AND FUNCTIONAL PROPERTIES

Extrusion cooking of raw materials is carried out at high pressure and high temperature conditions. As the raw material is conveyed in the barrel, the shear force is exerted by the rotating screw, and the barrel heating causes the food material to melt or plasticize (Van Zuilichem, 1992; Moscicki et al., 2007). The material is then conveyed under high pressure through a die or a series of dies; as a result, the product emerges as an expanded product. A schematic diagram of typical operation of a single and twin-screw extruder is shown in Figure 10.1 (Heldman et al., 1998).

Extrusion cooking is used for the preparation of many foods, ranging from the simplest expanded snacks to highly processed meat analogs. The following are the examples of popular products produced by extrusion cooking (Moscicki and Van Zuilichem, 2011):

- ready-to-eat extruded snacks, cereal flakes, and a wide variety of breakfast foods prepared from cereals;
- snack pellets, precooked pasta;
- bread crumbs, crispbread, emulsions, and pastes;
- chewing gum and sweets,

- aqua feed, pet food, calf-milk replacers, and feed concentrates;
- precooked flours, baby food, and functional components;
- texturized vegetable protein for meat analogs.

Extrusion conditions depend upon the type and amount of starches used in the processing. Temperature conditions used during extrusion conditions vary from 80–170°C (Gutkoski and El-Dash, 1999; Meng et al., 2010). The moisture content of raw material to be extruded varies and is generally found in the range of 10% to 30%, with a residence time of 30–90 s (Huber, 2001). Screw speed and feed rate used by different researchers for extrusion cooking varied depending upon the requirement and extruder used. It was noted from the work of different researchers that these processing conditions have a significant effect on the physical and functional properties (Table 10.1). The various physical and functional properties generally studied in extrudates are expansion ratio, bulk density, hardness, water solubility index (WSI), water absorption index (WAI), and color (Muthukumarappan and Karunanithy, 2012; Dar et al., 2014; Wani and Kumar, 2015a,b,c).

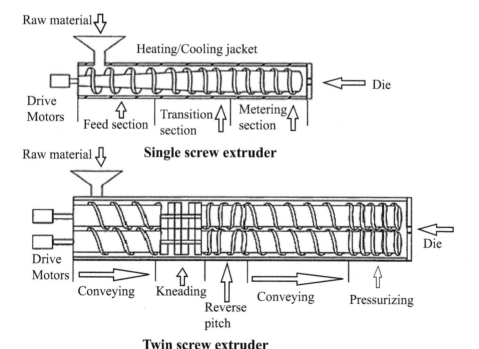

FIGURE 10.1 Typical operation of a single and twin screw extruder.

TABLE 10.1   Effect of Extrusion Cooking on Various Properties of Extrudates

Parameters (Expanded products)	Increases in			References
	Barrel Temperature	Screw Speed	Feed moisture	
Expansion ratio	Increases	Increases	Decreases	Oke et al., 2013; Wani and Kumar, 2016b
Bulk density	Decreases	Decreases	Increases	Koksel et al., 2004; Kothakota et al., 2013; Wani and Kumar, 2016b
Water absorption index	Decreases	Increases	Increases	Ding et al., 2005; Pelembe et al., 2002
Water solubility index	Increases	Increases	Decreases	Altan et al., 2008; Pathania et al., 2013; Wani and Kumar, 2016b
Hardness	Decreases	Decreases	Increases	Kumar et al., 2010; Oke et al., 2013
Specific mechanical energy	Decreases	Increases	No effect	Meng et al., 2010

10.2.1   EXPANSION RATIO

The degree of expansion is a significant property that depicts the quality of product and is directly related to the degree of roasting. It is a desirable characteristic and is also associated with other important factors such as crispness and WAI (Singh et al., 2007; Rodriguez-Miranda et al., 2011). In case of ready-to-eat extruded snacks, higher expansion is considered as the desirable property, which depend upon extrusion temperature, feed moisture, die diameter, screw speed, and feed rate (Seth et al., 2013).

Variation in barrel temperature, feed moisture content, and screw speed significantly affected the expansion ratio of all the extruded products. With increase in moisture content, the expansion ratio decreases, whereas it increases with screw speed and barrel temperature. Increase in the feed moisture content resulted in a substantial decrease in the expansion ratio as reported by number of researchers (Oke et al., 2013; Pathania et al., 2013; Wani and Kumar, 2016b). The reduced moisture content increases the drag force, which in turn exerts more pressure at the die, resulting in higher expansion ratio of the extruded product (Oke et al., 2013). An increase in the expansion ratio at higher barrel temperature and screw speed was reported by a number of researchers (Kothakota et al., 2013; Wani and Kumar, 2016b). The increased expansion ratio at higher temperature may be due to the gelatinization of starch and structure strengthening (Ainsworth et al., 2007). The higher expansion ratio due to higher screw speed may be due to high mechanical shear (Ding et al., 2006).

10.2.2 BULK DENSITY

Bulk density is one of the main important properties of extruded products. Lower bulk density, a desirable characteristic of extruded products, has been found to be obtained at low feed moisture and high barrel temperature. The effect of the process parameters of extrusion (temperature, moisture content, and screw speed) on bulk density of the extruded product was found to be inversely related to the expansion ratio (Pathania et al., 2013). Rise in barrel temperature decreased the bulk density. As the temperature rises, a higher potential energy leads the super-heated water to flash-off from extrudates as the product leaves the die (Koksel et al., 2004). With the increase in barrel temperatures, extrudates emerging from the die loses extra moisture and become lighter in weight and consequently causes a sharp reduction in the bulk density.

Screw speed was found to have a decreasing effect on bulk density (Kothakota et al., 2013). As reported by Hagenimana et al. (2006), the lower screw speed led to an increase in the bulk density of texturized rice, which may be due to a decrease in starch gelatinization. Higher screw speed causes increase in the shear force, which results higher pressure inside the extruder barrel, which in turn can cause greater expansion in extrudate and decrease bulk density. Sahagun and Harper (1980) have found similar trends using corn/soybean feed. Bulk density is mainly dependent on screw speed and temperature. However, several researchers have reported that the bulk density increased with increasing feed moisture (Meng et al., 2010), which has been directly co-related to the increase in weight with an increase in the amount of moisture. According to Ding et al. (2005), an increase in bulk density with moisture content may be due to change in molecular structure of amylopectin of the feed material as a result reduction in melt elasticity, thereby decreasing the expansion ratio but increasing the bulk density of extrudate.

10.2.3   WATER ABSORPTION INDEX (WAI)

The WAI is used to determine the volume occupied by the flour after swelling in excess of water, which keeps the integrity of flour in aqueous dispersion. It is defined as an index of starch gelatinization (Singh et al., 2007). The increase in WAI with screw speed may be due to high mechanical shear and higher expansion as a result of gelatinization. WAI decreases with the increase in temperature probably due to increased dextrinization at relatively higher temperature. A decrease in WAI with the rise in temperature may be due to degradation or decomposition of starch (Pelembe et al., 2002). In a study, Ding et al. (2006) found that WAI decreases with increase in temperature, if dextrinization or starch melting overcomes the gelatinization process.

Increase in feed moisture increases WAI. At higher moisture content, the viscosity of starch remains low, thereby allowing the starch molecules to move freely and thus enhancing the penetration of heat as a result of greater gelatinization (Lawton and Handerson, 1972). Increased feed moisture content and screw speed significantly increased the WAI of texturized rice, whereas higher barrel temperature was observed to cause reduction in WAI. Higher barrel temperature reduces WAI, probably due to increase in starch degradation (Hagenimana et al., 2006).

10.2.4   WATER SOLUBILITY INDEX (WSI)

WSI is an operational parameter that measures the ability of a powder to dissolve in water and is defined as the volume of sediments in milliliter after centrifuging. WSI is used as an indicator of degradation of molecular components and determines the amount of polysaccharides that are soluble to be released from the starch granules after excess of water has been supplemented to it (Van den Einde et al., 2003). Researchers found an increase in WSI with an increase in screw speed and temperature and decrease in WSI with an increase in moisture (Mezreb et al., 2003; Altan et al., 2008; Pathania et al., 2013; Wani and Kumar, 2016b). The increase in screw speed induced a sharp increase in specific mechanical energy (SME) that leads to high mechanical shear as a result of degradation of macromolecules; therefore, the molecular weight of starch granules decreased and hence WSI increased (Mezreb et al., 2003). Similarly, barrel temperature affected the WSI significantly. The starch degradation at increased temperature causes greater shearing action to the feed inside the barrel (Seth et al., 2013). Higher barrel temperature increases WSI due to increased solubility of starch molecules (Sobukola et al., 2012). Ding et al. (2005) reported that the higher lateral expansion is attributed to higher WSI. Decrease in WSI with an increase in moisture may be due to a decrease in lateral expansion due to plasticization of melt (Ding et al., 2005). Higher feed moisture during extrusion may reduce shearing action and thus may decrease WSI.

10.2.5 TEXTURE

Texture of the snacks is also associated with the expansion and cell structure of the product and hence can also be represented in terms of hardness. The hardness of cereals increased with higher moisture content for each extrusion temperature. Hardness is inversely related to the expansion of the extruded product (Wani and Kumar, 2016b). Higher feed moisture content increased hardness significantly, which may be due to reduction in expansion of the extruded product (Wani and Kumar, 2016b). A number of researchers found a positive correlation between hardness and moisture content (Ding et al., 2005; Kumar et al., 2010; Wani and Kumar, 2016b). Studies have shown decreased hardness with increase in barrel temperature (Kumar et al., 2010; Meng et al., 2010; Wani and Kumar, 2016b). The decrease in hardness with an increase in temperature has been reported due to higher expansion at greater temperatures. According to Ding et al. (2005), increasing temperature reduces melt viscosity of the raw mixture, but at the same time, it also increases the vapor pressure of water, which favors the growth of bubble; as the bubble grows, the expansion ratio increases, thus producing a low-density product and reducing the hardness of the extrudate. A significant positive correlation between bulk density and hardness was observed, as both hardness and bulk density decreased with an increase in temperature. A lower value of bulk density in a product naturally offers lower hardness (Altan et al., 2008). Hardness decreased with an increase in screw speed. The effect of screw speed on hardness might be due to its effect on the expansion of extrudates.

10.2.6   SPECIFIC MECHANICAL ENERGY (SME)

SME is used as a system parameter to model extrudate characteristics. Total input of mechanical energy given to the extruded ingredients plays a significant role for the conversion of starch. Increase in specific mechanical energy generally leads to higher degree of gelatinization of starch and finished product expansion. Therefore, higher SME is the desired parameter for the expanded products.

Increased specific mechanical energy has been reported with an increase in screw speed and decreased barrel temperature. However, no effective correlation between SME and feed moisture is reported (Meng et al., 2010). The parameter that affects the rheology of the food melt during extrusion will influence accordingly the SME. The reduction in SME with an increase in barrel temperature at the die has been found due to the alterations in melt viscosity (Meng et al., 2010). Increased barrel temperature would cause a reduction in the melt viscosity and as a result decrease in SME. A positive correlation between SME and screw speed has been observed. Similar results are reported by several researchers. On contrary to this, an increase in feed moisture lowers the melt viscosity, but hardly any change was observed in SME. Relatively narrow moisture range of the feed can be one of the possible reason (Meng et al., 2010).

Researchers have suggested that the feed material rich in protein may lead to decreased melt viscosity and product expansion (Onwulata and Konstance, 2006). When whey protein concentrate (WPC) was incorporated to a corn meal, a greater reduction in paste viscosity was observed during the extrusion. Moreover, Della Valle et al. (1997) reported that viscous behavior greatly depends upon the starch content, and melt viscosity was decreased with a decrease in starch content. Conclusively, a higher amount of protein and lesser starch in feed mixture may cause lower melt viscosity, which may result in lower SME (Meng et al., 2010). Similarly, a significant decrease in SME, was enhanced from the addition of dairy proteins (Onwulata et al., 2001).

10.3   EFFECTS OF EXTRUSION COOKING ON NUTRITIONAL PROPERTIES OF EXTRUDATES

One of the most demanding and challenging field in the area of food processing is health and nutrition. Maintenance and increase in the nutritional value during processing has remained an important area of research. The degradation of nutritional properties by high temperature is a serious problem in conventional processing of foods. Extrusion cooking is preferred over other techniques because of the continuous process with high temperature and short time, significant nutrient retention, and higher productivity.

Effects of extrusion cooking on nutritional and antinutritional properties are shown in Table 10.2. Positive effects consist of destruction of antinutritional factors, starch gelatinization, lesser oxidation, and enhanced soluble dietary fiber. Conversely, Maillard reaction between sugars and proteins decreases the nutritional quality of the protein and sugar, depending upon the composition of raw materials and process parameters. Vitamins that are sensitive to heat are lost in different amounts. Alteration in carbohydrates, amino acid profile, proteins, minerals, vitamins, dietary fiber, and some non-nutrient healthful components of food may either be advantageous or detrimental (Singh et al., 2007). Proper selection of ingredients and careful control of process conditions are therefore important to achieve nutritionally balanced extruded product.

TABLE 10.2 Effect of Extrusion Cooking on Nutritional and Antinutritional Parameters of Extruded Product

Parameters	Effect of extrusion cooking	References
Carbohydrate	Sugar losses (about 2–20% sucrose); Decrease in raffinose and stachyose; Increase in the starch digestibility	Borejszo and Khan, 1992; Camire et al., 1990; Wang and Klopfenstein, 1993
Total dietary fibre	Increased	Ralet et al., 1993; Vasanthan et al., 2002; Wani and Kumar, 2016d
Protein	Increases protein digestibility	Alonso et al., 2000b; Camire et al., 1990
Fat	No effect	Wicklund and Magnus, 1997
Minerals	Increase in the availability of minerals	Alonso et al., 2001; Wani and Kumar, 2016d
Vitamins	Reduces heat sensitive vitamins. Vitamins D and K are comparatively stable.	Andersson and Hedlund, 1990; Killeit, 1994; Singh et al., 2007
Antinutritionals	Destruction/annihilation of antinutritional factors, particularly haemagglutinins, trypsin inhibitors, phytates and tannins.	Alonso et al., 2000a

10.3.1 PROTEIN

Nutritional quality of protein is dependent upon the availability, amount, and digestibility of essential amino acids (EAAs). Among the parameters, digestibility is the most important determinant of quality in adult persons (FAO, 1985). Digestibility value of protein in extruded products is comparatively higher than that of the unextruded one.

Extrusion conditions have shown a significant effect on the protein quality of the extruded product. Among various extrusion conditions, the feed has shown the maximum influence on the digestibility of protein, followed by barrel temperature, as observed in the extrusion process of fish–wheat flour mixture. Increase in digestibility of the extrudates by 2–4% was observed by increasing the ratio of fish to wheat three times (Bhattacharya et al., 1988; Camire et al., 1990). It was observed that rise in the extrusion temperature (100–140°C) increases the rate of inactivation of protease inhibitors in wheat flour, which can directly result in greater protein digestibility. High extrusion temperature at 140°C does not show any negative influence on the digestibility of protein that may be due to the lesser retention time of feed material in the extruder (Bhattacharya et al., 1988). Protein digestibility of extruded corn-gluten can be increased with an increase in screw speed. The increase in shear forces causes denaturation of proteins more easily in the extruder, thereby facilitating change in functionality (Bhattacharya and Hanna, 1985). Several other researchers have reported that extrusion cooking causes destruction of antinutritional factors, particularly haemagglutinins, trypsin inhibitors, phytates and tannins, all of which inhibit protein digestibility (Alonso et al., 2000b).

By keeping the temperature constant, the inactivation of enzymes increases with rise in moisture content and retention time (Singh et al., 2000). An increase in product retention and moisture content probably decreased the effect of heat, which caused relatively lesser rate of inactivation of trypsin inhibitor and as a result yielded to lesser expansion. Lorenz and Jansen (1980) reported the maximum protein efficiency ratio at a temperature of 143°C and moisture content of 15–30% with residence time of 0.5–2 min. Higher extrusion temperature, lower feed moisture content, and longer residence time can be considered as key parameters for the destruction of antinutritional factors.

10.3.2 CARBOHYDRATE

Carbohydrates vary from simple sugars to fiber and complex starch. The process of extrusion showed a significant effect on various carbohydrates. A number of researchers have shown sugar losses during the extrusion process. In the study of extrusion of protein-enriched biscuits, about 2–20% of sucrose was lost at 170–210°C and 13% feed moisture (Camire et al., 1990). The conversion of sucrose into reducing sugars, fructose, and glucose and Maillard reactions could be a major reason for the loss of sucrose.

Oligosaccharides particularly stachyose and raffinose can cause flatulence and thus impair the nutritional utilization of legumes (Omueti and Morton, 1996). A decrease in stachyose and raffinose was observed in extrudates prepared from high-starch fractions of pinto beans (Borejszo and Khan, 1992). Lower content of both stachyose and raffinose was also observed in extruded snacks in comparison to that in unextruded soy grits and flour (Omueti and Morton, 1996). Therefore, the destruction of these flatulence-causing oligosaccharides in extruded legume products might improve the nutritional quality.

Starch is one of the main important carbohydrates and is the storage form of energy for plants. The major sources of starch in the human diet are corn, wheat, and rice and are also the main raw materials used for the development of extruded products. Studies have shown significant effect of extrusion on starch. The digestion of the ungelatinized starch cannot be easily carried out by humans and other monogastric species; therefore, this can easily be carried out using the extrusion technique.

Extrusion to a greater extent increased the availability of starch to enzymes of wheat bran and whole flour (Wang and Klopfenstein, 1993). Decreased feed moisture (16% vs. 20%) and temperature (160°C vs. 185°C) significantly increased the average starch molecular weight in wheat flour (Politz et al., 1994). According to Gautam and Choudhoury (1999), the breakdown of starch depends upon the screw configuration. The degradation/digestion of the starch can be dominated by extrusion to produce dextrin/free glucose for fermentation or syrups. High shear conditions are necessary for the maximum conversion of starch to glucose. The production of glucose from starch has been studied in number of food items such as cassava, barley, and corn (Van Zuilichem et al., 1990).

10.3.3 DIETARY FIBER

Extrusion cooking has been found to affect the dietary fiber content significantly. Increased dietary fiber content by extrusion process was reported by several researchers (Ralet et al., 1993; Wani and Kumar, 2016d). However, some contradictory findings have also been reported. There are factors that affect the fiber solubility, such as acid and alkaline treatment of the feed prior to extrusion (Ning et al., 1991). Soluble fiber of pea hull was increased over 10% through extrusion (Ralet et al., 1993). Extrusion cooking of wheat flour and whole-wheat meal at 161–180°C, 150–200 rpm screw speed, and 15% feed moisture revealed nonsignificant changes in dietary fiber content (Varo et al., 1983). No change in dietary fiber content was observed when wheat was extruded under lower conditions of temperature screw speed and feed moisture, except the solubility of fiber, which was slightly more soluble (Siljestrom et al., 1986).

Increase in dietary fiber content has been reported due to glucans, which is present in soluble as well as insoluble dietary fiber. Extrusion cooking increased the total dietary fiber of barley flours, which may be due an increase in soluble dietary fiber. However, for regular barley flour, the rise in both total dietary fibers as well as soluble dietary fiber was reported (Vasanthan et al., 2002). Transformation in dietary fiber during extrusion may be due to change of insoluble to soluble dietary fiber and formation of resistant starch along with formation of other enzyme-resistant indigestible glucans by transglycosidation.

10.3.4 LIPID

The raw material used for extrusion cooking contains native lipids. The most commonly used raw material for extrusion purposes are cereals like rice, wheat, and corn. These are typically low in lipid content. Generally, extrusion is not advisable for high-fat materials, particularly in case of puffed products, because lipid levels above 5–6% decreases the performance of extruder (Camire, 2000). This is because lipid causes a reduction in the torque within the barrel, and the expansion of the final product is also reduced because adequate pressure is not developed during extrusion process due to release of lipids from the physical disruption of cell walls and high temperature.

Meanwhile, a lower level of lipid (<5%) enables steady extrusion and development of the texture.

The fatty acid content of lipids obtained from corn meal samples were found similar before and after extrusion (Guzman et al., 1992). Similarly, nonsignificant variations in the distribution of palmitic and stearic acids have been reported for raw oat flour and its extruded product (Wicklund and Magnus, 1997).

10.3.5 VITAMINS

Vitamin content decreases by extrusion conditions of higher temperature, specific energy inputs, screw speed, die diameter, and feed rate. A substantial degradation can take place depending upon the vitamin. Because the structure and composition of vitamins vary, their stability may differ during the extrusion. The degree of degradation of vitamins during food processing depends on a number of parameters such as moisture, screw speed, temperature, and time. The protection of vitamins during the extrusion process is possible to a certain extent by reducing temperature and shear.

Vitamins D and K (lipid-soluble vitamins) are comparatively stable. Some of the vitamins such as E and A and their associated compounds such as tocopherols and carotenoids, respectively, are not stable under high temperature (Killeit, 1994). The losses of β-carotene during the extrusion process are mainly due to heat. High barrel temperatures decrease all trans-β-carotene in wheat flour by greater than 50% (Guzman-Tello and Cheftel, 1990). The effect of HTST extrusion cooking on vitamin stability was assessed by Pham and Del Rosario (1986) and Guzman-Tello and Cheftel (1987). Vitamin C (ascorbic acid) is sensitive to heat, and it is reduced when extruded at higher temperature and low moisture (10%) (Andersson and Hedlund, 1990). When ascorbic acid was added at a level of 0.4–1.0% to cassava starch in order to enhance the conversion of starch, about 50% retention was observed (Sriburi and Hill, 2000).

10.3.6 MINERALS

Minerals are uniformly distributed in foodstuffs. Phosphates, as a source of phosphorus, are commonly supplemented/added in processing, and minerals such as calcium and iron are usually supplemented/added to foodstuffs for enhancing the nutritional value (Camire et al., 1990). Some of the metals, mainly calcium, copper, iron and magnesium, act as catalysts for various enzymes. Minerals such as calcium and iron are necessary for bones and prevention of anemia (Camire et al., 1990). Fortification or enrichment of the extruded product with minerals can be done depending on the product and requirement of the masses.

In spite of greater significance of minerals for health, the stability of minerals during extrusion has not widely studied (Camire et al., 1990). Extrusion reduces the factors that can inhibit absorption, thereby improving the absorption of minerals. Insoluble complexes of phytate with minerals affect the absorption of minerals (Alonso et al., 2001). The process of extrusion has been found to hydrolyze the phytate complex, thus making phosphate molecules available for digestion. Extrusion cooking caused phytate hydrolysis in extruded products prepared using peas and kidney beans, which indicates better mineral availability after extrusion processing (Alonso et al., 2001). Extrusion causes a reduction of 13–35% of the phytate content of a wheat bran-starch-gluten mix. The availability of minerals in the extruded product prepared using fenugreek oats and pea was higher than that in raw materials (Wani and Kumar, 2016d).

The inhibitory effect of natural polyphenols on mineral absorption has been reported. Tannins may chelate the minerals and form insoluble complexes in the gastrointestinal tract, thus reducing their bioavailability. Mineral absorption can be increased during extrusion by the destruction of these polyphenolic compounds during processing. The variation in polyphenolic compounds after extrusion may be due to binding of phenolics and organic materials (Alonso et al., 2001).

The fiber components might alter mineral absorption. The presence of lignin, cellulose, and some hemicelluloses may impede mineral absorption by affecting the movement of the digestive tract. The chelating ability during the extrusion process can be changed from the reorganization of the dietary fiber components. Besides, phytate and other antinutritional factors present in food may bind with fiber molecules, and this can modify the availability of minerals (Alonso et al., 2000a,b).

Fortification of minerals prior to extrusion may pose other challenges. Phenolic compounds form complexes with iron that are dark in color and thus diminish the sensory acceptability of foods. Kapanidis and Lee (1996) reported that ferrous sulfate heptahydrate can be used as an appropriate source of iron in a simulated rice product without changing of the color of the product.

10.4 CONCLUDING REMARKS

A number of ready-to-eat product have been developed by extrusion cooking. Extrusion cooking has a significant effect on nutritional attributes of the extruded product. Beneficial effects on nutritional properties ranged from increased digestibility to the preparation of nutritionally balanced or enriched foods/feeds at low cost. Extrusion causes destruction of antinutritional factors, increase in soluble dietary fibers, and increase in the availability of minerals.

Few studies have shown the effects of extrusion cooking on the structural and pasting properties of extruded product; hence, this area needs to be explored and requires focus in research. Further research can be directed on the relationship between textural properties and product structure.

KEYWORDS

- cooking
- extrusion
- functional
- nutritional
- physical
- properties

REFERENCES

Ainsworth, P., Ibanoglu. S., Plunkett, A., Ibanoglu, E., & Stojceska, V., (2007). Effect of brewers spent grain addition and screw speed on the selected physical and nutritional properties of an extruded snack. *J. Food Eng.*, *81*,702–709.

Alonso, R., Aguirre, A., & Marzo, F., (2000a). Effects of extrusion and traditional processing methods on antinutrients and in vitro digestibility of protein and starch in feba and kindey beans. *Food Chem.*, *68*, 159–165.

Alonso, R., Orue, E., Zabalza, M. J., Grant, G., & Marzo, F., (2000b). Effect of extrusion cooking on structure and functional properties of pea and kindey bean proteins. *J. Sci. Food Agric.*, *80*, 397–403.

Alonso, R., Rubio, L. A., Muzquiz, M., & Marzo, F., (2001). The effect of extrusion cooking on mineral bioavailability in pea and kidney bean seed meals. *Animal Feed Sci. Tech.*, *94*, 1–13.

Altan, A., McCarthy, K. L., & Maskan, M., (2008). Evaluation of snack foods from barley–tomato pomace blends by extrusion processing. *J. Food Eng.*, *84*, 231–242.

Andersson, Y., & Hedlund, B., (1990). Extruded wheat flour: correlation between processing and product quality parameters. *Food Qual. Pref.*, *2*, 201–216.

Bhandari, B., D'Arcy, B., & Young, G., (2001). Flavour retention during high temperature short time extrusion cooking process: A review. *Int. J. Food Sci. Technol.*, *36*, 453–461.

Bhattacharya, M., & Hanna, M. A., (1985). Extrusion processing of wet corn gluten meal. *J. Food Sci.*, *50*, 1508–1509.

Bhattacharya, S., Das, H., & Bose, A. N., (1988). Effect of extrusion process variables on in vitro protein digestibility of fish-wheat flour blends. *Food Chem.*, *28*, 225–231.

Borejszo, Z., & Khan, K., (1992). Reduction of flatulence-causing sugars by high temperature extrusion of pinto bean high starch fractions. *J. Food Sci.*, *57*, 771–722 & 777.

Brennan, M. A., Derbyshire, E., Tiwari, B. K., & Brennan, C. S., (2013). Ready-to-eat snack products: the role of extrusion technology in developing consumer acceptable and nutritious snacks. *Int. J. Food Sci. Technol.*, *48*, 893–902.

Camire, M. E., (2000). Chemical and nutritional changes in food during extrusion. In: *Extruders in Food Applications;* Riaz, M. N., ed.; Boca Raton, FL: CRC Press; pp. 127–147.

Camire, M. E., Camire, A. L., & Krumhar, K., (1990). Chemical and nutritional changes. *Crit. Rev. Food Sci. Nutr.*, *29*, 35–57.

Carvalho, C. W. P., & Mitchelle, J. R., (2000). Effect of sugar on the extrusion of maize grits and wheat flour. *Int. J. Food Sci. Technol.*, *35*, 569–576.

Castells, M., Marin, S., Sanchis, V., & Ramos, A. J., (2005). Fate of mycotoxins in cereals during extrusion cooking: a review. *Food Add. Contam.*, *22*, 150–157.

Dar, A. H., Sharma, H. K., & Kumar, N., (2014). effect of extrusion temperature on the microstructure, textural and functional attributes of carrot pomace-based extrudates. *J. Food Proc. Preserv.*, *38*(1), 212–222.

Della Valle, G., Vergnes, B., Colonna, P., & Patria, A., (1997). Relations between rheological properties of molten starches and their expansion behavior in extrusion. *J. Food Eng.*, *31*, 277–296.

Ding, Q. B., Ainsworth, P., Plunkett, A., Tucker, G., & Marson, H., (2006). The effect of extrusion conditions on the functional and physical properties of wheat-based expanded snacks. *J. Food Eng.*, *73*, 142–148.

Ding, Q. B., Ainsworth, P., Tucker, G., & Marson, H., (2005). The effect of extrusion conditions on the physicochemical properties and sensory characteristics of rice-expanded snacks. *J. Food Eng.*, *66*, 283–289.

FAO/WHO/UNU; Energy and protein requirements, (1985). Tech. Rep. Series 724, Expert Consultation. Geneva: World Health Organisation.

Fellows, P., (2000). *Food Processing Technology: Principles and Practice.* Cambridge: Woodhead Publishing Ltd, pp. 177–182.

Gautam, A., & Choudhoury, G. S., (1999). Screw configuration effects on starch breakdown during twin screw extrusion of rice flour. *J. Food proc. Preserv.*, *23*, 355–375.

Gutkoski, L. C., & El-Dash, A. A., (1999). Effect of extrusion process variables on physical and chemical properties of extruded oat products. *Plant Food Hum. Nutr.*, *54*, 315–325.

Guzman, L. B., Lee, T. C., & Chichester, C. O., (1992). Lipid Binding During Extrusion Cooking. In: *Food Extrusion Science and Technology,* Kokini, J. L., Ho, C-T, Karwe, M. V., Ed.; Marcel Dekker: New York, USA, pp. 427–436

Guzman-Tello, R., & Cheftel, J. C., (1987). Thiamine destruction during extrusion cooking as an indicator of the intensity of thermal processing. *Int. J. Food Sci. Technol.*, *22*, 549–562.

Guzman-Tello, R., & Cheftel, J. C., (1990). Colour loss during extrusion cooking of beta carotene-wheat flour mixes as an indicator of the intensity of thermal and oxidative processing. *Int. J. Food Sci. Technol.*, *25*, 420–434.

Hagenimana, A., Ding, X. L., & Fang, T., (2006). Evaluation of rice flour modified by extrusion cooking. *J. Cereal Sci.*, *43*, 38–46.

Havck, B. W., & Huber, G. R., (1989). Single screw vs twin screw extrusion. *The Amer. Assoc. Cereal Chemist.*, *34*, 930–939

Heldman, D. R. et al., (1998). Food extrusion. In: *Principles of Food Processing*; Heldman D. R., & Hartel, R. W., (ed.), Springer, New York, pp. 253–283.

Hirth, M., Leiter, A., Beck, S. M., & Schuchmann, H. P., (2014). Effect of extrusion cooking process parameters on the retention of bilberry anthocyanins in starch based food. *J. Food Eng.*, *125*, 139–146.

Huber, G., (2001). Snack foods from cooking extruders, *In Snack Food Processing*. Lusas, E. W., & Rooney, L. W., (ed.), CRC Press, Baca Raton, FL. pp. 315–367.

Kapanidis, A. N., & Lee, T. C., (1996). Novel method for the production of color-compatible rice through extrusion. *J. Agric. Food Chem.*, *44*, 522–525.

Killeit, U., (1994). Vitamin retention in extrusion cooking. *Food Chem.*, *49*, 149–155.

Koksel, H., Ryu, G. H., Basman, A., Demiralp, H., & Ng, P. K. W., (2004). Effects of extrusion variables on the properties of waxy hulless barley extrudates. *Nahrung.*, *48*, 19–24.

Kothakota, A., Jindal, N., Thimmaiah, B., (2013). A study on evaluation and characterization of extruded product by using various by-products. *Afric. J. Food Sci.*, *7*(12), 485–497.

Kumar, N., Sarkar, B. C., & Sharma, H. K., (2010). Development and characterization of extruded product of carrot pomace, rice flour and pulse powder. *Afri. J. Food Sci.*, *4*(11), 703–717.

Lawton, B. T., & Handerson, B. A., (1972). The effects of extruder variables on the gelatinization of corn starch. *Can. J. Chem. Eng.*, *50*, 168–172.

Lorenz, K., & Jansen, G. R., (1980). Nutrient stability of full-fat soy flour and corn-soy blends produced by low-cost extrusion. *Cereal Food World*, *25*, 161–172.

Meng, X., Threinen, D., Hansen, M., & Driedger, D., (2010). Effects of extrusion conditions on system parameters and physical properties of a chickpea flour-based snack. *Food Res. Int.*, *43*, 650–658.

Mezreb, K., Goullieux, A., Ralainirina, R., & Queneudec, M., (2003). Application of image analysis to measure screw speed influence on physical properties of corn and wheat extrudates. *J. Food Eng.*, *57*, 145–152.

Moscicki, L., Mitrus, M., & Wojtowicz, A., (2007). Extrusion techniques in the agro-food industry, PWRiL, Warsaw, p. 223.

Moscicki, L., & Van Zuilichem, D. J., (2011). Extrusion-Cooking and Related Technique. In: *Applications, Theory and Sustainability*. Moscicki, L. (ed.), Wiley-VCH Verlag GmbH & Co. KGaA, Weinheim ISBN: 978-3-527-32888-8.

Muthukumarappan, K., & Karunanithy, C., (2012). Extrusion cooking process In: *Handbook of Food Process Design,* Ahemed, J., & Rahman, M. S., (ed.), First edition, Blackwell Publishing Ltd,; vol. 1, pp. 710–742.

Ning, L., Villota, R., & Artz, W. E., (1991). Modification of corn fiber through chemical treatments in combination with twin-screw extrusion. *Cereal Chem., 68,* 632–636.

Oke, M. O., Awonorin, S. O., Sanni, L. O., Asiedu, R., & Aiyedun, P. O., (2013). Effect of extrusion variables on extrudates properties of water yam flour – A response surface analysis. *J. Food Proc. Preserv., 37*(5), 1–18.

Omueti, O., & Morton, I. D., (1996). Development by extrusion of soyabari snack sticks: a nutritionally improved soy-maize product based on the Nigerian snack (kokoro). *Int. J. Food Sci. Nutr., 47,* 5–13.

Onwulata, C. I., & Konstance, R. P., (2006). Extruded corn meal and whey protein concentrate: Effect of particle size. *J. Food Proc. Preserv., 30*(4), 475–487.

Onwulata, C. I., Konstance, R. P., Smith, P. W., & Holsinger, V. H., (2001). Coextrusion of dietary fiber and milk proteins in expanded corn products. *LWT – Food Sci. Technol., 34*(7), 679–687.

Pathania, S., Singh, B., Sharma, S., Sharma, V., & Singla, S., (2013). Optimization of extrusion processing conditions for preparation of an instant grain base for use in weaning foods. *Int. J. Eng. Res. App., 3*(3), 1040–1049.

Pelembe, L. A. M., Erasmus, C., & Taylor, J. R. N., (2002). Development of a protein-rich composite sorghum-cowpea instant porridge by extrusion cooking process. *LWT – Food Sci. Technol., 35,* 120–127.

Pham, C. B., & Del Rosario, R. R., (1986). Studies on the development of texturized vegetable products by the extrusion process. III. Effects of processing variables on thiamine retention. *J. Food Technol., 21,* 569–576.

Politz, M. L., Timpa, J. D., & Wasserman, B. P., (1994). Quantitative measurement of extrusion-induced starch fragmentation products in maize flour using non aqueous automated gel-permeation chromatography. *Cereal Chem., 71,* 532–536.

Ralet, M. C., Della Valle, G., & Thibault, J. F., (1993). Raw and extruded fiber from pea hulls. Part 1: composition and physicochemical properties. *Carbohydr. Polymer, 20,* 17–23.

Rodriguez-Miranda, J., Ruiz-Lopez, I. I., Herman-Lara, E., Martinez-Sanchez, C. E., Delgado-Licon, E., & Vivar-Vera, M. A., (2011). Development of extruded snacks using taro (*colocasia esculenta*) and nixtamalized maize (zea mays) flour blends. *LWT – Food Sci. Technol., 44*(3), 673–680.

Sahagun, J. F., & Harper, J. M., (1980). Effects of screw restrictions on the performance of an autogenous extruder. *J. Food Proc. Eng., 3*(4), 199.

Seth, D., Badwaik, L. S., & Ganapathy, V., (2013). Effect of feed composition, moisture content and extrusion temperature on extrudate characteristics of yam-corn-rice based snack food. *J. Food Sci. Technol.,* DOI: 10.1007/s13197-013-1181-x.

Siljestrom, M., Westerlund, E., Bjorck, I., Holm, J., Asp, N. G., & Theander, O., (1986). The effects of various thermal processes on dietary fibre and starch content of whole grain wheat and white flour. *J. Cereal Sci., 4,* 315–323.

Singh, B., Sekhon, K. S., & Singh, N., (2007). Effects of moisture, temperature and level of pea grits on extrusion behavior and product characteristics of rice. *Food Chem., 100,*198–202.

Singh, D., Chauhan, G. S., Suresh, I., & Tyagi, S. M., (2000). Nutritional quality of extruded snakes developed from composite of rice broken and wheat bran. *Int. J. Food Prop., 3,* 421–431.

Singh, S., Gamlath, S., & Wakeling, L., (2007). Nutritional aspects of food extrusion: A review. *Int. J. Food Sci. Technol.*, *42*, 916–929.

Sobukola, O. P., Babajide, J. M., & Ogunsade, O., (2012). Effect of brewer spent grain addition and extrusion parameters on some properties of extruded yam starch-based pasta. *J. Food Proc. Preserv.*, doi:10.1111/j.1745-4549.2012.00711.x.

Sriburi, P., & Hill, S. E., (2000). Extrusion of cassava starch with either variations in ascorbic acid concentration or pH. *Int. Food Sci. Technol.*, *35*, 141–154.

Van den Einde, R. M., Van der Goot, A. J., & Boom, R. M., (2003). Understanding molecular weight reduction of starch during heating-shearing process. *J. Food Sci.*, *68*, 396–2904.

Van Zuilichem, D. J., (1992). *Extrusion Cooking.* Craft or Science? PhD thesis, Department of Food Science, Food and Bioprocess Engineering Group, Agricultural University Wageningen, The Netherlands.

Van Zuilichem, D. J., Van Roekel, G. J., Stolp, W., & Van't Riet, K., (1990). Modelling of the enzymatic conversion of cracked corn by twin-screw extrusion cooking. *J. Food Eng.*, *12*, 13–28.

Varo, P., Liane, R., & Koivistoinen, P., (1983). Effect of head treatment on dietary fibre: interlaboratory study. *J. Assoc. Official Anal. Chem.*, *66*, 933–938.

Vasanthan, T., Gaosong, J., Yeung, J., & Li, J., (2002). Dietary fibre profile of barley flour as affected by extrusion cooking. *Food Chem.*, *77,* 35–40.

Wang, W. M., & Klopfenstein, C. F., (1993). Effects of twin-screw extrusion on the nutritional quality of wheat, barley, oats. *Cereal Chem.*, *70*, 712–725.

Wani, S. A., & Kumar, P., 2016a Effect of incorporation levels of oat and green pea flour on the properties of an extruded product and their optimization. *Acta. Alimen.*, *45*(1), 28–35.

Wani, S. A., & Kumar, P., 2016b Development and parameter optimization of health promising extrudate based on fenugreek oat and pea. *Food Biosci.*, *14*(1), 34–40.

Wani, S. A., & Kumar, P., 2016c Fenugreek enriched extruded product: optimization of ingredients using response surface methodology. *Int. Food Res. J.*, *23*(1), 18–25.

Wani, S. A., & Kumar, P., 2016d Effect on nutritional, antioxidant and microstructural characteristics of nutritionally enriched snacks by extrusion cooking. *J. Food Proc. Preserv.*, *40*(2), 166–173.

Wicklund, T., & Magnus, E. M., (1997). Effect of extrusion cooking on extractable lipids and fatty acid composition in sifted oat flour. *Cereal Chem.*, *74*, 326–329.

CHAPTER 11

PROCESSING AND VALUE ADDITION OF CUSTARD APPLE

NAVNEET KUMAR,[1] HARISH KUMAR SHARMA,[2]
C. KHODIFAD BHARGAVBHAI,[1] and MANJEET PREM[3]

[1] *Department of Agricultural Process Engineering, College of Agricultural Engineering and Technology, Anand Agricultural University, Godhra (Gujarat), India*

[2] *Department of Food Engineering and Technology, Sant Longowal Institute of Engineering and Technology, Longowal, Sangrur–148106, Punjab, India*

[3] *Department of Farm Machinery and Power Engineering, College of Agricultural Engineering and Technology, Anand Agricultural University, Godhra (Gujarat), India*

CONTENTS

11.1 INTRODUCTION

Custard apple is one of the delicious fruits enjoyed by many for table purposes (Figure 11.1). Its pleasant flavor, mild aroma, and sweet taste have a

FIGURE 11.1   Custard apple *(Annona squamosa)* fruit.

widespread acceptance. It is found in wild and cultivated throughout India; it grows gregariously and widely in the hilly tracts and waste lands and has become completely naturalized in several districts of Gujarat (Middle, North Gujarat, and Saurashtra), Andhra Pradesh, Punjab, Rajasthan, Uttar Pradesh, Madhya Pradesh, Bihar, West Bengal, Assam, Maharashtra, Karnataka, Kerala, and Tamil Nadu. It is a native of South America and West Indies. Custard apples are usually consumed as a dessert fruit. This fruit is a rich source of starch, when firm, but as the amount of sugar increases from the degradation of starch, it softens. The sugars that are mainly present in custard apples are glucose and fructose (80–90%). Compared with other fruits, the custard apple fruit contains a significant amount of vitamin C, thiamine, magnesium, potassium, and dietary fiber. The calorific value is 300–450 kJ per 100 g and is nearly double that of apple, peach, and orange. However, despite of high sugar content, the glycemic index of custard apple is low (54), and the glycemic load is moderate (10.2) (Brand-Miller et al., 2003).

Annona fruits are climacteric, relatively soft, and are characterized by high levels of ethylene (100–300 μL kg^{-1} h^{-1}) during ripening. These fruits should be handled with care to minimize bruising (Pareek et al., 2011). The fruits show typical climateric peak (104.23 mg CO$_2$ kg^{-1} h^{-1}) on the second day of storage in fruits harvested at the matured stage under ambient condition. Lower peaks at 56.32 mg CO$_2$ kg^{-1} h^{-1} and 27.42 mg CO$_2$ kg^{-1} h^{-1} were also reported at 20°C and 15°C, respectively (Mallikarjuna et al., 2012). The recommended temperature for storage of custard apple is 8–12°C for storage up to 1–2 weeks with a minimum loss of quality (Brown and Scott, 1985). The fruits can be stored in polythene bags and paper bags along with wheat straw for least reduction in fruit weight and higher values of vitamin C during ripening, respectively (Chaudhry et al., 1985). Compositional and functional parameters of custard apple pulp are shown in Table 11.1.

The average weight of the fruit is 326 g and it possesses a very sweet flavour along with very good eating and transport quality (Liu, 2000). The seeded berries are larger and have higher sugar content (Mazumdar, 1977). The custard apple juice contains protein (4.48%), fat (1.56%), crude fiber (7.53%), carbohydrate (10.52%), food energy (74.04 Kcal), invert sugar (161.84 mg), fructose (167.27 mg/100 g), hydrated maltose (268.13 mg/100 g), vitamin A (16.63 µg/100 g RE), vitamin C (43.38 mg/100 g), total titrable acidity (0.027%), total solids (27.25%), and soluble solids (10.00%) (Amoo et al., 2008). The rind portion of the fruit also contains lactose, sucrose, galactose, and glucose (Chandraju et al., 2012). Alpha-pinene, beta-pinene, limonene, bornyl acetate, and germacrene D are the major compounds from about 53 compounds identified by simultaneous steam distillation-solvent extraction (Pino and Rosado, 1990).

Custard apple is also a source of the medicinal and industrial products. It is also used as an antioxidant, antidiabetic, hepatoprotective, cytotoxic, geno-toxic, antitumor, and antilice agent. The fruit is also used in various recipes, viz., jam, nectars, and ice creams. There is a demand to utilize the nutritional potential of custard apple and to develop various new value-added products, which will also reduce the postharvest losses of this perishable food.

11.2   PROCESSING AND VALUE ADDITION

Development of various value-added products like jam, fruit-flavored yoghurt, fruit drinks, and syrups. with fruit extracts from passion fruits like custard apple was suggested by Hoyos (1980). Singh et al. (2009) also stressed on value addition of forest produce for nutrition and stated the importance of some Indian fruits like custard apple for improving nutrition. A summary of research work reported is presented in Table 11.2.

11.2.1   CUSTARD APPLE PULP

Gamage et al. (1997) used minimal processing for custard apple pulp and treated it with 0.1–0.5% ascorbic acid. The ascorbic acid-treated pulp was of acceptable quality throughout the storage period of 4 weeks at 0°C. Shashirekha et al. (2008) studied the changes in volatile compounds of fruit

TABLE 11.1 Composition and Functional Parameters of 100 g of Edible Pulp of Custard Apple (*Annona squamosa*) Fruits

Components	Amount	Components	Amount
Yield of edible parts, %	36.7 ± 1.15	**Lipids**	
Water, g/100g	73.0 ± 1.10	Total saturates, mg/100g	48
Total soluble solids, °Brix	21.10 ± 0.13	Total mono unsaturates, mg/100g	114
Lightness (L*)	65.40 ± 0.93	Total poly unsaturates, mg/100g	40
Hue	95.07 ± 0.47	Trans, mg/100g	Nil
Proteins, g/100g	1.68 ± 0.8	Cholesterol, mg/100g	Nil
Lipids, g/100g	0.4 ± 0.3	**Dietary fibre**	
Carbohydrates, g/100g	19.6 ± 1.0	Total dietary fibre, g/100 g dry weight	26.15 ± 0.46
Fiber, g/100g	1.4 ± 0.6	Soluble fiber, g/100 g dry weight	6.91 ± 0.01
Total acidity, g/100g	0.1	Insoluble fiber, g/100 g dry weight	19.24 ± 0.47
Ash, g/100g	0.7 ± 0.1	**Bioactive compounds**	
Energy, calories/100g	96 ± 10	Vitamin C, g/100g fresh weight	50.79 ± 4.92
Calcium, mg/100g	26.2 ± 6.0	Total anthocyanins, mg TA/100g fresh weight	0.47 ± 0.02
Phosphorous, mg/100g	42 ± 14	Total phenolic compounds, mg of GAE/100 g fresh weight	207.60 ± 17.85
Iron, mg/100g	0.8 ± 0.5	Total flavonoids, mg of quercetin/100 g fresh weight	200.92 ± 3.83
Magnesium, mg/100g	21	Total carotenoids, mg of b-carotene/100 g fresh weight	1.44 ± 0.11

TABLE 11.1 (Continued)

Components	Amount	Components	Amount
Potassium, mg/100g	247	**Antioxidant activity**	
Sodium, mg/100g	9	Trolox equivalent antioxidant capacity (TFAC) – ABTS, μm/100 g FW)	646.25 ± 12.15
Zinc, mg/100g	0.10	Trolox equivalent antioxidant capacity (TFAC) – DPPH, μm/100 g FW)	368.55 ± 8.33
Titratable acidity, g of citric acid/100 g FW	0.36 ± 0.098	VCEAC – ABTS, mg/100 g FW	188.64 ± 3.57
		VCEAC – DPPH, mg/100 g FW	79.43 ± 1.80
Vitamins			
Vitamin, A, mg	0.005 ± 0.001	Vitamin B_5, mg	0.9 ± 0.3
Vitamin B_1, mg	0.1 ± 0.01	Vitamin B_6, mg	0.2000
Vitamin B_2, mg	0.113	Vitamin B_{12}, mg	0.13 ± 0.05
Vitamin B_3, mg	0.883	Ascorbic acid, mg	37.38 ± 4.62

(*Source:* Pareek et al., 2011; Moo-Huchin et al., 2014; USDA, 2016).

TABLE 11.2 Processing and Value Addition of Custard Apple

1	Products	Various value added products of custard apple like jam, fruit-flavored yoghurt, fruit drinks and syrups etc. were suggested (Hoyos, 1980, Kadam, 2001)
2	Taste enhancer	Polyphenol oxidase enzyme extracted from custard apple pulp was used for taste enhancement in cocoa nibs (Lima et al., 2001)
3	Jam	Jam prepared with 50% custard apple pulp obtained high sensory score and found fit for consumption till 4 months (Singh et al., 2006)
4	Flavored pulp and dehydration	Frozen and stored (for 12 months) did not differ in flavor spectrum, Heating at 55°C, 85°C (pasteurization) for 20 min each increased the flavor spectrum, and spray dried samples with whole milk powders exhibited fruit flavor. (Shashirekha et al., 2008)
5	Fruit fermentations	Technologies for pectolytic enzymes, fining agents and parameters for fruit fermentations were reviewed (Muniz et al., 2008)
		Fruit wine (alcohol percentage in distillate –8.2%) using custard apple were prepared (Deshpande et al., 2010)
		Custard apple wine prepared using *Saccharomyces cerevisiae* (NCIM 3282) showed free radicle scavenging activity towards DPPH, DMPD and FRAP (Jagtap and Bapat, 2015)
6	Toffee	Prepared using 55% custard apple pulp achieved maximum sensory score (Mundhe et al., 2008)
7	Milk shake	Buffalo milk & custard apple pulp ratio – 90:10 was prepared (Poul et al., 2009, 2010)
8	Pulp	Minimal processing by addition 0.1%–0.5% ascorbic acid and storage temperature of 0°C temperature for 4 weeks (Gamage et al., 1997)
		Inactivation of Polyphenol Oxidase in custard apple pulp by addition of 2000 ppm of ascorbic acid was recommended (Pawar et al., 2010)
		Viscosity behavior of pulp was studied and improved stability by addition of Xanthan gum were obtained (Santos et al., 2014)
		Pulp stored for a period of 6 months by addition of 1500 ppm of potassium metabisulphite (Sravanthi et al., 2014)
		Drying of custard apple pulp was recommended at 60°C in hot air dryer. (Kumar et al., 2015)

TABLE 11.2 (Continued)

9	Ice-cream	Ice cream proportion: 15% Custard apple pulp and 15% sugar was recommended (Yadav et al., 2010)
		Ice cream proportion: 15% pulp, 15% sugar, 10% fat in different combination, ascorbic acid 0.3% was more acceptable (Pawar et al., 2011)
10	RTS beverage	Custard apple and lime (3:2) with 15% TSS and 0.2% acidity was recommended (Pilania et al., 2010)
		Effect of chemical preservative, potassium metabisulphite on microbial and sensory quality of RTS beverage (Markam and Singh 2012)
11	Squash	Keeping quality and acceptability of prepared squash through sensory evaluation during 32 days storage was studied (Patil et al., 2011)
12	Nectar	Custard apple (20% pulp) was used (Shrivastava et al., 2013)
13	Binding agent in tablets	Custard apple pulp powder had two folds increase in the disintegration time in tablets in comparison to Polyvinylpyrrolidone (Thube et al., 2011)
14	Powder	Freeze drying at –40°C temperature and 20% Maltodextrin level was performed (Bharadiya et al., 2010)

pulp of custard apple, as influenced by the conditions of processing. Sweet and pleasant-flavored pulp from mature ripe fruits was subjected to treatments such as freezing and storage (for 12 months), heated to 55°C (critical temperature) and 85°C (pasteurization temperature) for 20 min each, and spray dried with skim and whole milk powders. Terpenes such as alpha-pinene, beta-pinene, linalool, germacrene-d, and spathulenol, esters like sec-butyl butanoate, and methyl linolenate, along with benzyl alcohol and two oxygenated sesquiterpenes were found to be the major volatiles of the fresh pulp. Heating the fresh pulp at 55°C and 85°C produced better flavor, and volatile compounds were relatively more at 85°C. Poul et al. (2009) found that custard apple milk shake prepared from 90 parts of buffalo milk and 10 parts of custard apple pulp was most acceptable and ranked between "liked very much" to "liked extremely." Pawar et al. (2010) heated custard apple pulp and observed that a linear increase in temperatures exhibited accelerated inhibition of polyphenol oxidase (PPO) activity, leading to 100% inhibition at 83°C temperature for 2 min. The complete inhibition of the PPO activity was also observed at the temperature of 82°C for 5 min. However, heat treatment decreased the consumer acceptability of pulp in terms of sensorial characteristics. Addition of 2000 ppm of ascorbic acid without heat treatment showed highest sensorial properties without discoloration compared to heat-treated samples. Santos et al. (2014) evaluated the viscosity behavior of custard apple pulp at dilution of 1:1 and by addition of xanthan gum at concentrations 0.10% and 0.15% under different rotation speeds (6, 12, 20, 30, 50, 60, and 100 rpm) and temperatures (10°C, 20°C, 30°C, 40°C and 50°C) and reported improved stability of the samples with the addition of xanthan gum. All samples were classified as non-Newtonian fluid, with pseudoplastic behavior. Sravanthi et al. (2014) extracted the pulp from custard apples and stored for a period of 6 months by adding 1500 ppm of potassium metabisulfite. Stored product at lower temperature was better in quality than the sample stored at higher temperatures.

11.2.2 CUSTARD APPLE JAM AND TOFFEE

The jam prepared using 50% custard apple had better organoleptic score (Singh et al., 2006) and was fit for consumption up to 4 months of storage. The amount of total soluble solids (TSS), total sugars, and reducing sugars increased continuously during the storage period for 150 days, while the amount of ascorbic acid and nonreducing sugars decreased. Mundhe et al. (2008) found that toffee prepared using 55% custard apple pulp scored the maximum overall acceptability. This toffee was presented as an ideal supplement to the diet of young children.

11.2.3 MILK SHAKE AND ICE CREAM

A method of custard apple milk shake was reported by Poul et al. (2010) who recommended the composition of 90:10 for milk and custard apple, respectively (Figure 11.2). Ice cream prepared with incorporation of 15% custard apple pulp and 15% sugar level had an overall acceptability score of

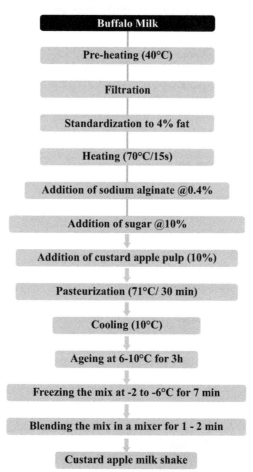

FIGURE 11.2 Preparation of custard apple milk shake (adapted from Poul et al., 2009).

8.05 (Yadav et al., 2010) (Figure 11.3). Pawar et al. (2011) prepared low fat custard apple ice cream from 15% custard apple pulp, 15% sugar, 10% fat, and ascorbic acid in different combinations and reported that 0.3% level of ascorbic acid was most acceptable and rated between "liked very much" to "liked extremely" for all sensory attributes.

11.2.4   CUSTARD APPLE BEVERAGES

Ready-to-serve beverage of custard apple and lime was prepared by Pilania et al. (2010), and they reported that the blended juice of custard apple and

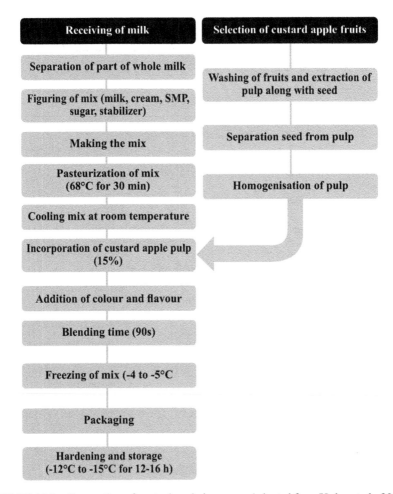

FIGURE 11.3   Preparation of custard apple ice cream (adapted from Yadav et al., 2010).

lime (3:2) with 15% TSS and 0.2% acidity was the best with respect to color (off white), taste, overall acceptance, and ascorbic acid content (Figure 11.4). Patil et al. (2011) reported that the maximum increase in the TSS level of the custard apple squash was observed during a storage period of 32 days. The acidity and ascorbic acid and sugar content of the custard apple squash increased with storage, while the tannin content and organoleptic score decreased as the storage length increased. Markam and Singh (2012) evaluated the effect of chemical preservative (potassium metabisulfite) on the microbial and sensory quality of the custard apple ready-to-serve (RTS) beverage stored at ambient temperature (28–32 °C) for 180 days with an interval of 1 month. The mean score of taste panel for color, flavor, appear-

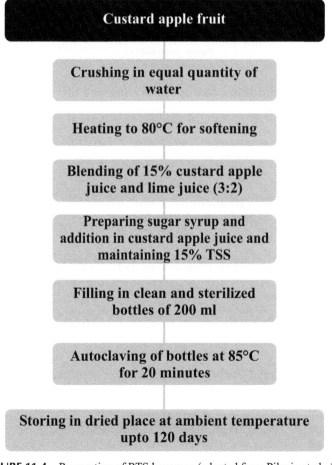

FIGURE 11.4   Preparation of RTS beverage (adapted from Pilania et al., 2010).

ance and overall acceptance significantly ($p < 0.05$) decreased, whereas the microbial contamination was maximum at 30 days of storage. Shrivastava et al. (2013) worked out on finding the best recipe for custard apple beverage with a maximum shelf—life and reported that the nectar prepared with the treatment (20% pulp, 0.3% acidity, 25% TSS) contained highest acidity, TSS, total sugars, and reducing sugars with a moderate amount of non-reducing sugars and TSS:acid ratio; thus, this composition was found to be suitable for the preparation of custard apple nectar at the commercial scale (Figure 11.5).

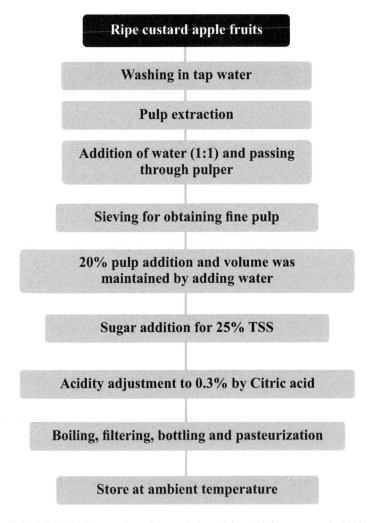

FIGURE 11.5   Preparation of nectar (adapted from Shrivastava et al., 2013).

11.2.5 FERMENTED PRODUCTS

Lima et al. (2001) used polyphenol oxidase enzyme extracted from ripe custard apple pulp on the cocoa (*Theobroma cacao* L.) nibs for taste improvement. Treatment of nonautoclaved unfatted cocoa nibs with enzyme for 210 min resulted in 15%, 15%, 10%, and 18% reduction in total phenol, tannin, flavan-3-ois, and anthocyanins, respectively. Nibs of autoclaved unfatted cocoa had 25%, 26%, 23%, and 51% reduction of total phenols, tannins, flavan-3-ois, and anthocyanins after enzyme treatment. Muniz et al. (2008) also reviewed the technologies used in the process (pectolytic enzymes, fining agents) as well as the parameters involved in custard apple fruit fermentation. Deshpande et al. (2010) processed custard apple and prepared nutrient-rich fermented products as the alcohol percentage of the distillate prepared from custard apple was 8.2%, which was within the range of the values of alcohol percentage of the fruit wines (8% to 13%). Jagtap and Bapat (2015) evaluated the potential of custard apple in the production of a fermented beverage using the yeast *Saccharomyces cerevisiae* (NCIM 3282) and assessed the antioxidant capacity, total phenolic content, and DNA damage-protecting activity of custard apple fruit wine (CAFW). It was revealed that CAFW was not only effective in scavenging DPPH- and DMPD (N,N-dimethyl-p-phenylenediamine dihydrochloride) free radicals but also had good ferric-reducing antioxidant capacity. Three hydroxybenzoic acids (gallic acid, protocatechuic acid, and gentisic acid) and two hydroxycinnamic acids (caffeic acid and p-coumaric acid) were also identified.

11.2.6 CUSTARD APPLE POWDER

Bharadiya et al. (2010) obtained custard apple powder using freeze drying. The set point temperature was −40°C and 20% maltodextrin level were found to be the best on basis of physical, bio-chemical and organoleptic parameters (Figure 11.6). Thube et al. (2011) prepared and studied the effect of custard apple pulp powder (CAPP) as an excipient on the properties of acetaminophen tablet, and the disintegration test showed that the tablets containing CAPP in the presence of polyvinyl pyrrolidone (PVP) as a binder had two-fold increase in the disintegration time. Moreover, the disintegration time for the tablets prepared by replacing PVP with CAPP in the presence of

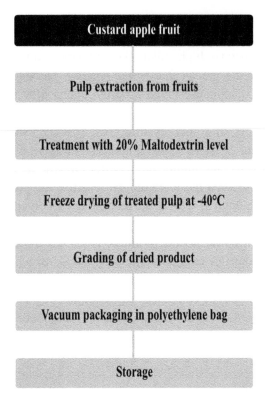

FIGURE 11.6   Drying of custard apple pulp (adapted from Bharadiya et al., 2010).

starch as a disintegrant was reduced drastically. Dissolution studies further confirmed the enhancement of binding potential of PVP and disintegrating potential of starch by CAPP. Performance of the binder and the disintegrant in the tablet was modified by the presence of CAPP. Kumar et al. (2015) extracted custard apple pulp and dried it at 60–80°C in a force convection dryer; they recommended drying of custard apple pulp at 60°C with 1 cm thickness based on sensory evaluation (Figure 11.7). The average drying rates for custard apple were 0.18, 0.20, 0.23, 0.24, and 0.25 g water/g dry matter/min for temperatures of 60°C, 65°C, 70°C, 75°C, and 80°C, respectively. Drying kinetics for thin layer drying of ripe custard apple were also fitted in Two-term model appropriately. The estimated values for effective diffusivities for the custard apple pulp drying were 2.13×10^{-5}, 2.54×10^{-5}, 2.94×10^{-5}, 3.25×10^{-5}, and 3.45×10^{-5} m^2/s at the temperatures of 60°C, 65°C, 70°C, 75°C, and 80°C. The activation energy was 31.92 kJ/mol.

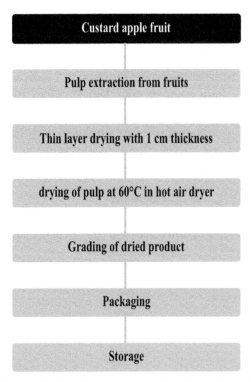

FIGURE 11.7   Thin layer drying of custard apple pulp (adapted from Kumar et al., 2015).

11.3   CASE STUDY: COLOR CHANGES DURING RIPENING

Color is an important attribute because it is usually considered the first property that the consumer observes. The color usually changes with the heat processing of food; however, the retention of food color may be used to predict the quality deterioration of food (Shin and Bowmilk, 1995). The change in color during storage was also reported by Kumar et al. (2011). A study on color was conducted, and the color of custard apple was assessed using a Miolata Chromameter (CR–400) every day till ripening. The CIE Lab color space is based on the concept that colors can be considered as combinations of red and yellow, red and blue, green and yellow, and green and blue. To determine the exact combination of colors of a product, the coordinates of a three-dimensional color space were assigned, and L^*, a^*, and b^* values were recorded. The L^* value gives a measure of lightness of the product color from 100 for perfect white and 0 for black. The redness/

greenness and yellowness/blueness are denoted by $a*$ and $b*$ values, respectively. The total color difference, $\Delta E*$, was also calculated, which accounts for the differences between the $L*$, $a*$, and $b*$ of the stored samples in comparison to the fresh samples. The chroma, $C*$, indicates the purity of a color or its freedom from white or gray. Hue is one of the main properties (called color appearance parameters) of a color; it is defined as "the degree to which a stimulus can be described as similar to or different from stimuli that are described as red, green, blue, and yellow." The difference in hue angle, $\Delta H*$, is the difference between the stored samples and fresh samples (HunterLab, 1996).

$$\textit{Total colour difference,}\, \Delta E^* = \sqrt{\Delta L^{*2} + \Delta a^{*2} + \Delta b^{*2}} \tag{1}$$

$$\textit{Chroma,}\, C^* = \sqrt{a^{*2} + b^{*2}} \tag{2}$$

$$\textit{Difference in hue angle,}\, \Delta H^* = \sqrt{\Delta E^{*2} - \Delta L^{*2} - \Delta C^{*2}} \tag{3}$$

11.3.1   KINETIC CHANGES DURING STORAGE

The use of zero- and first-order equations for the loss in food quality during storage was reported by Singh (2000). The experimental data obtained from the storage study were fitted to these equations. Changes in color $L*$, $a*$, $b*$, ΔE, $C*$, and $\Delta H*$ values of custard apples as a result of storage were investigated using the following zero- and first-order kinetics as shown in equations (4) and (5), respectively.

$$C = C_0 \pm k_0 t \tag{4}$$

$$\ln C = \ln C_0 \pm k_1 t \tag{5}$$

where C is the measured value of response, C_0 is the initial value of the corresponding response, t is the storage time, and K_0 and K_1 are the reaction rate constants for zero and first order, respectively. (+) and (−) signs represent the increase and decrease in the corresponding quality response of food, respectively.

The goodness of fit of the selected mathematical models to the experimental data was evaluated with the correlation coefficient (R^2), the reduced chi-square (χ^2), and the root mean square error (RMSE). R^2 is a measure of the amount of variation around the mean that is explained by the model. Reduced chi-square (χ^2) is the mean square of the deviations between the experimental and predicted values for the models and was used to determine the goodness of fit. The RMSE gives the deviation between the predicted and experimental data, and it is required to reach zero value. The goodness of fit will be better, if R^2 values are higher and χ^2 and RMSE values are lower (Kumar et al., 2011).

The reduced chi-square (χ^2) and the RMSE were calculated using the following expressions:

$$\chi^2 = \frac{\sum_{i=1}^{N} (MR_{Exp,i} - MR_{Pre,i})^2}{(N - Z)} \tag{6}$$

$$RMSE = \sqrt{\frac{1}{N} \sum_{i=1}^{N} (MR_{Pre,i} - MR_{Exp,i})^2} \tag{7}$$

where $MR_{Exp,i}$ is the ith experimental moisture ratio, $MR_{Pre,i}$ is the ith predicted moisture ratio, N is the number of observations, and z is the number of constants. In this study, the regression analysis was performed using MS Office 2007, Redmond, Washington.

The change in color during ripening was observed with the help of the Chroma meter, and the observations of different color parameters were recorded. It can be observed from Table 11.3 that L^*, a^*, b^*, ΔE, C^*, and ΔH^* values varied from 43.11 to 47.88, -8.62 to -5.45, 20.06 to 24.17, 1.32 to 6.31, 20.48 to 25.66, and 0.12 to 1.22, respectively, for the samples kept under refrigerated conditions. Color, L^*, value of the custard apple fruit decreased with an increase in the storage period during ripening, indicating an increase in darkness during ripening. The color, a^*, value also increased with the increase in the storage period, which is evident due to attaining the maturity level during storage. The color, b^*, values also decreased with the increase in the storage period, indicating a decrease in yellowness of the product.

TABLE 11.3 Observation and Prediction of Change in Color Parameters of the Samples in Refrigerated Storage Using Zero- and First-Order Models

Color parameters	Time (days)				
	0	1	2	3	4
Observed values					
L* value	47.88	47.23	45.75	45.05	43.11
a* value	−7.59	−8.62	−7.23	−5.45	−5.57
b* value	23.67	24.17	22.95	20.71	20.06
ΔE* value	-	1.32	2.28	4.62	6.31
C* value	24.86	25.66	24.06	21.42	20.82
ΔH* value	-	0.81	0.12	1.22	0.90
Predicted values (Zero order)					
L* value	48.15	46.98	45.81	44.63	43.46
a* value	−8.33	−7.61	−6.89	−6.17	−5.45
b* value	24.45	23.38	22.31	21.25	20.18
ΔE value	-	1.03	2.76	4.50	6.23
C* value	25.83	24.60	23.36	22.13	20.90
ΔH* value	-	0.56	0.70	0.83	0.97
Predicted values (First order)					
L* value	48.19	48.16	48.14	48.11	48.08
a* value	−8.41	−8.31	−8.20	−8.09	−7.98
b* value	24.52	24.47	24.42	24.37	24.33
ΔE* value	-	1.33	1.87	2.41	2.95
C* value	25.91	25.86	25.81	25.75	25.70
ΔH* value	-	0.56	0.82	1.08	1.33

It can also be observed from Table 11.4 that R^2 values for L*, b*, ΔE*, and C* for first-order models are above 0.95, indicating the suitability of both the models for the prediction of color, which may be linked to the maturity indices of the fruit in refrigerated storage. The zero-order model also have higher R^2 values and for L* ΔE*, indicating model suitability.

L*, a*, b*, ΔE, C*, and ΔH* values varied from 42.83 to 47.88, −7.49 to −1.43, 18.62 to 23.67, 1.89 to 9.43, 18.68 to 24.86, and 0.72 to 5.02, respectively, for the samples kept in plastic crates at 25–30°C and 60–65% relative humidity (Table 11.5). A similar decrease in L* and b* and an increase in a* values were observed as in refrigerated conditions; however, the change in values is more than that observed in refrigerated conditions.

TABLE 11.4 Coefficients and Statistical Parameters for Zero- and First-Order Models in Color Parameters of the Samples in Refrigerated Storage

	Zero order model					First order model				
	Ko	Co	R^2	χ^2	RMSE	K_1	C_1	R^2	χ^2	RMSE
L* value	−1.17	48.15	0.97	0.15	0.30	−0.03	48.19	0.96	13.58	2.85
a* value	0.72	−8.33	0.70	0.74	0.67	−0.11	8.41	0.70	4.83	1.70
b* value	−1.07	24.45	0.85	0.64	0.62	−0.05	24.52	0.95	11.52	2.63
ΔE value	1.73	−0.70	0.98	0.11	0.26	0.54	0.791	0.99	5.43	1.80
C* value	−1.23	25.83	0.83	1.03	0.79	−0.05	25.915	0.99	15.61	3.06
ΔH* value	0.13	0.43	0.14	0.18	0.33	0.26	0.303	0.18	0.25	0.39

TABLE 11.5 Observation and Prediction of Change in Color Parameters of the Samples in Plastic Crates by Using Zero- and First-Order Models

Color parameters	Time (days)				
	0	1	2	3	4
Observed values					
L* value	47.88	46.25	45.98	45.69	42.83
a* value	−7.59	−6.74	−5.59	−5.28	−1.43
b* value	23.67	23.35	22.85	22.39	18.62
ΔE value	–	1.87	2.88	3.43	9.43
C* value	24.86	24.30	23.52	23.00	18.68
ΔH* value	–	0.72	1.70	1.89	5.02
Predicted values (Zero order)					
L* value	47.86	46.79	45.73	44.66	43.60
a* value	−8.08	−6.70	−5.32	−3.95	−2.57
b* value	24.39	23.28	22.18	21.07	19.97
ΔE value	–	0.91	3.24	5.56	7.89
C* value	25.61	24.24	22.87	21.51	20.14
ΔH* value	–	0.37	1.68	2.99	4.30
Predicted values (First order)					
L* value	47.90	47.87	47.85	47.83	47.80
a* value	−9.51	−9.15	−8.79	−8.43	−8.07
b* value	24.52	24.47	24.42	24.37	24.32
ΔE value	–	1.53	2.04	2.54	3.05
C* value	25.80	25.73	25.67	25.61	25.55
ΔH* value	–	1.01	1.61	2.20	2.79

Colorfulness is the visual sensation according to which the perceived color of an area appears to be more or less chromatic. The C^* value was retained more in case of refrigerated samples than in the samples stored in trays. The changes in the ΔE^* value and ΔH^* value were more in samples stored at room temperature.

It can also be observed from Table 11.6 that R^2 values for a^*, b^*, ΔE^*, and C^* values for first-order models were above 0.92, indicating the suitability of both the models for the prediction of color, which may be associated with the ripening stage of the fruit stored in refrigerated storage.

11.4 CONCLUDING REMARKS

Custard apple fruits are generally stored in crates/ cartons for ripening at 25–30°C and 60–65% relative humidity. The ripening of the fruit can be delayed under refrigerating conditions by 2 days. The decrease in L^* and b^* and an increase in a^* values were observed under refrigerated and ambient conditions. The C^* value was retained more in the case of refrigerated samples than in the samples stored in crates. The ΔE^* value and ΔH^* value were also more in samples stored at room temperature, indicating more overall changes and stimulus value. The life of custard apple pulp can be enhanced by the addition of ascorbic acid without heat treatment. The pulp can be stored by the addition of 1500 ppm of potassium metabisulfite up to 6 months under cold storage. The custard apple was also successfully incorporated in various value-added products, viz., ice cream (@15% pulp), toffee (@55% pulp), milk shake (@10% pulp), RTS beverage (@60% pulp), jam (@50% pulp), and nectar (@20% pulp). The shelf-life of RTS beverage stored at

TABLE 11.6 Coefficients and Statistical Parameters for Zero- and First-Order Models in Color Parameters of the Samples in Plastic Crates

	Zero order model					First order model				
	K_0	C_0	R^2	χ^2	RMSE	K_1	C_1	R^2	χ^2	RMSE
L^* value	−1.07	47.86	0.85	0.67	0.63	−0.02	47.90	0.84	11.80	2.66
a^* value	1.38	−8.08	0.85	1.13	0.82	−0.36	9.51	0.92	24.63	3.84
b^* value	−1.11	24.39	0.73	1.51	0.95	−0.05	24.52	1.00	13.60	2.86
ΔE^* value	2.33	−1.41	0.77	2.65	1.26	0.50	1.031	0.95	14.13	2.91
C^* value	−1.37	25.61	0.78	1.79	1.04	−0.06	25.798	0.96	20.52	3.51
ΔH^* value	1.31	−0.94	0.82	0.62	0.61	0.59	0.419	0.85	1.73	1.02

ambient temperature (28–32°C) was 180 days with potassium metabisulfite. The development of process technology for custard apple powder followed by its utilization should also be explored.

KEYWORDS

- custard apple
- processing
- value addition

REFERENCES

Amoo, I. A., Emenike, A. E., & Akpambang, V. O. E., (2008). Compositional evaluation of *Annona cherimoya* (custard apple) fruit. *Trends Applied Sci. Res., 3*(2), 216–220.

Bharadiya, N. N., Memon, K. N., & Cholera, S. P., (2010). Preparation of custard apple powder. *A Dissertation Report Submitted to The Faculty of Agricultural Engineering and Technology, CAET, JAU,* Junagadh.

Brand-Miller, J., Foster-Powell, F., & Colaguira, S., (2003). *The New Glucose Revolution.* Publisher: Headline, ISBN: 9780733616440.

Brown, B. I., & Scott, K. J., (1985). Cool storage conditions for custard apple fruit (*Annona atemoya Hort.*). Singapore *J. Primary Indust., 13,* 23–31.

Chandraju, S., Mythily, R., & Kumar, C. S. C., (2012). Qualitative chromatographic analysis of sugars present in non-edible rind portion of custard apple (*Annona squamosa* L.). *J. Chem. Pharma. Res., 4*(2), 1312–1318.

Chaudhry, A. S., Singh, G. N., & Singh, A. R., (1985). Effect of wrapping materials and ripening media on the physico-chemical compositions of custard apple (*Annona squamosa Linn.*). *Indian J. Agric. Res., 19*(2), 90–92.

Deshpande, K. G., Kshirsaoar, A. B., Kesralikar, M., Narwade, P., Gample, S., & Pujari, V., (2010). Rare fruits as a source of nutrient rich fermented products. *Asian J. Bio. Sci., 4*(2), 300–303.

Gamage, T. V., Yuen, C. M. C., & Wills, R. B. H., (1997). Minimal processing of custard apple (*Annona atemoya*) pulp. *J. Food Process Preserv., 21*(4), 289–301.

Hoyos, P., (1980). Fruit mixture. German Patent 800464003.

HunterLab, (1996). CIE Lab Color scale, Applications Note, Technical Services Department, Hunter Associates Laboratory Inc., Virginia, *8*(7), 1–4.

Jagtap, U. B., & Bapat, V. A., (2015). Phenolic composition and antioxidant capacity of wine prepared from custard apple (*Annona squamosa L.*) fruits. *J. Food Process Preserv., 39*(2), 175–182

Kadam, S. S., (2001). New products from the arid and semi-arid fruits. *Indian J. Hort., 58*(1/2), 170–177.

Kumar, N., Sarkar, B. C., & Sharma, H. K., (2011). Colour kinetics and storage characteristics of carrot, pulse and rice byproduct based extrudates, *British Food J., 114*(9), 1279–1296.

Kumar, S., Kumar, N., Seth, N., & Jethva, K. R., (2015). A study on physical properties and drying characteristics of custard apple, Paper presented at "National Seminar on Emerging Trends in Food Quality and Safety" at AAU, Anand during 15th-16th October.

Lima, E. D. P. D. A., Pastore, G. M., Barbery, S. D. F., Garcia, N. H. P., Brito, E. S. D., & Lima, C. A. D. A., (2001). Obtaining and use of polyphenoloxidase enzyme extracted from ripe custard apple (*Annona squamosa L.*) pulp on the cocoa (*Theobroma cacao L.*) nibs in taste improvement." *Revista Brasileira de Fruticultura., 23*(3), 709–713.

Liu, S., (2000). The performance of African Pride custard apple cultivar. *China Fruits., 3*, 29–30.

Mallikarjuna, T., Ranganna, B., & Dronachari, M., (2012). Respiration studies of custard apple (*Annona squamosa L.*) fruit under different maturity levels at ambient and refrigerated condition. *J. Dairying Foods Home Sci., 31*(1), 25–27.

Markam, R., & Singh, V., (2012). Studies on microbial and sensory quality of custard apple RTS beverage. *Asian J. Hort., 7*(2), 460–464.

Mazumdar, B. C., (1977). Differences between seeded and seedless berries of custard apple (*Anona squamosa L.*) *Plant Sci., 103*.

Moo-Huchin, V. M., Estrada-Mota, I., Estrada-Leon, R., Cuevas-Glory, L., Ortiz-Vazquez, E., Vargas, M. L. V., Betancur-Ancona, D., & Sauri-Duch, E., (2014). Determination of some physicochemical characteristics, bioactive compounds and antioxidant activity of tropical fruits from Yucatan, Mexico, *Food Chem., 152*, 508–515.

Mundhe, S. A., Kshirsagar, R. B., Kulkarni, D. N., & Patil, B. M., (2008). Studies on utilization of custard apple pulp in toffee. *Indian J. Nutri. Diet., 45*(11), 472–478.

Muniz, C. R., Borges, M. D. F., & Freire, F. D. C. O., (2008). Tropical and subtropical fruit fermented beverages, *Microbial Biotechnol Hort., 3*, 35–69.

Pareek, S., Yahia, E. M., Pareek, O. P., & Kaushik, R. A., (2011). Postharvest physiology and technology of Annona fruits, *Food Res Int., 44*, 1741–1751.

Patil, S. M., Raut, V. U., Pushpendra, K., & Pankaj, L., (2011). Standardization of recipes for production of custard apple squash. *Progressive Agriculture., 11*(2), 472–474.

Pawar, S. L., Karanjkar, L. M., & Poul, S. P., (2011). Sensory evaluation of low fat custard apple ice-cream. *J. Dairying, Foods Home Sci., 30*(1), 32–34.

Pawar, V. N., Kardile, W. G., & Hashmi, S. I., (2010). Studies on inactivation profile of polyphenol oxidase (PPO) from custard apple (*Annona squamosa L.*) pulp by heat treatment and its effects of sensorial quality. *Food Sci. Res. J., 1*(2), 137–141.

Pilania, S., Dashora, L. K., & Singh, V., (2010). Standardization of recipe and juice extraction method for preparation of ready-to-serve beverage from custard apple (*Annona squamosa L.*), Int. J. Process Post Harvest Technol., 1*(2), 65–72.

Pino, J. A., & Rosado, A., (1999). Volatile constituents of custard apple (*Annona atemoya*). *J. Essential Oil Res., 11*(3), 303–305.

Poul, S. P., Sontakke, A. T., Munde, S. S., & Adangale, A. B., (2010). Composition and economics of custard apple milk shake, *Asian J. Animal Sci., 4*(2), 139–142.

Poul, S. P., Sontakke, A. T., Munde, S. S., Adangale, A. B., & Jadhav, P. B., (2009). Process standardization for custard apple milk shake. *J. Dairying, Foods Home Sci., 28*(3/4), 202–205.

Santos, L. N. B. D., Centenaro, B. M., & Ohata, S. M., (2014). Influence of temperature and rotation speed on addition hydrocolloid on the apparent viscosity behavior of custard apple pulp (*Annona squamosa L.*), *Revista Brasileira de Produtos Agroindustriais.*, *16*(3), 299–312.

Shashirekha, M. N., Baskaran, R., Rao, L. J., Vijayalakshmi, M. R., & Rajarathnam, S., (2008). Influence of processing conditions on flavor compounds of custard apple (*Annona squamosa L.*) *LWT Food. Sci. Technol. 41*(2), 236–243.

Shin, S., & Bhowmilk, S. R., (1995). Thermal kinetics of colour changes in pigeon pea puree. *J. Food Eng.*, *24*(1), 77–86.

Shrivastava, R., Dubey, S., Dwivedi, A. P., Pandey, C. S., & Banafar, R. N. S., (2013). Effect of recipe treatment and storage period on biochemical composition of custard apple (*Annona squamosa L.*) nectar. *Progressive Hort.*, *45*(1), 110–114.

Singh, D., Chaudhary, M., Chauhan, P. S., Prahalad, V. C., & Kavita, A., (2009). Value Addition to forest Produce for Nutrition and Livelihood, *The Indian Forester.*, *135*(9), 1271–1284.

Singh, P., Patel, D., Agrawal, S., & Panigrahi, H., (2006). Value addition of custard apple through jam for year round availability. *Proceedings of the National Symposium on Production, Utilization and Export of Underutilized Fruits with Commercial Potentialities.* Kalyani, Nadia, West Bengal, India.: Mohanpur, India: Bidhan Chandra Krishi Viswavidyalaya., 262–265.

Singh, R. P., (2000). Scientific principles of shelf-life evaluation. In: *Shelf life of Foods*, Man, C. M. D., & Jones, A. A. (eds.). Maryland, USA, Aspen Publishers Inc., 3–22.

Sravanthi, T., Waghrey, K., & Daddam, J. R., (2014). Studies on preservation and processing of custard apple (*Annona squamosa L.*) pulp. *Int. J. Plant, Animal Environm. Sci.*, *4*(3), 676–682.

Thube, R., Purohit, S., & Gothoskar, A., (2011). Study of effect of Custard Apple Pulp Powder as an excipient on the properties of acetaminophen tablet. *World Applied Sci. J.*, *12*(3), 364–371

USDA, (2016). Basic Report 09321, Sugar apples, National Nutrient Database for Standard Reference, Release 28, Accessed on November 26, 2016. 07:34 EST.

Yadav, C. M., Karanjkar, L. M., & Kashid, U. B., (2010). Effect of assimilation of custard apple (*Annona squamosa*) pulp on chemical quality and cost of ice-cream. *J. Dairying Foods Home Sci.*, *29*(2), 86–91.

CHAPTER 12

PEARL MILLET: FLOUR AND STARCH PROPERTIES

KAWALJIT SINGH SANDHU,[1] ANIL KUMAR SIROHA,[1] MANINDER KAUR,[2] and SNEH PUNIA[1]

[1]Department of Food Science and Technology, Chaudhary Devi Lal University, Sirsa, India, Tel: +91-1666-247124, Fax: +91-1666-248123, E-mail: kawsandhu@rediffmail.com

[2]Department of Food Science and Technology, Guru Nanak Dev University, Amritsar, India

CONTENTS

12.1   MILLET

The term "millet" is broadly used to describe various small-seeded grasses that are harvested for food or feed. They can be classified within the genera

Brachiaria, Digitaria, Echinochloa, Eleusine, Panicum, Paspalum, Pennisetum, Setaria, and Sorghum (Yang et al., 2012). Species of millets include the following: browntop (*Brachiariaramose*), barnyard (*Echinochloacrus galli*), finger (*Eleusine coaracana*), proso (*Panicum miliaceum*), kodo (*Paspalum scrobiculatum*), little (*Panicum sumatrense*), pearl (*Pennisetum glaucum*), and foxtail (*Setaria italica*) millets. Amongst these millet species, pearl millet is the most widely cultivated one.

Pearl millet is a dual-purpose crop used for feed and fodder and grown in arid and semi-arid regions of the world. It is a widely cultivated cereal in the world, after rice, wheat, and sorghum. It is mainly grown in Asia and Africa. According to FAO (2016), India ranks first in the annual production of millet, i.e., 1,14,20,000 tons in 2014, which was followed by Niger with an annual production of 33,21,753 tons. In India, pearl millet is mainly grown in Rajasthan, Maharashtra, Gujarat, Uttar Pradesh, and Haryana, and together, these states comprise an area containing more than 90% of pearl millet. Nambiar et al. (2011) reported that the consumption of pearl millet is reported to have health benefits. It is also recommended for the treatment of constipation, weight loss, anemia, diabetes, celiac disease, diarrhea, and ulcers due to the presence of high fiber.

12.2 GRAIN

Pearl millet grains are smaller than cereal grains such as maize and sorghum. Due to their size, they pack closely together leaving little air space. The diameter and thickness of the grains range from 1.7 to 3.7 mm and 1.4 to 1.8 mm, respectively (Badau et al., 2002). The density of grains ranges from 1.1 to 1.3 g/ml, while grain hardness varies from 30.3 to 68.0. Color of pearl millet grains varies from off white to yellow, slate gray, brown and purple (Taylor, 2004). The pearl millet grain comprises about 75% endosperm, 17% germ, and 8% pericarp (Serna-Saldivar and Rooney, 1995).

12.3 FLOUR

Pearl millet is milled into flour for its use in various food products. The flour is mainly used for making *chapatti* and *khichri* in India. Various workers have reported proximate composition of pearl millet. Badau et al. (2002) reported protein, ash, crude fiber, and soluble carbohydrate contents in the range from 9.6% to 12.5%, 1.1% to 1.5%, 1.2% to 4.4%, and 71.7% to 76.3%, respectively. Abdalla et al. (1998) also studied the proximate composition of pearl millet grain and reported ash, fat, protein, and starch content in the range between l.6–2.4%, 2.7–7.1%, 8.5–15.1%, and 58–70%, respectively. Sade et al. (2009) reported 14.0% crude protein, 5.7% fat, 2.1% ash, 2.0% crude fiber, and 76.3% carbohydrate content in the pearl millet. Ali et al. (2003) reported fat and ash content in the range from 6.7% to 7.8% and 2.1% to 2.3%, respectively, for pearl millet cultivars. The protein, fat, ash, fiber, and carbohydrate content of pearl millet flour observed in our laboratory was 9.7%, 6.5%, 1.65%, 3.1%, and 72.5%, respectively (Table 12.1). Water absorption capacity (WAC), oil absorption capacity (OAC), foaming capacity, and emulsion activity were observed as 173%, 105%, 18%, and 45%, respectively, for pearl millet flour (Table 12.1).

Different processing methods are employed to improve the functional properties of pearl millet flours. Sade et al. (2009) observed the effect of germination and fermentation on functional properties of pearl millet flours. These processing techniques improved the functional properties of flours. The carbohydrate content decreased during fermentation, while it increased during germination and roasting. Processing significantly increased the WAC, OAC, least gelation concentration, and bulk density (BD) of flours. Ali et al. (2003) studied the effects of fermentation on in vitro protein digestibility of two pearl millet genotypes. The results indicated that there was a gradual reduction in pH with an increase in time, a marginal change in the protein content, and a significant ($P<0.05$) increase in *in vitro* protein digestibility for both the cultivars.

TABLE 12.1   Proximate Composition and Functional Properties of Pearl Millet Flour

Properties	Value* (%)
Ash	1.65
Fat	6.5
Protein	9.7
Fiber	3.1
Carbohydrates	72.5
Water absorption capacity	173
Oil absorption capacity	105
Foaming capacity	18
Emulsion activity	45

*Values are means of triplicate determinations.

The decrease in the phytic acid content was observed with an increase in germination time. No effect was observed in the manganese content with germination time. Phytic acid reduction and an increase in extractable minerals had a good correlation with an increase in germination time (Badau et al., 2005). Lestienne et al. (2007) studied abrasive decortications on the losses of nutrients such as starch, lipids, proteins, iron, and zinc as well as phytase activity and antinutritional compounds of grains from two pearl millet cultivars. Phytase activity and some antinutritional parameters decreased with decortication, but no such effect was observed for lipid and protein content. Falade et al. (2013) studied the effect of γ-irradiation on seed color and functional and pasting properties of pearl millet cultivars. Color, BD, and WAC were not significantly affected by γ-irradiation. The OAC of the cultivar SOSAT was not significantly affected, but γ-irradiation had a significant effect on the cultivar ZATIV. Peak viscosity (PV), trough viscosity (TV), final viscosity (FV), and setback viscosity (SV) decreased significantly ($p<0.05$) with an increased dose of γ-irradiation.

The functional properties of pearl millet flour have been reported to improve by the addition of different flours to it. After supplementation of soybean flour (5%, 10%, and 15%) to pearl millet flour, BD, WAC, OAC, and emulsifying and foaming properties were improved (Ali et al., 2012). Properties of agidi (millet product) from different blends of pearl millet and bambara groundnut flour were studied (Zakari et al., 2010). The results showed that protein, fat, and carbohydrate contents were increased by the addition of bambara groundnut flour. BD increased from 0.32 to 0.44 g/ml with an increase in bambara flour.

12.4   SHELF-LIFE OF FLOUR

Grain of pearl millet is small as compared to other cereals but has larger germ size. Hence, it comprises a higher amount of triglycerides, which are rich in unsaturated fatty acids (Lai and Varriano-Marston, 1980a). Thus, the flour of pearl millet is prone to rancidity within few days (Lai and Varriano-Marston, 1980b). These changes due to rancidity are more predominant under high moisture and oxygen exposure during the storage. Lipase enzyme, which is mainly in the aleurone layer, pericarp, and germ, is responsible for the hydrolysis of triglycerides in the pearl millet grain, which results in an off odor and taste (Galliard, 1999). Thermal treatment can be used to prevent the deterioration of triglycerides of flour.

Shelf-life of pearl millet flour is less than those of flours of various cereal grains. However, its nutritional and functional properties are comparable to those of the cereals. Thus, it can be a good substitute for cereal flours. The shelf life of pearl millet flours can be increased by treating the flours with a variety of processing techniques. Kapoor and Kapoor (1990) evaluated different treatments such as addition of antioxidants (butylated hydroxytoluene, butylated hydroxyanisole, and ascorbic acid), thermal treatment, defatting, and salting on shelf-life of flour and analyzed the properties after 10, 20, and 30 days of storage. The amount of peroxides was lower in antioxidant-treated flour than in others. The lowest acid value and peroxide value were observed for defatted flour. Arora et al. (2002) used a dry oven at 100°C to heat pearl millet grain for 2 h and evaluated the extent of hydrolytic rancidity of the resulting flour during 1 month of storage at ambient conditions. Fat acidity and free fatty acid (FFA) increased during storage; both these parameters increased by less than two-fold for heat-treated grain flour and by at least four-fold for the untreated grain flour.

The effect of hydrothermal treatment (moisture, 30% ± 2% and steaming at 1.05 kg/m^2 for 0, 10, 15, 20 and 25 min) on the whole and pearled pearl millet grains revealed that the flour from steamed before pearling (SBP, 20 min) and steamed after pearling (PBS, 15 min) grains showed no lipase activity and with acceptable physical, functional, and pasting properties (Yadav et al., 2012a). A significant ($p<0.05$) increase in BD and functional properties and a decrease in pasting properties of treated flour samples were observed with an increase in the duration of steaming. Flour samples (SBP, 20 min; PBS, 15 min) were found acceptable even after 50 days when stored in polyethylene pouches (75 μm) at ambient conditions (15–35°C). Nantanga et al. (2008) reported that fat acidity of flour from the untreated grain increased from 0.11 to 3.73 g KOH/kg during 3 months of storage, whereas no significant increase ($p<0.05$) was observed for wet thermal-treated samples. The shelf-life of wet thermal-treated samples was more as than that of untreated grain.

Meera et al. (2011) studied the storage stability of flour upon thermal treatment of sorghum grains. Storage stability was positively correlated with temperature. Moist heat exposure of grains for 15 min retarded hydrolytic rancidity significantly during the storage for 6 months. It retarded the development of rancidity irrespective of the morphological variations and the lipid content. Yadav et al. (2012b) reported that microwave heating of pearl millet grains decreased lipase activity significantly ($p<0.05$) with an increase in the moisture level from 12% to 18% and the maximum reduction was observed at 18% moisture level for 100 s. A significantly ($p<0.05$) lower change in the FFA value during storage up to 30 days was observed in flour from the grains, treated for 80 s at 18% moisture level as compared to that for the control flour.

Tiwari et al. (2014) studied flour quality during storage by pearling, thermal treatment, and fermentation of pearl millet flour. A reduction in antinutritional factors, fat acidity, and FFA was observed after the pre-treatments, and these parameters thereafter increased with an increase in the storage period. Heat treatment was the best pre-milling treatment. Chavan and Kachare (1994) used 0.05M HCl solution to lower the pH to inactivate the lipase enzyme. During the staorage of the flour in cloth bags at ambient conditions, fat acidity of the untreated flour was increased by about 6 folds, whereas fat acidity of acid soaked-grain flour increased by about 1.5 folds.

12.5   EFFECT OF DIFFERENT METHODS OF PROCESSING ON ANTIOXIDANT PROPERTIES

Mild heat-treated foods offer better health benefits. Processing methods are applied to cereals to improve their nutritional values, nutrient availability, texture, inactivation of toxic substances, and enzyme inhibitors (Bakr and Gowish, 1991). Thermal processing techniques are primarily employed for improving palatability, stability, and safety of the food products (Wang et al., 2014).

Germination, steaming, and roasting had a significant effect on the nutraceutical and antioxidant properties of little millet (*Panicum sumatrense*) (Pradeep and Guha, 2011). The reduction in total phenolic content during cooking could be due to either destruction of phenolic compounds or other chemical rearrangements of phenolic compounds, such as formation of new insoluble components with other organic substances (Satwadhar et al., 1981). The type of grains, the nature and location of phenolic compounds in these grains, and the severity and duration of heat treatments are responsible for variation in phenolic compounds in cereal grains during thermal treatments (Towo et al., 2003; Chandrasekara et al., 2012).

Ragaee et al. (2006) studied the antioxidant activity of selected cereals (hard wheat, soft wheat, barley, sorghum, rye, and millet). The highest antioxidant activity was reported for sorghum followed by pearl millet. Total phenols, DPPH scavenging capacity, and ABTS+ scavenging activity of pearl millet were 1387 µg/g (gallic acid equivalent), 23.83 µmol/g, and 21.4 µmol/g, respectively.

Many processing methods have been adopted to improve the functional and nutraceutical properties of pearl millet. Antioxidant properties of flour from different thermally treated grains (cooking and toasting) are shown in Table 12.2.

Toasting is a mild heat treatment, and it increases the antioxidant properties, whereas the reverse effect was observed in cooking. Chandrasekara et al. (2012) reported that dehulling and cooking affected total phenolic content (TPC), radical scavenging, and antioxidant activities (AOA) of the grains. The AOA of extracts followed the order hull > whole grain > dehulled grain > cooked dehulled grain. Pradeep and Guha (2011) studied the effect of germination, steaming, and roasting on the antioxidant properties of little millet and concluded that TPC, flavonoids, and tannin contents of treated little millet increased by 21.2, 25.5, and 18.9 mg/100 g, respectively. TPC, total flavonoids content, AOA, and the Fe^{2+} chelating activity of the roasted millet extract were the highest. Alvarez-Jubete et al. (2010) studied the effect of sprouting and baking on antioxidant properties and observed that TPC and AOA increased after sprouting but decreased after baking.

Toasting is a mild heat treatment as it releases the bound antioxidant compounds, which results in an increase in antioxidant properties. Fares and Menga (2012) studied the effect of toasting on carbohydrate profile and antioxidant properties of chickpea flour and observed that the overall nutritional profile of the flour was improved after the toasting process. Sandhu et al. (2015a) also reported that the toasting of oats resulted in a significant ($p<0.05$) decrease in physical parameters, hunter color $L*$ value, and TFC.

TABLE 12.2 Antioxidant Properties of Control, Toasted, And Cooked Pearl Millet Flour

Properties	Control flour	Toasted flour	Cooked flour
AOA (%)	37.7	42.3$_{\uparrow12.2}$	28.5$_{\downarrow24.4}$
TPC (µg GAE/g)	2668	2795$_{\uparrow4.7}$	1754$_{\downarrow34.2}$
TFC(µg CE/g)	2484	2569$_{\uparrow3.4}$	569$_{\downarrow77.0}$
Metal chelating activity (%)	20.3	28.2$_{\uparrow38.9}$	32.3$_{\uparrow59.1}$

Values are means of triplicate determinations.

However, it increased WAC, OAC, hunter color $a*$ and $b*$ values, TPC, AOA and metal chelating values (MCA).

Gujral et al. (2011) evaluated that the color values $a*$ and $b*$ and the non-enzymatic browning index significantly increased upon roasting. TPC and TFC decreased significantly ($p<0.05$), while an increase in DPPH radical scavenging activity, reducing power, and β-glucan was observed in roasted groats. Sharma and Gujral (2011) also reported similar results for barley cultivars during the roasting and cooking process. Gallegos-Infante et al. (2010) reported that TPC increased as compared to the native counterpart during cooking and roasting of barley, but it decreased after germination. Roasted, cooked, germinated, and unprocessed barley grains showed good antioxidant activity. These results were contrary to the results reported by Gujral et al. (2011) and Sharma and Gujral (2011).

Sharma et al. (2012) studied the effect of extrusion cooking on antioxidant properties of grit of different hulled barley cultivars, and a significant increase in AOA and decrease in TPC and TFC were observed. The increase in feed moisture and temperature increased the MCA, whereas the reducing power decreased as compared to control samples.

Cardoso et al. (2014) reported the effects of dry and wet heat on the antioxidant profile of sorghum flours. 3-Deoxyanthocyanidins and TPC were stable to dry heat, and both were reduced when wet heat was employed. Processing with dry heat increased vitamin E content and its retention, and reduced the carotenoid content. The AOA in dry heated flour either remained constant or increased, while it decreased after wet heat treatment. The grains milled before processing in oven and microwave ovens have more vitamin E and less carotenoids than the flour milled after processing. The wet heat processing decreased antioxidant compounds, except carotenoids, which showed an increase. Gujral et al. (2013) reported that the addition of oat flours to wheat increased TPC but decreased the AOA, while both were decreased upon baking. The reducing power of the oat-blended flour was higher than that of the wheat flours and increased after baking.

12.6 STARCH

Starch is the major component (55–60%) of the pearl millet grain (Hoover et al., 1996). It is the major carbohydrate of plants, which is deposited as insoluble and semi-crystalline granules in storage tissues (roots, tubers, grains) and occurs comparatively lesser in most vegetative tissues of plants (Copeland et al., 2009). Starch granules are composed of two alpha-glucans, named amylose and amylopectin, which account for around 98–99% of the dry weight. Amylose content (AC) of starches is specified as waxy, 0–2%; very low, 5–12%; low, 12–20%; intermediate, 20–25%; and high 25–33% (Juliano, 1992). AC of starch changes with the botanical source and climatic and soil conditions during grain development (Morrison et al., 1984; Yano et al., 1985; Morrison and Azudin, 1987). Amylose is a linear/less branched polymer of α-D glucose units linked through α-D-(1→4) glycosidic linkages, while amylopectin is a highly branched polymer of the same linked with α-(1→6) and α-D-(1→4) glycosidic linkages (Karim et al., 2000; Hoover, 2001).

Amylopectin, with an average molecular weight of 10^7-10^9 (Aberle et al., 1994), is composed of linear chains of (1→4)-α-D-glucose residue connected through (1→6)-α-linkages. Morrison and Karkalas (1990) have shown that there are three typical types of amylopectin: (1) normal amylopectin, which contains very long chains with frequent branching, (2) amorphous amylopectin with similarly extended A and B chains that elute from gel permeation chromatography (GPC) columns with amylose, and (3) high-molecular-weight amylopectin with A and B chains of 5–15 glucose residues that are longer than normal.

Amylopectin has stabilizing effects, whereas amylose forms a gel. It also possesses strong tendency to form complexes with lipids and other components. Many studies reported that starch biosynthesis, starch content, molecular structure, and the percentage of amylose and amylopectin can be affected by mutations (Shannon and Garwood, 1984). Choi et al. (2004) reported AC of the three starches (amaranth, pearl millet, and waxy sorghum) in the range between 3.2–6.0%, with the highest being observed for pearl millet starch. Annor et al. (2014) reported physical and molecular properties of starch of pearl, foxtail, proso, and finger millet. Their AC ranged from 28.6% to 33.9%, with finger and pearl millets showing higher proportions of long amylose chain than foxtail and proso millets.

Swelling power (SP) measures the water-holding capacity (WHC) upon heating in water, followed by cooling, and centrifuging, while solubility (SB) reflects the degree of dissolution during the starch swelling procedure. AC, SP, and SB of starch from pearl millet were observed as 14.2%, 14.1 g/g, and 10.4%, respectively (Table 12.3).

TABLE 12.3 Physicochemical and Pasting Properties of Pearl Millet Starch

Properties	Values
Amylose content (%)	14.2
Swelling power (g/g)	14.1
Solubility (%)	10.4
Peak viscosity (cP)	1068
Breakdown viscosity (cP)	506
Setback viscosity (cP)	351
Pasting temperature (°C)	81.2

Values are means of triplicate determinations.

SP and SB provide measures of the magnitude of interaction between starch chains within the amorphous and crystalline domains. The degree of this interaction is affected by the amylose:amylopectin ratio, the characteristics of amylose and amylopectin in terms of molecular weight/distribution, and the branching degree and chain length of amylose and amylopectin (Kaur et al., 2007). Upon hydrothermal treatment in presence of excess water, starch molecules lost their crystallinity, and an increase in SB and SP was observed due to hydrogen bonding of water molecules to the exposed hydroxyl groups of amylose and amylopectin (Tester and Karkalas, 1996; Singh et al., 2003). SP of starch has been reported to depend on the WHC of starch molecules by hydrogen bonding (Lee and Osman, 1991). The granules become more susceptible to shear disintegration on swelling and released more soluble material. Sandhya and Bhattacharya (1989) indicated that starch granules with low AC, being less rigid, swell freely when heated. The starch granules with higher AC, on the other hand, being better reinforced and thus more rigid, probably swell less freely. However, amylose and lipid interaction also inhibit swelling of starch granules (Tester and Morrison, 1990). Balasubramanian et al. (2014) reported SP and SB of 3.80 g/g and 16.09%, respectively, at 60°C for pearl millet starch. Hoover et al. (1996) reported that SP and SB values in the range from 18.0 to 28.6 and 9.45% to 10.40%, respectively, at 90°C for pearl millet starch.

Cooking properties of starches during heating and cooling cycles can be understood through pasting properties. When starch is heated in an aqueous system, the loss of crystalline region occurs and starch surface increases. Various factors such as amylose leaching, amylose-lipid complex formation, and granule swelling affect the hot paste viscosity of starches (Lu et al., 1997). Breakdown viscosity (BV) is the measure of starch paste resistance to heat and shear, while SV exhibited the tendency of starch pastes to retrograde, which is an index of starch retrogradation (Asante et al., 2013). Starch exhibits unique viscosity behavior with change of temperature, concentration, and shear rate (Nurul et al., 1999). The pasting behavior is helpful for understanding the textural change or retrogradation potency of the applied products. Many researchers have reported the pasting characteristics of pearl millet starches (Balasubramanian et al., 2014; Choi et al., 2004; Wu et al., 2014). Pasting properties such as PV, BV, SV, and pasting temperature (PT) of pearl millet starch studied in our laboratory have been reported in Table 12.3, and the representative graph is shown in Figure 12.1.

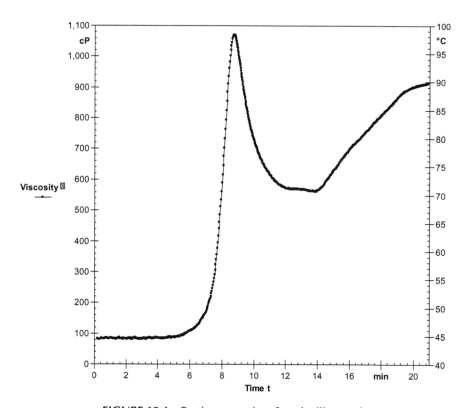

FIGURE 12.1   Pasting properties of pearl millet starch.

Choi et al. (2004) reported the values for PV, BV, and SV for pearl millet starch of 202.8, 107.3, and 20.8 RVU, respectively, which were higher than those of the amaranth starch and lower than those of the sorghum starch. PT of amaranth, waxy sorghum, and waxy millet starches were 75.7°C, 73.3°C, and 75.2°C, respectively. Hoover et al. (1996) reported values of PV, FV, and PT in the range of 167 to 180, 273 to 430 (BU), and 89.3°C to 90°C, respectively, for pearl millet starch. In another study, PV, TV, BV, SV, FV, and PT values were 2708, 1124, 1583, 3229, and 4353 cP and 84.15°C, respectively, for pearl millet starch (Balasubramanian et al., 2014). X-ray diffraction (XRD) is used to study the presence and characteristics of the crystalline structure of the starch granules. Starch granules possess a semi-crystalline structure corresponding to different polymorphic forms, and based on this, starch can be classified into three types, namely A, B, and C (Buleon et al., 1998). Arrangement of the amylopectin helices are supposed to be responsible for starch crystallinity, whereas amylose is associated with amorphous regions (Zobel, 1988; Singh et al., 2006). According to Zobel (1964), type-A starches show strong peaks at 2θ of 15.3°, 17.1°, 18.2°, and 23.5°, while for type-B starches, strong bands appear at 5.6°, 14.4°, 17.2°, 22.2°, and 24°. For type-C starches, the signals are stronger at 5.6°, 15.3°, 17.3°, and 23.5° (2θ). A type diffraction pattern is the characteristic of cereal starches like maize and wheat. B-type diffraction pattern is common for tuber starches like potato and canna (Watcharatewinkul et al., 2009), while C-type is common for legume starches (Gernat et al., 1990; Sandhu and Lim, 2008). Pearl millet starches show type-A crystallinity similar to cereal starches (Hoover et al., 1996; Annor et al., 2014; Florence and Urooj, 2015). Higher readily digestible starch content in the pearl millet starch could be due to A-type X-ray diffraction pattern of starch, which is more susceptible to enzymatic hydrolysis and also due to elimination of structural barrier to amylase hydrolysis (Florence and Urooj, 2015).

Many researchers have used rheological methods to determine the viscoelasticity of the starch pastes (Kaur et al., 2004; Sandhu et al., 2004; Kaur and Singh, 2015; Jan et al., 2016). The most common method of studying the viscoelastic properties of starch is by using a dynamic rheometer. The continuous assessment of the starch suspensions during temperature and frequency sweep tests can be done by using a dynamic rheometer. Rheological properties of pearl millet starch suspension during heating and frequency sweep measurements of starch paste are shown in Figures 12.2 and 12.3, respectively.

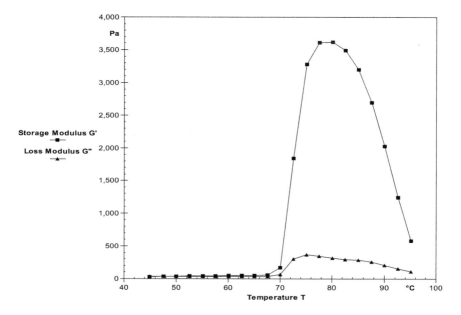

FIGURE 12.2 Rheological properties of starch during heating.

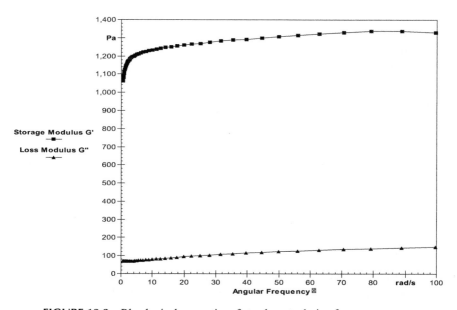

FIGURE 12.3 Rheological properties of starch paste during frequency sweep.

The starch gel is characterized with storage modulus (G'), loss modulus (G''), and phase angle (tan δ). G' is a parameter of the energy stored in the material and recovered from it per cycle, while G'' is a compute of the energy dissipated or lost per cycle of sinusoidal deformation (Ferry, 1980). Another parameter indicating the physical behavior of a system is tan δ, which is the ratio of the energy lost to the energy stored for each cycle. The G' of starch progressively increases at a certain temperature (TG') to a maximum (peak G') and then drops with continued heating in a dynamic rheometer. The initial increase in G' and G'' is caused by the starch granules swelling progressively and finally becoming closely packed (Eliasson, 1986). Excessive heating of starch suspension causes a decrease in G', indicating that the gel structure is ruptured due to prolonged heating (Tsai et al., 1997). This may be due to the melting of the residual crystalline region in the gelatinized starch granules, which deforms and loosens the particles (Eliasson, 1986). The rheological properties are influenced by the starch concentration, amylose content temperature, heating rate, and mechanical treatments. Furthermore, the addition and presence of other components (lipids, proteins, sugars, salts, etc.) also influence these properties (Eliasson and Gudmundsson, 1996).

Morphological characteristics of starches from different plant sources vary with the genotype and biological origin. The size and shape of starch granules varied with biological origin (Svegmark and Hermansson, 1993). Starch granules range in size (1 to 100 μm diameter) and shape (polygonal, spherical, lenticular) and can differ with the content, structure, and organization of the amylose and amylopectin molecules, the branching pattern of amylopectin, and the level of crystallinity (Lindeboom et al., 2004). The physicochemical properties such as AC, percent light transmittance, SP, and WAC significantly correlate with the average granule size of the starches (Zhou et al., 1998; Singh and Singh, 2001; Kaur et al., 2004). Light microscopy and scanning electron microscopy (SEM) can be used to study the morphological properties of starch granules (French, 1984; Fannon et al., 1992; Jane et al., 1994). SEM is used to study the granule morphology more accurately than light microscopy. High resolution of SEM also provides a more comprehensive approach on granule surface features and granule morphology (Chmelik, 2001). Choi et al. (2004) reported polygonal or slightly round shape of waxy millet and waxy sorghum starch granules with visible pores on their surface. The millet starch granules have been reported to vary from 2.5 to 24 μm in size and were mainly polygonal with a few spherical ones (Annor et al., 2014).

12.7 STARCH MODIFICATION

Native starches have limitations such as low shear resistance, thermal resistance, thermal decomposition, and higher tendency toward retrogradation, which limit their use in industrial food applications. Therefore, starch used in the food industry is often modified to overcome undesirable changes in product texture and appearance caused by retrogradation or breakdown of starch during processing and storage (Van Hung and Morita, 2005). For specific applications, starch is modified to alter its physical and chemical features to improve its functional characteristics (Hermansson and Svegmark, 1996). Starch modification can be achieved through chemical derivatization, decomposition, or hydrothermal treatment of starch (Singh et al., 2007). SEM is widely used to determine structural changes caused by chemical modifications (Kim et al., 1992; Kaur et al., 2004). Various researchers have reported morphological properties of modified starches (Koo et al., 2010; Arueya and Oyewale, 2015; Sandhu et al., 2015b). In a study in our laboratory, a slight change in morphological properties occurs after the physical and chemical modifications. SEM of native and modified starches (crosslinked, octenyl succinic anhydride (OSA)-modified, and heat moisture treatment (HMT)-modified) are shown in Figure 12.4.

In enzymatic modification, the molecular weight of native starch is decreased by the action of hydrolytic enzymes (Le et al., 2009). This provides derivatives with varying viscosity, gel consistency, thermal stability, and sweetness. In such modifications, pre-gelatinized starches were subjected to enzymatic degradation resulting in various products (Alexander, 1992). The commonly used enzymes are α and β amylases, isoamylase, and pullulanase. The enzyme α-amylase attacks on α (1, 4)-linkages of the starch and produces maltodextrins and lower DE dextrins. Isoamylases and pullulanase attack at specific sites such as 1, 6-linkages in the starch and give high DE syrups (Duffner et al., 2000).

12.7.1 MODIFICATION WITH CROSSLINKING AGENTS

Native starches are mostly crosslinked with various crosslinking agents such as sodium trimetaphosphate, sodium tripolyphosphate, epichlorohydrin, and phosphoryl chloride (Ratnayake and Jackson, 2008). Hirsch and

Modifications

Physical	**Chemical**	**Enzymatic**
Heat moisture treatment	Esterification	α amylase
Annealing	Cross-linking	β-amylase
Steam treatment	Acid thinned	Iso-amylase
Extrusion	Oxidation	Pullulanase
Gamma irradiation	Etherification	

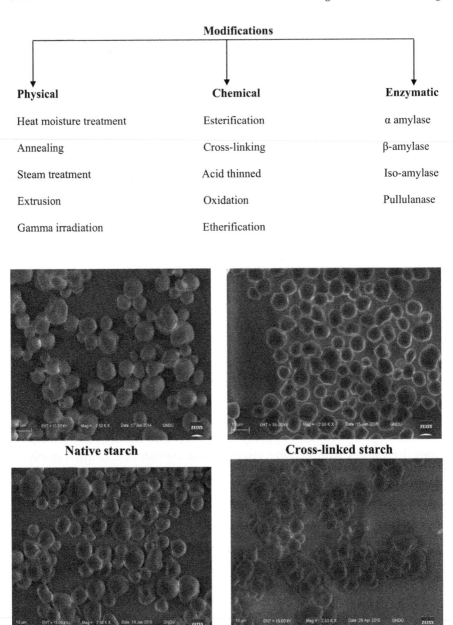

Native starch

Cross-linked starch

OSA modified starch

HMT starch

FIGURE 12.4   Morphological properties of native and modified starches.

Kokini (2002) reported that phosphoryl chloride imparts greater viscosity than sodium trimetaphosphate- and epichlorohydrin-treated granules. Starch source, crosslinking reagent concentration and composition, pH, reaction time, and temperature affect the crosslinking of starch (Lim and Seib, 1993; Chung et al., 2004).

12.7.2 *MODIFICATION WITH OCTENYL SUCCINIC ANHYDRIDE (OSA)*

Starch modification with octenyl succinic anhydride (OSA) was patented by Caldwell et al. (1953). In many countries, the modification of starch with 3% OSA on dry basis of starch is permitted (CFR, 2001). OSA starches stabilize the oil–water interface of an emulsion. The glucose part of starch binds the water and lipophilic, and the octenyl part binds the oil; hence, the separation of the oil and water phase is prevented (Murphy, 2000). These starches are commonly utilized in a large range of oil-in-water emulsions for food and beverage, biomedical, and other industrial products (Jeon et al., 1999; Park et al., 2004).

12.7.3 *MODIFICATION WITH HEAT MOISTURE TREATMENT (HMT)*

Chemical modification of starch is widely used, but interest in the physical modification of starch has also increased in food applications (Zavareze and Dias, 2011). The physical modification of starch has been gaining wider acceptance because of no by-products in the modified starch. Consideration of physically modified starches as natural and safe ingredients provide it an advantage over other modification as their addition in food are not limited by food legislations (BeMiller, 1997). The term HMT is employed where low moisture levels (<35% w/w) are applied, whereas annealing relates to high moisture treatment (<40% w/w). Both modifications took place below gelatinization temperature and have been shown to cause structural changes of starch chains to a varying extent (Hoover, 2010). Different methods have been described by various researchers to modify the starches by HMT (Hormdok and Noomhorm, 2007; Hung et al., 2015; Juansang et al., 2015; Lee and Moon, 2015).

12.7.4   MODIFICATION WITH ACETIC ANHYDRIDE

Chemical modification involving acetylation is a widely used method for starch modification for improvement in functional or physicochemical properties and to increase their applications in a variety of food products. Acetylated starch is prepared commonly with acetic anhydride and vinyl acetate. According to "The Joint FAO/WHO Expert Committee on Food Additives (JECFA)," starch acetate is defined as a starch ester prepared by addition of acetic anhydride or vinyl acetate under alkaline conditions. The addition of esterification reagents resulted in the substitution of hydroxyl groups with acetyl moieties. However, according to Food and Drug Administration, starch acetate should not contain more than 2.5% acetyl groups (Codex, 1996; JECFA, 2001). The European Union (EU) assigned E1420 to starch acetates under the category of food additives.

12.8   CONCLUDING REMARKS

Pearl millet is a drought tolerant crop and has low production cost. It has starch content comparable to other starch sources (corn, wheat, and potato). It can be a cost-effective substitute for corn starch. Applications of pearl millet starch can be enhanced by starch modification. Fat content of pearl millet is higher than that of other cereal grains, which makes the flour lesser stable. By improving the storage stability of the flour, it can be incorporated in wheat flour to meet the demand of increasing population. Due to low cost and high production, there is an urgent need for more research on pearl millet starch for its various properties so that its further applications in various food products can be increased.

KEYWORDS

- antioxidant activity
- dynamic rheology
- morphology
- pearl millet

- **starch modification**
- **x-ray diffraction**

REFERENCES

Abdalla, A. A., El Tinay, A. H., Mohamed, B. E., & Abdalla, A. H., (1998). Proximate composition, starch, phytate and mineral contents of 10 pearl millet genotypes. *Food Chem., 63*, 243–246.

Aberle, T., Burchard, W., & Radosta, S., (1994). Conformation, contributions of amylose and amylopectin to the structural properties of starches from various sources. *Starch, 46*, 329–335.

Alexander, R. J., (1992). Maltodextrins: *Production Properties and Applications*. In Starch hydrolysis products, worldwide technology production and applications; Scenck, F. W., & Hebeda, R. A., Ed., pp. 233–276.

Ali Maha, A. M., Tinay Abdullahi, H. El., & Abdalla, A. H., (2003). Effect of fermentation on the *in vitro* protein digestibility of pearl millet. *Food Chem., 80*, 51–54.

Ali Maha, A. M., Tinay Abdullahi, H. El., Elkhalifa Elmoneim, O., Mallasy, L. O., & Babiker, E. E., (2012). Effect of different supplementation levels of soybean flour on pearl millet functional properties. *Food Nutr. Sci., 3*, 1–6.

Alvarez-Jubete, L., Wijngaard, H., Arendt, E. K., & Gallagher, E., (2010). Polyphenol composition and *in vitro* antioxidant activity of amaranth, quinoa buckwheat and wheat as affected by sprouting and baking. *Food Chem., 119*, 770–778.

Annor, G. A., Marcone, M., Bertoft, E., & Seetharaman, K., (2014). Physical and molecular characterization of millet starches. *Cereal Chem., 91*, 286–292.

Arora, P., Sehgal, S., & Kawtra, A., (2002). The role of dry heat treatment in improving the shelf life of pearl millet flour. *J. Nutri. Health, 16*, 331–336.

Arueya, G. L., & Oyewale, T. M., (2015). Effect of varying degrees of succinylation on the functional and morphological properties of starch from acha (*Digitaria Exilis Kippis Stapf*). *Food Chem., 177*, 258–266.

Asante, M. D., Offei, S. K., Gracen, V., Adu-Dapaah, H., Danquah, E. Y., & Bryant, R., (2013). Starch physicochemical properties of rice accessions and their association with molecular markers. *Starch, 65*, 1022–1028.

Badau, M. H., Nkama, I., & Ajalla, C. O., (2002). Physicochemical characteristics of pearl millet cultivars grown in Northern Nigeria. *Int. J. Food Prop., 5*, 37–47.

Badau, M. H., Nkama, I., & Jideani, I. A., (2005). Phytic acid content and hydrochloric acid extractability of minerals in pearl millet as affected by germination time and cultivar. *Food Chem., 92*, 425–435.

Bakr, A. A., & Gowish, R. A., (1991). Nutritional evaluation and cooking quality of dry cowpea (*Vigna signensis* L.) grown under various agricultural conditions. I. Effect of soaking and cooking on the chemical composition and nutritional quality of cooked seeds. *J. Food Sci. Tech., 28*, 312–316.

Balasubramanian, S., Sharma, R., Kaur, J., & Bhardwaj, N., (2014). Characterization of modified pearl millet (*Pennisetum typhoides*) starch. *J. Food Sci. Tech., 51*, 294–300.

Bemiller, J. N., (1997). Starch modification: Challenges and prospects. *Starch, 49,*127–131.

Buleon, A., Colonna, P., Planchot, V., & Ball, S., (1998). Starch granules: Structure and bio-synthesis. *Int. J. of Biol. Macromol., 23,* 85–112.

Caldwell, C. G., & Wurzburg, O. B., (1953). Polysaccharide derivatives of substituted dicarboxylic acids. *National Starch Products Inc.* USPTO, US 1953/2661349.

Cardoso Morais, L., Montini, T. A., Pinheiro, S. S., Helena Maria, P., Martino Hercia, S. D., & Moreira, A. B., (2014). Effects of processing with dry heat and wet heat on the antioxidant profile of sorghum. *Food Chem., 152,* 210–217.

CFR (Code of Federal Regulation), (2001). Food starch modified. *Food Additives permitted in food for human Consumption.* Govt. Printing office, Washington, USA.21/1/172/172.892.

Chandrasekara, A., Naczk, M., & Shahidi, F., (2012). Effect of processing on the antioxidant activity of millet grains. *Food Chem., 133,* 1–9.

Chavan, J. K., & Kachare, D. P., (1994). Effect of seed treatment on lipolytic deterioration of pearl millet flour during storage. *J. Food Sci. Tech., 31,* 80–81.

Chmelik, J., (2001). Comparison of size characterization of barley starch granules determined by electron and optical microscopy, low angle laser light scattering and gravitational field-flow fractionation. *J. Inst. Brew., 107,* 11–17.

Choi, H., Kim, W., & Shin, M., (2004). Properties of Korean amaranth starch compared to waxy millet and waxy sorghum starches. *Starch/Stärke, 56,* 469–477.

Chung, H. J., Woo, K. S., & Lim, S. T., (2004). Glass transition and enthalpy relaxation of cross-linked corn starches. *Carbohyd. Polym., 55,* 9–15.

Codex F. C., (1996). *Committee on Food Chemicals Codex.* Food and Nutrition Board, Institute of Medicine, National Academy of Sciences. Published: National Academy Press, Washington, DC.

Copeland, L., Blazek, J., Salman, H., & Tang, M. C., (2009). Form and functionality of starch. Food Hydrocolloid., 23, 1527–1534.

Duffner, F., Bertoldo, C., Andersen, J. T., Wagner, K., & Antranikian, G., (2000). A new thermoactive pullulanase from Desulfurococcus mucosus: cloning, sequencing, purification, and characterization of the recombinant enzyme after expression in *Bacillus subtilis. J. Bacteriol., 182,* 6331–6338.

Eliasson, A. C., & Gudmundsson, M., (1996). *Starch:* Physicochemical and functional aspects. In: *Carbohydrates in Food.* Eliasson, A. C., Ed., Marcel Dekker, New York, pp. 431–503.

Eliasson, A. C., (1986). Viscoelastic behavior during the gelatinization of starch I. Comparison of wheat, maize, potato and waxy-barley starches. *J. Texture Stud., 17,* 253–265.

Falade, K. O., & Kolawole, T. A., (2013). Effect of γ-Irradiation on color, functional and physicochemical properties of pearl millet (*Pennisetum glaucum* L. R. Br.) cultivars. *Food Bioprocess Tech., 6,* 2429–2438.

Fannon, J. E., Hauber, R. J., & BeMiller, J. N., (1992). Surface pores of starch granules. *Cereal Chem., 69,* 284–288.

FAO. Food and Agricultural Organization of the United Nations. http://faostat.fao.org/beta/en/#data/QC (accessed 27.10.2016)

Fares, C., & Menga, V., (2012). Effects of toasting on the carbohydrate profile and antioxidant properties of chickpea (*Cicer arietinum* L.) flour added to durum wheat pasta. *Food Chem., 131,* 1140–1148.

Ferry, J. D., (1980). *Viscoelastic Properties of Polymers 3rd Ed.*, J. Wiley and Sons, New York.

Florence, S. P., & Urooj, A., (2015). Isolation and characterization of starch from pearl millet (*Pennisetum typhoidium*) flours. *Int. J. Food Prop., 18*, 2675–2687.

French, D., (1984). In: *Starch Chemistry and Technology*, 2nd Edn. Whistler, R. L., BeMiller, J. N. Paschall E. F., Ed., Academic Press, Orlando, USA, pp. 183.

Gallegos-Infante, J. A., Rocha-Guzman, N. E., Gonzalez-Laredo, R. F., & Pulido-Alonso, J., (2010). Effect of processing on the antioxidant properties of extracts from Mexican barley (*Hordeum vulgare*) cultivar. *Food Chem., 119*, 903–906.

Galliard, T., (1999). Rancidity in cereal products. In: *Rancidity in Foods*, Allen, J. C., Hamilton, R. J., ed., Aspen Publishers, Gaithersburg, Maryland, pp. 140–156.

Gernat, C., Radosta, S., Damaschun, G., & Schierbaum, F., (1990). Supramolecular structure of legume starches revealed by X-ray scattering. *Starch/Stärke, 42*, 175–178.

Gujral, H. S., Sharma, P., & Singh, R., (2011). Effect of sand roasting on beta glucan extractability, physicochemical and antioxidant properties of oats. *LWT – Food Sci. Tech., 44*, 2223–2230.

Gujral, H. S., Sharma, P., Gupta, N., & Wani, A. A., (2013). Antioxidant properties of legumes and their fractions as affected by cooking. *Food Sci. Biotech, 22*, 187–194.

Hermansson, A. M., & Svegmark, K., (1996). Developments in the understanding of starch functionality. *Trends Food Sci. Tech., 7*, 345–353.

Hirsch, J. B., & Kokini, J. E., (2002). Understanding the mechanism of cross-linking agents (POCl$_3$, STMP, and EPI) through swelling behavior and pasting properties of cross-linked waxy maize starches. *Cereal Chem., 79*, 102–107.

Hoover, R., (2001). Composition, molecular structure, and physico-chemical properties of tuber and root starches: A review. *Carbohyd. Polym., 45*, 253–267.

Hoover, R., (2010). The impact of heat–moisture treatment on molecular structures and properties of starches isolated from different botanical sources. *Crit. Rev. Food Sci. Nutri., 50*, 835–847.

Hoover, R., Swamidas, G., Kok, L. S., & Vasanthan, T., (1996). Composition and physico-chemical properties of starch from pearl millet grain. *Food Chem., 56*, 355–367.

Hormdok, R., & Noomhorm, A., (2007). Hydrothermal treatments of rice starch for improvement of rice noodle quality. *LWT – Food Sci. Tech., 40*, 1723–1731.

Hung, P. V., Vien, N. L., & Phi Nguyen, T. L., (2015). Resistant starch improvement of rice starches under a combination of acid and heat-moisture treatments. *Food Chem., 191*, 67–73.

Jan, R., Saxena, D. C., & Singh, S., (2016). Pasting, thermal, morphological, rheological and structural characteristics of chenopodium (*Chenopodium album*) starch. *LWT – Food Sci. Tech., 66*, 267–274.

Jane, J. L., Kasemsuwan, T., Leas, S., Zobel, H., Robyt, J. F., (1994). Anthology of starch granule morphology by scanning electron microscopy. *Starch/Stärke, 46*, 121–129.

JECFA (2001) Starches; Summary of evaluations performed by the Joint FAO/WHO expert committee on food additives.

Jeon, Y., Lowell, A. V., & Gross, R. A., (1999). Studies of starch esterification : reactions with alkenyl succinates in aqueous slurry systems. *Starch/Stärke, 51*, 90–93.

Juansang, J., Puttanlek, C., Rungsardthong, V., Puncha-Arnon, S., Jiranuntakul, W., & Uttapap, D., (2015). Pasting properties of heat-moisture treated canna starches using different plasticizers during treatment. *Carbohyd. Polym., 122*, 152–159.

Juliano, B. O., (1992). Structure, chemistry, and function of the rice grain and its fractions. *Cereal Food World, 37*, 772–774.

Kapoor, R., & Kapoor, A. C., (1990). Effect of different treatments on keeping quality of pearl millet flour. *Food Chem., 35*, 277–286.

Kaur, L., Singh, J., McCarthy, O. J., & Singh, H., (2007). Physico-chemical, rheological and structural properties of fractionated potato starches. *J. Food Eng., 82*, 383–394.

Kaur, L., Singh, N., & Singh, J., (2004). Factors influencing the properties of hydroxypropylated potato starches. *Carbohyd Polym., 55*, 211–223.

Kaur, M., & Singh, S., (2015). Physicochemical, morphological, pasting and rheological properties of tamarind (*Tamarindus indica* L.) kernel starch. *Int. J. Food Prop.,* DOI: 10.1080/10942912.2015.1121495.

Karim, A. A., Norziah, M. H., & Seow, C. C., (2000). Methods for the study of starch retrogradation. *Food Chem., 71*, 9–36.

Kim, H. R., Hermansson, A. M., & Eriksson, C. E., (1992). Structural characteristics of hydroxypropyl potato starch granules depending on their molar substitution. *Starch/Stärke, 44*, 111–116.

Koo, S. H., Lee, K. Y., Lee, H. G., (2010). Effect of cross-linking on the physicochemical and physiological properties of corn starch. *Food Hydrocolloid., 24*, 619–625.

Lai, C. C., & Varriano-Marston, E., 1980a Lipid content and fatty acid composition of free and bound lipids in pearl millets. *Cereal Chem., 57*, 271–274.

Lai, C. C., & Varriano-Marston, E., 1980b Changes in pearl millet meal during storage. *Cereal Chem., 57*, 275–277.

Le, Q. T., Lee, C. K., Kim, Y. W., Lee, S. J., Zhang, R., Withers, S. G. et al., (2009). Amylolytically-resistant tapioca starch modified by combined treatment of branching enzyme and maltogenic amylase. *Carbohyd. Polym., 75*, 9–14.

Lee, C. J., & Moon, T. W., (2015). Structural characteristics of slowly digestible starch and resistant starch isolated from heat–moisture treated waxy potato starch. *Carbohyd. Polym., 125*, 200–205.

Lee, Y. E., & Osman, E. M., (1991). Correlation of morphological changes of rice starch granules with rheological properties during heating in excess water. *J. Korean Agric. Chem. Soc., 34*, 379–385.

Lestienne, I., Buisson, M., Lullien-Pellerin, V., Christian, P., & Treche, S., (2007). Losses of nutrients and anti-nutritional factors during abrasive decortication of two pearl millet cultivars (*Pennisetum glaucum*). *Food Chem., 100*, 1316–1323.

Lim, S. T., & Seib, P. A., (1993). Location of phosphate esters in a wheat starch phosphate by ^{31}P-nuclear magnetic resonance spectroscopy. *Cereal Chem., 70*, 145–152.

Lindeboom, N., Chang, P. R., & Tyler, R. T., (2004). Analytical, biochemical and physicochemical aspects of starch granule size, with emphasis on small granule starches: A review. *Starch/Stärke, 56*, 89–99.

Lu, S., Chen, L. N., & Lii, C. Y., (1997). Correlations between the fine structure, physicochemical properties and retrogradation of amylopectins from Taiwan rice varieties. *Cereal Chem., 74*, 34–39.

Meera, M. S., Bhashyam, M. K., & Ali, S. Z., (2011). Effect of heat treatment of sorghum grains on storage stability of flour. *LWT – Food Sci. Tech., 44*, 2199–2204.

Morrison, W. R., & Azudin, M. N., (1987). Variation in the amylose and lipid contents and some physical properties of rice starches. *J. Cereal Sci., 5*, 35–44.

Morrison, W. R., & Karkalas, J., (1990). Methods in Plant Biochemistry. In: *Starch, vol. 2.* Academic Press, New York, pp. 323–352.

Morrison, W. R., Milligan, T. P., & Azudin, M. N., (1984). A relationship between the amylose and lipid contents of starches from diploid cereals. *J. Cereal Sci., 2*, 257–271.

Murphy, P., (2000). Starch. In: *Handbook of Hydrocolloids*. Phillips, G. O., Williams, P. A., Ed., Boca Raton, FL: CRC Press, pp 41–65.

Nambiar, V. S., Dhaduk, J. J., Sareen, N., Shahu, T., & Desai, R., (2011). Potential Functional implications of pearl millet (*Pennisetum glaucum*) in health and disease. *J. Appl. Pharmaceut. Sci., 1*, 62–67.

Nantanga Komeine, K. M., Seetharaman, K., Kock. H. L., & Taylor, J. R. N., (2008). Thermal treatments to partially pre-cook and improve the shelf-life of whole pearlmillet flour. *J. Sci. Food Agric., 88,* 1892–1899.

Nurul, M., Azemi, B. M., & Manan, D. M. A., (1999). Rheological behavior of sago (*Metroxylon sagu*) starch paste. *Food Chem., 64*, 501–505.

Park, S., Chung, M., & Yoo, B., (2004). Effect of octenyl succinylation on rheological properties of corn starch pastes. *Starch/Stärke, 56*, 399–406.

Pradeep, S. R., & Guha, M., (2011). Effect of processing methods on the nutraceutical and antioxidant properties of little millet (*Panicum sumatrense*) extracts. *Food Chem., 126,* 1643–1647.

Ragaee, S., & Abdel-Aal, E. M., (2006). Pasting properties of starch and protein in selected cereals and quality of their food products. *Food Chem., 95*, 9–18.

Ratnayake, W. S., & Jackson, D. S., (2008). Phase transition of cross-linked and hydroxyl propylated corn (*Zea mays* L.) starches. *LWT – Food Sci. Tech., 41*, 346–358.

Sade, F. O., (2009). Proximate, antinutritional factors and functional properties of processed pearl millet (*Pennisetum glaucum*). *J. Food Tech., 7*, 92–97.

Sandhu, K. S., & Lim, S. T., (2008). Digestibility of legume starches as influenced by their physical and structural properties. *Carbohyd. Polym., 71*, 245–252.

Sandhu, K. S., Godara, P., Kaur, M., & Punia, S., (2015a). Effect of toasting on physical, functional and antioxidant properties of flour from oat (*Avena sativa* L.) cultivars. *Journal of the Saudi Society of Agricultural Sciences*, http://dx.doi.org/10.1016/j.jssas.2015.06.004.

Sandhu, K. S., Sharma, L., & Kaur, M., (2015b). Effect of granule size on physicochemical, morphological, thermal and pasting properties of native and 2-octenyl-1-ylsuccinylated potato starch prepared by dry heating under different pH conditions. *LWT – Food Sci. Tech., 61*, 224–230.

Sandhu, K. S., Singh, N., & Kaur, M., (2004). Characteristics of the different corn types and their grain fractions: physicochemical, thermal, morphological, and rheological properties of starches. *J. Food Eng., 64,* 119–127.

Sandhya Rani, M. R., & Bhattacharaya, K. R., (1989). Rheology of rice-flour pastes: effect of variety, concentration, and temperature and time of cooking. *J. Texture Stud., 20,* 127–137.

Satwadhar, P. N., Kadam, S. S., & Salunkhe, D. K., (1981). Effects of germination and cooking on polyphenols and *in vitro* protein digestibility of horse gram and moth bean. *Plant Foods Hum. Nutri., 31,* 71–76.

Serna-Saldivar, S., & Rooney, L. W., (1995). Structure and Chemistry of Sorghum and Millet. In: *Sorghum and Millet: Chemistry and Technology*. Dendy, D. A. V. (ed.), American Association of Cereal Chemist, St Pual MN, pp. 69–124.

Shannon, J. C., & Garwood, D. L., (1984). Genetics and physiology of starch development. In: *Starch: Chemistry and Technology*. Whistler, R. L., BeMiller, J. N., Paschall, E. F., (ed.). Academic Press, Orlando, pp. 25–86.

Sharma, P., & Gujral, H. S., (2011). Effect of sand roasting and microwave cooking on antioxidant activity of barley. *Food Res. Int., 44*, 235–240.

Sharma, P., Gujral, H. S., & Singh, B., (2012). Antioxidant activity of barley as affected by extrusion cooking. *Food Chem., 131,* 1406–1413.

Singh, J., & Singh, N., (2001). Studies on the morphological, thermal and rheological properties of starch separated from some Indian potato cultivars. *Food Chem., 75,* 67–77.

Singh, J., Kaur, L., & McCarthy, O. J., (2007). Factors influencing the physico-chemical, morphological, thermal and rheological properties of some chemically modified starches for food applications-A review. *Food Hydrocoll., 21,* 1–22.

Singh, J., McCarthy, O., & Singh, H., (2006). Physico-chemical and morphological characteristics of new Zealand Taewa (*Maori potato*) starches. *Carbohyd. Polym., 64,* 569–581.

Singh, N., Singh, J., Kaur, L., Sodhi, N. S., & Gill, B. S., (2003). Morphological, thermal and rheological properties of starches from different botanical sources. *Food Chem., 81,* 219–231.

Svegmark, K., & Hermansson, A. M., (1993). Microstructure and rheological properties of composites of potato starch granules and amylose: a comparison of observed and predicted structure. *Food Strucure, 12,* 181–193.

Taylor, J. R. N., (2004). *Millet: Pearl.* In: *Encyclopedia of Grain Science.* vol. *2,* ed.; Wrigley, C., Corke, H., Walker, C. E., Elsevier, London, pp. 253–261.

Tester, R. F., & Karkalas, J., (1996). Swelling and gelatinization of oat starches. *Cereal Chem., 73,* 271–277.

Tester, R. F., & Morrison, W. R., (1990). Swelling and gelatinization of cereal starches. I. Effect of amylopectin, amylose and lipids. *Cereal Chem., 67,* 551–557.

Tiwari, A., Jha, S. K., Pal, R. K., Sethi, S., & Krishan, L., (2014). Effect of pre-milling treatments on storage stability of pearl millet flour. *J. Food Process Pres., 38,* 1215–1223.

Towo, E. E., Svanberg, U., & Ndossi, G. D., (2003). Effect of grain pre-treatment on different extractable phenolic groups in cereals and legumes commonly consumed in Tanzania. *J. Sci. Food Agric., 83,* 980–986.

Tsai, M. L., Li, C. F., & Lii, C. Y., (1997). Effects of granular structure on the pasting behavior of starches. *Cereal Chem., 74,* 750–757.

Van Hung, P., & Morita, N., (2005). Effect of granule sizes on physicochemical properties of cross-linked and acetylated wheat starches. *Starch/Stärke, 57,* 413–420.

Wang, T., He, F., & Chen, G., (2014). Improving bioaccessibility and bioavailability of phenolic compounds in cereal grains through processing technologies: A concise review. *J. Funct. Foods, 7,* 101–111.

Watcharatewinkul, Y., Puttanlek, C., Rungsardthong, V., & Uttapap, D., (2009). Pasting properties of a heat-moisture treated canna starch in relation to its structural characteristics. *Carbohyd. Polym. 75,* 505–511.

Wu, Y., Lin, Q., Cui, T., & Xiao, H., (2014). Structural and physical properties of starches isolated from six varieties of millet grown in China. *Int. J. Food Prop., 17,* 2344–2360.

Yadav, D. N., Kaur, J., Anand, T., & Singh, A. K., (2012a). Storage stability and pasting properties of hydrothermally treated pearl millet flour. *Int. J. Food Sci. Tech., 47,* 2532–2537.

Yadav, D. N., Anand, T., Kaur, J., & Singh, A. K., (2012b). Improved storage stability of pearl millet flour through microwave treatment. *Agr. Res., 1,* 399–404.

Yang, X., Wan, Z., Perry, L., Lu, H., Wang, Q., Zhao, C., Li, J., Xie, F., Yu, J., & Cui, T., (2012). Early millet use in northern china. *P. Natl. Acad. Sci. USA, 109,* 3726–3730.

Yano, M., Okuno, K., & Fuwa, H., (1985). Effect of environmental temperature at the milky state on amylose content and fine structure of amylopectin of waxy and non-waxy endosperm starches of rice (*Oryza sativa* L.). *Agr. Biol. Chem., 49,* 373-379.

Zakari, U. M., Hassan, A., & Abbo, E. S., (2010). Physico-chemical and sensory properties of "Agidi" from pearl-millet (*Pennisetum glaucum*) and bambara groundnut (*Vigna subterranean*) flour blends. *Afr. J. Food Sci., 4,* 662–667.

Zavareze, E. R., & Dias, A. R. G., (2011). Impact of heat–moisture treatment and annealing in starches: A review. *Carbohyd. Polym., 83,* 317–328.

Zhou, M., Robards, K., Glennie-Holmes, M., & Helliwell, S., (1998). Structure and pasting properties of oat starch. *Cereal Chem., 75,* 273–281.

Zobel, H. F., (1988). Starch crystal transformations and their industrial importance. *Starch/Stärke, 40,* 1–7.

Zobel, H. F. X., (1964). Ray analysis of starches granules. In: *Carbohydrate Chemistry.* Whistler, R. L. (ed.), Academic Press, New York, pp. 109–113.

CHAPTER 13

AGRO INDUSTRIAL BYPRODUCTS UTILIZATION: FROM TRASH TO TREASURE

KIRAN BAINS and PUSHPA DHAMI

Department of Food and Nutrition, Punjab Agricultural University, Ludhiana, 141004, India

CONTENTS

13.1   INTRODUCTION

Agro-industrial waste wreaks havoc due to its inefficient dumping and excessive burning, causing huge environmental pollution (Afsar and Naser, 2008). Therefore, there has been an escalating interest on how to utilize agro-industrial waste products for human health, animal feed (Maneerat et al., 2015), and other potential uses. Agriculturalists and food technologists who previously treated these crop and food residues as trash and discarded them into the encompassing environs have now understood the importance of agro-based byproducts and have changed their minds. Now, strategies are being planned to utilize these useful by-products for ruminant feeds, poultry feed, and most importantly for human health.

The growing trend of fitness and good health has been well encapsulated in the fitness freaky burgeoning population. This has led to an increased consumer demands for healthy, nutritional, and functional foods. Agro-industrial byproducts, which are claimed to be enriched in bioactive compounds and nutraceuticals, have an imperative role to play for increasing the nutritional needs and food choices that would meet the needs of fitness loving consumers (Rudra et al., 2015). Adding value to farm byproducts becomes crucial for rural growth as well for making them self-sustainable.

By making the food available that was earlier considered as unusable, the problem of poverty and malnutrition can be addressed to a certain extent. Further, some earlier nonutilized products that are reportedly rich source of antioxidants can be utilized by people suffering from lifestyle disorders like diabetes, cardiovascular disease (CVD), obesity, and cancer. Research on these potent byproduct utilization has shown to effectively meet the criteria of human nutrition as well as livestock nutrition. The reuse of agro-industrial waste and its planned dumping is now paving way to the problems of environmental pollution and strengthening its sustainability.

13.2   AGRO-INDUSTRY BYPRODUCTS

Agro-industrial wastes such as wheat middlings (wheat bran and germ), wheat bran oil, rice bran, rice straw, and sugarcane bagasse are not only cost-effective, but they are also very rich sources of natural carbon (Singh et al., 2012). The importance of adequate composition of food ingredients for increased immunity and improved human health has shown the consumers a direction toward increased consumption of fruits and vegetables as well as foods with antioxidant properties. This growing interest of consumers for these foods has attracted the attention of suppliers who now tend to focus on reusing agro-industrial waste from healthy plant (Teixeria et al., 2014) and animal sources.

13.2.1   AGRO-INDUSTRY BYPRODUCTS OF PLANT ORIGIN

13.2.1.1   Rice Milling Byproducts

Rice bran, rice husk, and broken rice are important byproducts of paddy milling industry. Rice husk in its loose form is mostly used for energy production, while rice bran has a high nutritive value with appreciable amounts of protein, fat, and vitamins B and E. Ashraf et al. (2012) studied the relevance of rice bran and broken rice as a cost-effective and health-boosting substitute that could replace fat in soft-dough biscuits. Such low-in-fat biscuits can constitute the food basket of people at risk of obesity, hypertension, CVD, and diabetes. Iso-malto-oligosaccharides (IMO), produced from potato processing waste (PPW) and starch-rich broken rice (BR) (Basu and Prapulla, 2016), can be promoted as a prebiotic dietary fiber marked with efficacy in preventing colon cancer, constipation, diverticulosis, etc.

Rice bran oil (RBO) is widely consumed in many countries such as China, Japan, India, Indonesia, and Korea as a cooking oil (Ghosh, 2007). Pertaining to its high smoke point and delicate flavor, RBO has become an important part of cooking and is also used in salad dressing. Tocopherol, tocotrienol, oryzanol, and squalene are the key constituting elements of RBO that impart antioxidant properties to it. Rice straw is used to feed animals. Chaff is used as a fuel and as a bedding, adsorbent, or packing material. The rice husk ash (RHA), a byproduct of burnt rice husk, is difficult to dispose off, thus resulting into serious environmental pollution. This problem seeks specific attention from researchers who could offer possible solution related to its disposal including its reuse (Kumar et al., 2015).

Rice milling byproducts (hulls and bran) were evaluated and turned out to be effective to adsorb metal ions from aqueous solutions (Marshall et al., 1993). In order to produce briquettes that meet the needs of consumers, different combinations of raw materials (separate rice husk and bran) and methods are used to produce rice-husk briquettes using a multipiston press. Rice husk and bran can be transformed into briquettes, which can meet the needs of consumers in building homes and other infrastructures (Ndindeng et al., 2015). A flow diagram representing the utilization of some byproducts of rice is given in Figure 13.1.

13.2.1.2   Wheat Milling Byproducts

Wheat is the chief ingredient in various cereal-based products like breads, cookies, biscuits, crackers, pancakes, pastries, muffins, noodles, waffles, pastas, macroni, and pizza. Wheat milling industries approximately releases

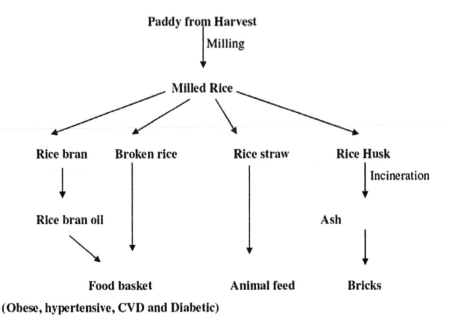

FIGURE 13.1   Flow diagram for the utilization of rice milling byproducts.

25-40% of byproducts such as wheat middlings (wheat germ and bran) and wheat bran oil, and these byproducts can be used for making composites for baked products targeted toward nutritional improvement, meat substitute, cosmetics, bioethanol production, succinic acid production, and nutraceutical products.

Wheat middlings (wheat bran and germ layers) account for 75% of the phytonutrients (Slavin, 2003) in the wheat kernel. Most phytochemicals in both pure and mixed forms impart potential antioxidant properties that provide a guard against lifestyle or metabolic disorder like diabetes, obesity, cancer, and CVD (Liang, 2007). Wheat bran from flour milling is an important livestock feed (generally high in fiber (cellulose), low in energy, and less digestible), while germ is valuable addition to feed concentrate (low in fiber, high in energy and protein, and highly digestible).

Wheat straw can be used as biomass for fuel and for mulching in the garden. The fiber derived from the young wheat stems can be used in paper manufacturing. A flow diagram representing the utilization of some byproducts of wheat is given in Figure 13.2.

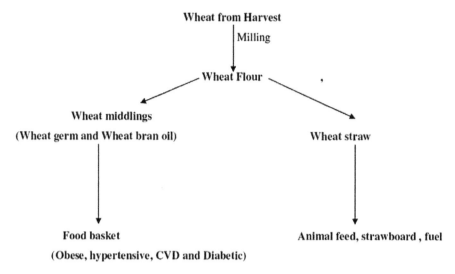

FIGURE 13.2   Flow diagram for the utilization of wheat milling byproducts.

13.2.1.3   Sugar Industry Byproducts

Sugar industry, the second largest agro-based industry in India, contributes in the socio-economic growth of the country. Beet molasses, black strap, bagasse, vinasse, and scum are the prominent byproducts of this industry. Beet molasses and black strap consist of increased amounts of sucrose and desirable substrates for biological hydrogen production. Photo-fermentation is a process where organic acids are converted to hydrogen, which is effectively used in a variety of feed stocks such as sugars (Keskin and Hallenbeck, 2012). Refining of byproduct streams and waste is an outcome of fermentative production of chemicals and biopolymers. This has become a notably important research area with great promising prospects for industrial applications (Koutinas et al., 2014).

Some important industrial byproduct and solid waste of sugar industry are sugar beet pulp (SBP) and fly ash (FA), which may be considered for the removal of zinc and copper from polluted water (Pehlivan et al., 2006). The adsorbent, derived from baggase fly ash, can absorb lead and chromium up to 96–98% (Gupta and Ali, 2004). Hakan and Gul (2010) also produced activated carbons from sugar beet bagasse by chemical activation used to remove nitrate from aqueous solutions.

In the production of ethanol, sugar beet and wheat grains yield vinasse and distiller's dried grains with soluble (DDGSs), respectively, which are used as animal feed. The ethanol production from wheat generates more greenhouse gases (GHGs) compared to sugar beet (Weinberg and Kaltschmitt, 2013). During sugar production, desugared sugar cane extract (DSE) is obtained from *Saccharum officinarum,* which has very strong antioxidant activity. In the food industry, fruit injuries during harvesting are directly linked with the tyrosinase-mediated browning effect; therefore, using tyrosinase inhibitors to control this effect is a crucial step in the manufacturing process of fruit pulp (Seo et al., 2003). The recycling of DSE may improve the economical value for the future development of natural antioxidants and/or tyrosinase inhibitors (Chung et al., 2011), which promise to pave way for treating oxidative stress-induced complications in humans. A flow diagram representing the utilization of some byproducts of sugarcane is given in Figure 13.3.

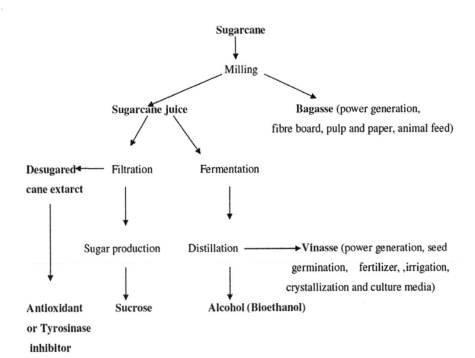

FIGURE 13.3   Flow diagram for the utilization of sugar cane byproducts.

13.2.1.4 Fruit and Vegetable Byproducts

13.2.1.4.1 Apple pomace

Apples after milling or pressing produce apple juice, which can be juiced or canned. The juice after fermentation can be used in making vinegar, ciderkin, and cider. The processing of apple fruit gives the byproduct apple pomace, which is equal to 25–35% of the processed fruit (Schieber et al., 2003) after milling and pressing of apples for cider, apple juice, or puree production (Kafilzadeh et al., 2008). It is biodegradable with high biochemical oxygen demand (BOD). Its richness in water and fermentable sugars results in quick spoilage, causing pollution. The use of apple pomace for feeding livestock offers a way to alleviate this problem (Crawshaw, 2004). Attributed to its high moisture and fermentable sugars, fresh apple pomace spoils readily and often calls for longer preservation using the dehydration method (Crawshaw, 2004; Shalini et al., 2010).

Alternative methods for disposing of apple pomace include composting, fuel production, and spraying it as mulch for landfilling or landspreading (Shalini et al., 2010). Apple pomace is a potent raw material of mulch or compost (Copas, 2004). Composting must be done in combination with a source of carbon to have an adequate C:N ratio. In UK, apple pomace has proven to be effective in suppressing weeds and offers to be a valuable substrate for grass seed germination in difficult areas like roadsides (Copas, 2004).

Due to improper harvesting and processing, culled, dropped and damaged apples (broken, injured during plucking, or unfit for packing) are available in plenty during the apple season and are sometimes utilized for feeding livestock (NDDB, 2012). Earlier, it was merely used as an animal feed, but now, pectin production from apple pomace is considered as one of the best utility approach from both ecological and economical dimensions.

Apple pomace amounts to 10–15% of pectin on a dry weight basis (Oreopoulou et al., 2007) and is recovered by acid extraction and precipitation. Pectin derived from apple pomace bears more prominent gelling properties than citrus pectin. Apple pomace is a key source of polyphenols, dietary fiber, and minerals (Figuerola et al., 2005; Sudha et al., 2007). After juice extraction, most of the polyphenols remain in the apple pomace which is reported to inhibit colon cancer in vitro (Veeriah et al., 2006).

Thus, apple pomace has multitude uses including fuel (ethanol production), direct burning, gasification, anaerobic digestion (methane generation), food (pomace jam, sauce, and confectionery products such as pomace powder for toffees), pectin production, citric acid production, fiber extraction, and livestock feed (Shalini et al., 2010). Apple pomace can also be utilized in extruded products (Kumar et al., 2013).

13.2.1.4.2   Grapes Pomace

Grapes are world's largest fruit produce. Nearly 80% of the total fruit crop is utilized in making wine with a turnover of more than 60 million tons annually worldwide (Teixeria et al., 2014). Grape juice is produced after crushing and blending grapes into a liquid, leaving the solid waste called grape pomace. The extracted juice is then fermented and made into wine or vinegar. Wine industry diffuses large quantities of wastes that increases the BOD and chemical oxygen demand (COD) that threatens the lives of flora and fauna of encompassing discharged areas (Louli et al., 2004). Therefore, discharging of winery waste has become a serious environmental problem that needs effective solutions other than use as fertilizers because the use of fertilizers/composts from grapes pomace results in an increased leaching of nitrogen in soils and oxygen decline due to the presence of tannins and other compounds (Bekhit et al., 2016). A clear and effective environmental management of these wastes and their utilization for other purposes, including ruminant and livestock feeding, is needed (Bekhit et al., 2016).

Byproducts from grape processing can be categorized into solid byproducts (stems, leaves, skins, seeds, and pulp), highly viscous byproducts (lees), and low viscous byproducts (wastewater) (Bekhit et al., 2016). Grape skins and seeds are composed of health-boosting polyphenols, which exhibit many functions such as antioxidant, anticarcinogen, anti-inflammatory, and antibacterial activities. Grape seeds are not only rich source of high-valued oil but also of phenolic compounds such as catechin, epicatechin, gallic acid, and procyanidins (Jasna et al., 2009). The seeds (pips) are sometimes extracted to prepare oil. Lees add textural creaminess and impart aromas/flavors to the wine. The treated wastewater from wine production can be utilized for irrigation.

Attributed to its richness in phenolic compounds, grape waste can be employed in the development of many products ranging from medical to food applications, decreasing the growth of spoilage and pathogenic micro-organisms, and inhibiting lipid oxidation (Mattos et al., 2016). Recent reports present a wide array of biological activities, e.g., prevention of cataract (Yamakoshi et al., 2002), radioprotective effects (Castillo et al., 2000), antihyperglycemic effects (Pinent et al., 2004), antihyperlipidemic effect (Bas et al., 2005), and effective antioxidant enzyme systems.

13.2.1.4.3 Pineapple Byproducts

Pineapple bags the 12th position among fruit crops worldwide and had a turn-over of more than 18 million tons in 2009 (FAO, 2011). The most important producers of canned pineapples are Asian countries including Philippines, Thailand, Malaysia, and Indonesia (Rohrbach et al., 2003). Pineapples can be consumed fresh, cooked, juiced, or preserved. Pineapple is used in desserts such as fruit salad, pizza toppings, yogurt, jam, sweets, and ice cream. The byproducts of pineapple amounts nearly 30–35% of the fresh fruit weight and find applications in soil improvements and livestock nutrition (Hepton et al., 2003).

Pineapple-processing operations pose a serious problem of acceptable and safe solid-waste disposal. Therefore, utilizing these byproducts as an animal feed can help to reduce the impact on environment. These wastes after recycling into animal feeds for ruminants are used as an organic fertil-izer for increased crop yield and present an example of crop-livestock inte-gration, which enhances economic competitiveness without threatening the environment (Liang, 2001).

The pineapple waste after extraction of juice is purified to derive bro-melain enzyme (Ketnawa et al., 2012), which helps in meat tenderization. Bromelain, papain, and ficin are some examples of proteolytic enzymes extracted from plant sources and are widely used as meat tenderizers world-wide (Naveena et al., 2004). Muscle proteins treated with bromelain extract from pineapple has effective tenderization on tough muscle foods without adversely affecting other quality parameters (Ketnawa et al., 2012).

Pineapple wastes (peels, residual pulp, and skin) contain different bioac-tive compounds, which can be beneficial in diabetes, obesity, CVD, and can-cer. Pineapple fruit residues have also been utilized as a valuable biosorbent to eliminate toxic heavy metals like cadmium, mercury, copper, lead, nickel and zinc (Senthilkumaar et al., 2000). The pineapple leaves are also utilized in making coarse textiles and threads in some Southeast Asian regions (Tran, 2006). Solid pineapple waste can be exploited to produce alcohol, methane and volatile fatty acids (Babel et al., 2004). They are also utilized for the production of organic acids such as citric acids that acidify and enhance flavor in substrates from pharmaceutical, food, and beverage industries.

13.2.1.4.4 Mango Byproducts

After bananas, mangos are the most widely produced tropical fruit crop (FAO, 2011). Mangoes are usually sweet, and the taste and texture of the flesh and pulp varies with cultivars and find applications in making juices, fruit bars, ice cream, chutneys, pickles, etc. Mango processing results in nearly 40-50% of byproducts, which are utilized to feed livestock (de la Cruz Medina et al., 2002; Sruamsiri et al., 2009). Mango processing is estimated to yield 150,000 and 400,000 tons of wastes worldwide, causing huge environmental pollution in the surroundings of the processing plants. Thus, utilizing mango wastes for livestock feeding becomes a wise way to reduce environmental loads (El-Kholy et al., 2008). After processing, mango biowastes (peels and seeds) are obtained bearing huge amounts of compounds with antioxidant property that can be reused to alleviate their environmental impact. These byproducts are potent sources of phenolic compounds (antioxidants) and pectins and are considered to be a cost-effective valuable food with nutraceutical characteristics (Berardini et al., 2005).

Soong and Barlow (2004) presented potent antioxidant activity of mango waste with relatively high phenolic contents such as tocopherols, campesterol, stigmasterol, and β-sitosterol. Nunez Selles (2005) and Schiber et al. (2003) reported that the antioxidant efficacy of the mango seed kernel was attributed to its high content of polyphenols, phytosterols, sesquiterpenoids, and microminerals like copper, selenium, and zinc. Mango seed kernels contain a low value protein but rich in essential amino acid contents with larger values of lysine, leucine, and valine (Kittiphoom, 2012). The major use of bioactive components, derived from mango byproducts are in food, pharmaceutical, nutraceutical, and cosmetic industries (Jahurul et al., 2015).

Fat extracted from mango seed has attracted considerable interest from researchers pertaining to its unique physicochemical characteristics, which are found to be similar to cocoa butter (Jahurul et al., 2014). Blends of mango seed fat and palm stearin could be a potential replacer of cocoa butter without significantly changing the physicochemical properties of the product.

The flour obtained from dried mango peel has found applications as a functional ingredient in various food products such as bread, biscuits, sponge cakes, noodles, and other bakery products (Aziz et al., 2012). A flow diagram representing the utilization of some byproducts of mango byproducts is given in Figure 13.4.

13.2.1.4.5 Citrus Byproducts

Citrus (*Citrus* spp.) fruits are one of the most important fruits produced worldwide (Crawshaw, 2004). Citrus fruits include important crops like oranges, lemons, grapefruits, and limes. The fruits are generally eaten fresh and are peeled and can be easily split into segments. Production of citrus juice from fruits is an integrated industry, and the waste generated has serious environmental impact, which is required to be assessed across the entire processing, from fruit cultivation to harvesting to the production to the extraction of juice and essential oils. Citrus seeds are pressed and the oil is extracted, leaving an important byproduct called citrus seed meal. Citrus seeds consist of approximately 20–37% oil, depending upon the cultivar and species (Waheed et al., 2009).

In treating leather and textiles, crude citrus oils are often used for the preparation of soaps and detergents. Some developing countries have regarded it as a new source of edible oil where oil shortage exists (Shahidi et al., 2005).

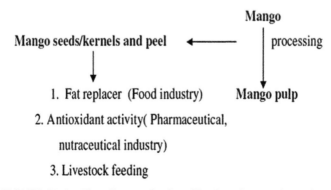

FIGURE 13.4   Flow diagram for the utilization of mango byproducts.

Citrus molasses is an important byproduct of citrus juice extraction. It is either reincorporated in the dried citrus pulp or sold to distilleries or added to grass silage or fed directly to animals (Grant, 2007). Pertaining to its high energy content and excellent digestibility in ruminant animals, dried citrus pulp is utilized as a substitute of cereals in ruminant feeds. The use of dried citrus pulp for animal feeding was found to be an effective way to decrease waste output. Anaerobic digestion of the dried citrus pulp can be carried out to produce biogas (Beccali et al., 2010). A flow diagram showing the utilization of some byproducts of apple, grape, and citrus fruits is given in Figure 13.5.

13.2.1.4.6   Onion Byproducts

According to FAO, onions are a widely grown crop worldwide and are cultivated in nearly 175 countries, and India is one of the leading producers of onion, producing nearly 194.01 million tons (FAO, 2014). Onions are often

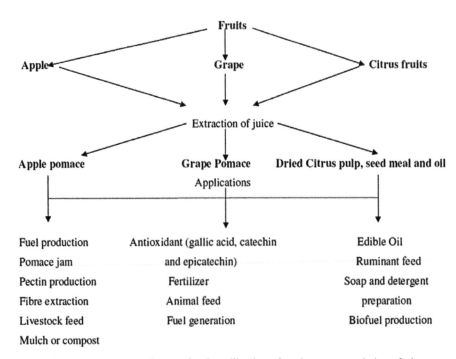

FIGURE 13.5   Flow diagram for the utilization of apple, grapes and citrus fruits.

served as cooked vegetable or a part of a prepared savory dish. These can be eaten as raw or sometimes used to make chutneys and pickles. Onions are widely available in fresh, canned, frozen, pickled, chopped, and caramelized forms. Onion wastes are one of the major agriculture waste materials with serious environmental implications and are the outcomes of industrially processed onions (Choi et al., 2015). Onion wastes may be a source of antioxidant and antibrowning bioactive ingredients.

Onion vinegar can be obtained from different fermentation systems, and it could be a novel valuable product derived from onion wastes and byproducts (Horiuchi et al., 2004; Gonzalez-Saiz et al., 2008). Quercetin present in onion waste has been researched for their extensive health benefits in treating osteoporosis, certain forms of cancer, and pulmonary and cardiovascular diseases and as antiaging agent. The peculiar ability of quercetin to scavenge free radicals such as hydroxyl radical and peroxynitrite is suggested to be the reason of these possible improved health conditions and status (Boots et al., 2008).

13.2.1.4.7 Carrot Byproducts

Carrot is a root vegetable widely available in various colors ranging from orange, purple, black, red, and white to yellow varieties. Carrots juice yield is reported to be 60–70%, and the pomace left is considered nutritious. Carrot pomace contains appreciable amount of carotene (Singh et al., 2006), vitamin B (Manjunatha et al., 2003), and dietary fiber (Bao and Chang, 1994). Solid waste is highly rich in insoluble fiber and potentially reduces cholesterol levels, and therefore, it can be extensively exploited as a key ingredient. Carrot pomace is a major byproduct produced during the industrial juice extraction process and is reportedly rich in carotenoids, phenolic compounds, and fiber which makes it an effective functional ingredient, aiming to improve the quality of foods and help to alleviate the environmental pollution (Ortega et al., 2013). Attempts have been made to utilize carrot pomace in foods such as cake, bread, and extrudates (Singh et al., 2006; Upadhyay et al., 2010; Kumar et al., 2010; Dar et al., 2013). The dried carrot pomace has ascorbic acid and carotene in the range of 13.53 to 22.95 mg and 9.87 to 11.57 mg per 100 g, respectively (Upadhyay et al., 2008).

13.2.2 AGRO-INDUSTRY WASTE OF ANIMAL ORIGIN

13.2.2.1 Dairy Industry Byproducts

Dairy industry processes milk from cows, buffaloes, goats, sheep, horse, and camels for human consumption. By employing heat and dehydration treatment to the raw milk received from farmers, dairy plants extend its marketable life. The waste water of dairy plant contains huge quantities of milk ingredients such as inorganic salts, casein, detergents, and sanitizers used for washing, which represents a value much higher than the specified limits of Bureau of Indian standard (BIS) for the discharge of industrial effluents. Whey is a serious pollutant with a very high BOD (30,000–50,000 mg/l) and COD (60,000–80,000 mg/l). Discarding of whey reportedly results in a potential loss of nutrients. Whey, as a byproduct, is now potentially utilized into whey powder, whey protein concentrate, whey protein isolate, and lactose on the commercial scale. The health-promoting properties of whey relates to its protein content, which mainly comprises α-lactalbumin, β-lactoglobulin, lactoperoxidase, and lactoferrin.

Whey proteins are a potent source of branched chain amino acids (BCAA) like leucine, isoleucine, and valine. Unlike other essential amino acids, BCAAs are effectively metabolized into the muscle tissue and are the very first amino acids, which get utilized during periods of exercise and endurance trainings (Sherwood et al., 2007). Pertaining to its high amount of proteins with enhanced nutritional value, whey beverages are a promising source of energy and nutrients for enduring athletes. The generation of whey from the smaller traders is a serious concern because of draining of whey. This problem can be resolved to the larger extent if the awareness could be created to use whey either for supplying to the industry in time or feeding the cattle.

13.2.2.2 Poultry Industry Byproducts

Poultry meat is one of the commonly eaten types of meat worldwide and accounts for 30% of the total meat production globally. Poultry is available in different forms such as fresh, frozen, joints (cuts), whole birds, bone-in or deboned, and seasoned and is consumed in different ways from raw to ready to cooked. Poultry meat production is predicted to grow at around 2.3% per year to nearly 134.5 million tons by 2023, thus making it one of the biggest meat sectors after 2020 (OECD/FAO, 2015). Poultry waste disposed off in the environment are potential carriers of bacteria, insects, vermin, and viruses, which in turn contaminate the water (leaching of nutrients and harmful microorganisms) and pollute air (toxic gases and odorants) (FAO, 2011). Therefore, transforming of poultry byproducts into animal feed is a wise step to alleviate the environmental pollution.

Poultry byproduct meal is effectively exploited in making the feed of terrestrial animals, which accompanies other byproducts like meat and bone meal, feather meal, and blood meal (Meeker et al., 2006). The nutrient composition of poultry byproducts or meal varies based on the substrate and the processing technique used (Watson, 2006). The hatchery byproduct meal from the processing of poultry hatchery wastes include infertile eggs, hard outer shells of hatched eggs, dead embryos, and dead chicks (Freeman, 2008; Al-Harthi et al., 2010). The disposal of hatchery wastes should be quick as it is an environmental concern. Therefore, processing of hatchery wastes to feed animals is a promising way to reduce the environmental pollution and related concerns (Freeman, 2008). By-product utilization from poultry industry is shown in Figure 13.6.

13.2.2.3 Fish Industry Byproducts

Fish and fish products are widely consumed globally and are considered the richest source of high-quality animal protein. Both fresh raw fish and fish trimmings are utilized to produce fish meal using different processing techniques such as cooking, pressing, drying, and milling (IFFO, 2006). Fish meal is a potent source of essential omega-3 fatty acids (EPA and DHA), highly valued digestible protein, and other essential vitamins and minerals (IFOMA, 2001). Omega-3 fatty acids from fish oil has multidimensional health benefits in treating cardiovascular, neurogenerative, and metabolic diseases. Fish protein concentrate is also a product obtained from fishery waste. Because of its higher processing cost, it is usually costly than fish meal.

The processing of fish industry produces wastewater that has high COD and BOD due to the excessive disposing of organic materials such as hair, blood, fat, flesh, and excreta. Processed wastewater from fish industry is found to have increased levels of nitrogen, phosphorus, and chlorine. It also contains some pathogens including *Campylobacter* and *Salmonella* (World

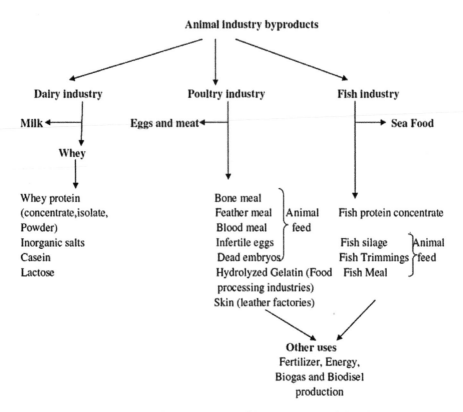

FIGURE 13.6   Flow diagram for the utilization of dairy, poultry and fish industry byproducts.

Bank, 2007). Production of fish silage offers an alternative step to fish meal, especially in location where a small amount of fishery waste is produced (Abdullah, 1983). A flow diagram representing the utilization of some byproducts of dairy, poultry, and fish industries is given in Figure 13.6.

13.3   CONCLUDING REMARKS

The processing of fruits, vegetables, oilseeds, and animal products produce a wide array of waste materials such as wheat bran, rice straw, bagasse, vinasse, peels, pomaces, stones, seeds, oilseed meals, bone meal, and hair. The disposal of these materials usually leads to an environmental problem, which is further aggravated by strict legal restrictions imposed by the countries. Therefore, the successful development of a byproduct utilization process, which can add value to the agro-waste and is economically viable on the commercial scale, is the need of the present day. Efforts are required to take value-added process to the industry for the utilization of byproducts from the agro-industry.

KEYWORDS

- **agro-industry waste**
- **antioxidants**
- **biogas**
- **byproduct**
- **fertilizer**
- **livestock feed**
- **pomace**

REFERENCES

Abdullah, J., (1983). Utilization of bycatches for the production of fish silage. *Philippine J. Vet. Anim. Sci., 9*(1–4), 296.

Afshar, M. A., & Naser, M. S., (2008). Nutritive Value of Some Agro-Industrial By-products for Ruminants: A review. *World J. of Zoo., 3*(2), 40–46.

Al-Harthi, M. A., El-Deek, A. A., Salah El-Din, M., & Alabdeen, A. A., (2010). A nutritional evaluation of hatchery by-product in the diets for laying hens. *Egypt. Poult. Sci., 30*(1), 339–351.

Ashraf, S. ; Saeed, S. M. G., Sayeed,S. A., Ali,R., Saeed, H., & Ahmed, M., (2012). Effect of fat-replacement through rice milling byproducts on the rheological and baking behavior of dough. *Afr. J. Agric. Res,, 7*(44), 5898–5904.

Aziz, N. A. A., Wong, L. M., Bhat, R., & Cheng, L. H., (2012). Evaluation of processed green and ripe mango peel and pulp flours (*Mangifera indica* var Chokanan) in term of chemical composition, antioxidant compounds and functional properties. *J. Sci. Food. Agr., 92*, 557–563.

Babel, S., Fukushi,K., & Sitanrassamee, B., (2004). Effect of acid speciation on solid waste liquefaction in an anaerobic acid digester. *Water Res., 38*, 2417–2423.

Bao, B., & Chang, K. C., (1994). Carrot pulp chemical composition, color, & water-holding capacity as affected by blanching. *J. Food Sci., 59*(6), 1159–1161.

Bas, D., Ferna´ndez-Larrea, J. M., & Blay, J., (2005). Grape seed procyanidins improve atherosclerotic risk index and induce liver CYP7A1 and SHP expression in healthy rats. *J. Fed. Am. Soc. Exp. Biol, 19*, 479–481.

Basu, A., & Prapulla, S. G., (2016). Sustainable utilization of food industry waste and by-products for the production of prebiotic isomaltooligosaccharides (IMO) Published in: *Biotechnology and Biochemical Engineering*, Springer Singapore, pp. 55–64.

Beccali, M., Cellura, M., Iudicello, M., & Mistretta, M., (2009). Resource consumption and environmental impacts of the agrofood sector: life cycle assessment of Italian citrus-based products. *J. Environ. Manage*, *43*(4), 707–724.

Bekhit, A. E. A., Cheng, V. J., Harrison, R., Jing, Y. Z., Bekhit, A. A., Ng, T. B., & Kong, L. M., (2016). Technological aspects of by-product utilization. In: *Valorization of Wine Making By-products*. Bordiga, M. (Ed.), CRC Press, 117–198.

Berardini, N., Knodler, M., Schieber, A., & Carle, R., (2005). Utilization of mango peels as a source of pectin and polyphenolics. *Innov. Food Sci. and Emerg. Technol.*, *6*(4), 442–452.

Boots, A. W., Haenen, G. R., & Bast, A., (2008). Health effects of quercetin: from antioxidant to nutraceutical. *Eur. J. Pharmacol.*, *13*(2–3), 325–37.

Castillo, J., Benavente-García, O., Lorente, J., Alcaraz, M., Redondo, A., Ortuño, A., & Del Rio J. A., (2000). Antioxidant activity and radioprotective effects against chromosomal damage induced in vivo by X-rays of flavan-3-ols (procyanidins) from grape seeds (Vitis vinifera): comparative study versus other phenolic and organic compounds. *J. Agric. Food Chem*, *48*, 1738–1745.

Choi, S., Eun, J. C., Jae-Hak, M., & Jong Bae, H., (2015). Onion skin waste as a valorization resource for the by-products quercetin and biosugar. *Food Chem*, *188*, 537–542.

Chung, Y. M., Wang, H. C., El-Shazly, M., Leu, Y. L., Cheng, M. C., Lee, C. L., Chang, F. R., & Wu, Y. C., (2011). Antioxidant and tyrosinase inhibitory constituents from a desugared sugar cane extract, a byproduct of sugar production. *J Agric Food Chem.*, *59*(17), 9219–25.

Copas, L., (2004). *Apple Pomace*. NACM Technical Report Miscellaneous, pp. 1–5.

Crawshaw, R., (2004). *Co-product Feeds: Animal Feeds from the Food and Drinks Industries*. Nottingham, U.K., University Press, Nothingham.

Dar, A. H., Kumar, N., & Sharma, H. K., (2013). Physical and micro structural changes in carrot pomace based extrudates. *Ital J Food Sci.*, *25*(3), 313–321.

De la Cruz Medina, J., & Garcia, H. S., (2002). Mango: Postharvest operations. In: Mejia, D., Lewis, B. InPho Post-Harvest Compendium. AGSI/FAO, *Instituto. Tecnologico. de Veracruz*, Rome.

El-Kholy, K. F., Solta, M. E., Rahman, S. A. E., Saidy, D. M., & Foda, D., (2008). Use of some agro-industrial by products in Nile Tilapia fish diets. *8th International Symposium on Tilapia in Aquaculture*. Aquafish Collaborative Research Support Program, American Tilapia Association.

FAO, (2011). FAOSTAT. Food and Agriculture Organization of the United Nations, Rome.

FAO, (2014). Major Food And Agricultural Commodities And Producers – Countries By Commodity". Food and Agriculture Organization of the United Nations. Rome, www.Fao.org.

Figuerola, F., Hurtado, M. L., Estevez, A. M., Chiffelle, I., & Asenjo, F., (2005). Fibre Concentrates from Apple Pomace and Citrus Peel as Potential Fibre Sources for Food Enrichment. *Food Chem.*, *91*, 395–401.

Freeman, S. R., (2008). Utilization of poultry byproducts as protein sources in ruminant diets. *PhD Thesis*, North Carolina State University, Raleigh, USA.

Ghosh, M., (2007). Review on recent trends in rice bran oil processing. *J. Am. Oil Chem. Soc*, *84*(4), 315–324.

González-Sáiz, J. M., Esteban-Díez, I., Rodríguez-Tecedor, S., & Pizarro, C., (2008). Valorization of onion waste and by-products: MCR-ALS applied to reveal the compositional profiles of alcoholic fermentations of onion juice monitored by near-infrared spectroscopy. *Biotechnol Bioeng, 101*(4), 776–87.

Grant, E., (2007). With corn prices soaring, maybe it's time to look at citrus-based feeds. Citrus World. *Angus. Journal*, pp. 234–238.

Gupta, V. K., & Ali, I., (2005). Removal of lead and chromium from wastewater using bagasse fly ash—a sugar industry waste. *J. Colloid Interface Sci., 271*(2), 321–328.

Hakan, D., & Gül, G., (2010). Removal of nitrate from aqueous solutions by activated carbon prepared from sugar beet bagasse. *Bioresour Technol, 101*, 1675–1680.

Hepton, A., & Hodgson, A. S., (2003). Processing. In: *The Pineapple: Botany, Production and Uses*. Bartholomew, D. P., Paul, R. E., Rohrbach, K. G., CABI Publishing, USA, Chapter 11, pp. 281.

Horiuchi, J., Tada, K., Kobayashi, M., Kanno, T., & Ebie, K., (2004). Biological approach for effective utilization of worthless onions – vinegar production and composting. *Resour. Conserv. Recy., 40*, 97–109.

IFFO, (2006). What is fishmeal and fish oil?, *International Fishmeal and Fish Oil Organisation*, The marine ingredients organization, London, UK.

IFOMA, (2001). Advantages of using fishmeal in animal feeds., Sociedad nacional de pesqueria, Lima, Peru.

Jahurul, M. H., Zaidul, I. S., Ghafoor, K., Al-Juhaimi, F. Y., Nyam, K. L., Norulaini, N. A., Sahena, F., & Mohd Omar, A. K., (2015). Mango (*Mangifera indica* L.) by-products and their valuable components: A review. *Food Chem., 15*(183), 173–80.

Jahurul, M. H. A., Zaidul, I. S. M., Norulaini, N. N. A., Sahena, F., Jaffri, J. M., & Omar, A. K., (2014). Supercritical carbon dioxide extraction and studies of mango seed kernel for cocoa butter analogy fats. *Cyta J Food, 12*(1), 97–103.

Jasna, S. D., (2009). By-products of fruits processing as a source of phytochemicals 8th Simposium "*Novel Technologies and Economics Development*," University of Niš, Faculty of Technology, Leskovac,, October 21–24.

Kafilzadeh, F., Tassoli, G., & Maleki, A., (2008). Kinetics of digestion and fermentation of apple pomace from juice and puree making. *Res. J. Biol. Sci., 3*(10), 1143–1146.

Keskin, T., & Hallenbeck, P. C., (2012). Hydrogen production from sugar industry wastes using single-stage photofermentation. *Bioresour. Technol., 112*, 131–6.

Ketnawa, S., Chaiwut, P., & Rawdkuen, S., (2012). Pineapple wastes: a potential source for bromelain extraction. *Food Bioprod. Process., 90*, 385–391.

Kittiphoom, S., (2012). Utilization of Mango seed. *Int. Food. Res. J., 19*(4), 1325–1335.

Koutinas, A. A., Vlysidis, A., Pleissner, D., Kopsahelis, N., Lopez Garcia, I., Kookos, I. K., Papanikolaou, S., Kwan, T. H., & Lin, C. S., (2014). Valorization of industrial waste and by-product streams via fermentation for the production of chemicals and biopolymers. *Chem Soc Rev., 43*(8), 2587–627.

Kumar, N., Sarkar, B. C., & Sharma, H. K., (2010). Development and characterization of extruded product using carrot pomace and rice flour. *Int. J. Food Engg., 6*(3), Article 7, DOI:10.2202/1556-3758.1824.

Kumar, R. R., Sharma H. K., & Kumar, N., (2013). Development and characterization of apple pomace and rice flour based extrudates. *Int. J. Post Harvest Technol and Innov, 3*(3), 285–303.

Kumar, A., Sengupta, B., Dasgupta, D., Mandal, T., & Datta, S., (2015). Recovery of value added products from rice husk ash to explore an economic way for recycle and reuse of agricultural waste. *Rev. Environ. Sci. Bio., 15*(1), 47–65.

Liang, J. B., (2001). The use of agricultural byproducts for sustainable livestock production – the Malaysian experience. In: *The Second Symposium on Sustainable Utilization of Agricultural byproducts for Animal Production, 26–27* July.

Liang, Y., (2007). *Wheat Antioxidants*, Department of Nutrition and Food Science the University of Maryland, John Wiley & Sons, Inc., publication.

Louli, V., Ragoussis, N., & Magoulas, K., (2004). Recovery of phenolic antioxidants from wine industry by-products. *Bioresour Technol.*, *92*(2), 201–208.

Maneerat, W., Prasanpanich, S., Tumwasorn, S., Laudadio, V., & Tufarelli, V., (2015). Evaluating agro-industrial by-products as dietary roughage source on growth performance of fattening steers. *Saudi J Biol Sci.*, *22*(5), 580–558.

Manjunatha, S. S., Mohan Kumar, B. L., & Das Gupta, D. K., (2003). Development and Evaluation of carrot kheer mix. *J Food Sci Tech*, *40*, 310–312.

Marshall, E. W., Elaine, T. C., & William, J. E., (1993). Use of rice milling byproducts (hulls & bran) to remove metal ions from aqueous solution. *J Environ Sci Health A Tox Hazard Subst Environ Eng*, *28*(9), 1977–92.

Meeker, D. L., & Hamilton, C. R., (2006). An overview of the rendering industry. In: *Essential Rendering. Meeker (Ed).* National Renderers Association,, Printed by Kirby Lithographic Company, Inc. Arlington, Virginia.

Mattos, G. N., Renata, V. T., Angela, A. L. F., & Lourdes, M. C. C., (2016). Grape by-products extracts against microbial proliferation and lipid oxidation: A review. *J. Sci. Food Agr.*, *97*, 1055–1064.

Naveena, B. M., Mendiratta, S. K., & Anjaneyulu, A. S. R., (2004). Tenderization of buffalo meat using plant proteases from cucumis trigonus roxb (kachri) and zingiber officinale roscoe (Ginger Rhizome). *Meat Sci*, *68*(3),363–369.

NDDB., (2012). Nutritive value of commonly available feeds and fodders in India., National Dairy Development Board, Animal Nutrition Group, Anand, India.

Ndindenga, S. A., Mbassib, J. E. G., Mbachamc, W. F., Manfula, J., Graham, S. A. J., Moreiraa, J., Dossou, J., & Futakuchi, K., (2015). Quality optimization in briquettes made from rice milling by- products. *Energy Sustain Dev*, *29*, 24–31.

Nunez-Selles, A. J., (2005). Antioxidant therapy: myth or reality. *J. Braz. Chem. Soc.*, *16*(4), 699–710.

OECD/Food and Agriculture Organization of the United Nations, (2015). OECD-FAO Agricultural Outlook., OECD Publishing, Paris. http://dx.doi.org/10.1787/agr_outlook-2015-en.

Oreopoulou, C. V., & Tzia, C., (2007). *Utilization of By-Products and Treatment of Waste in the Food*, W. Russ Eds., Springer, USA, pp. 209–232.

Ortega, M .H., Kissangou1, G., Hugo, N. M., María, E. S. P., & Alicia, O. M., (2013). microwave dried carrot pomace as a source of fiber and carotenoids. *J. Food Nutr. Sci.*, *4*, 1037–1046.

Pehlivan, E., Cetin, S., & Yanık, B. H., (2006). Equilibrium studies for the sorption of zinc and copper from aqueous solutions using sugar beet pulp and fly ash. *J. Hazard. Mater.*, *135*, 193–199.

Pinent, M., Blay, M., Blade, M. C., Salvado, M. J., Arola, L., & Arde'vol, A., (2004). Grape seed-derived procyanidins have an antihyperglycemic effect in streptozotocin-induced diabetic rats and insulinomimetic activity in insulin-sensitive cell lines. *J. Endocrinol*, *145*, 4985–4990.

Rohrbach, K. G., Leal, F., & Coppens, G., (2003). History, distribution and world production. In: *The Pineapple: Botany, Production and Uses.* Bartholomew, D. P., Paul, R. E., Rohrbach, K. G. (Eds), CABI Publishing.

Rudra, S. G., Nishad, J., Jakhar, N., & Kaur, C., (2015). Food industry waste: mine of nutraceuticals. *IJSET*, *4*(1), 205–229.

Schieber, A., Hilt, P., Streker, P., Endre, H. U., Rentschler, C., & Carlea, R., (2003). A new process for the combined recovery of pectin and phenolic compounds from apple pomace. *Innov. Food Sci. Emerg.*, *4*, 99–107.

Seo, S.Y., Sharma, V. K., & Sharma, N., (2003). Mushroom tyrosinase: Recent prospects. *J. Agric. Food Chem.*, *51*, 2837–2853.

Senthilkumaar, S., Bharathi, S., Nithyanandhi, D., & Subburam, V., (*2000*). Biosorption of toxic heavy metals from aqueous solutions. *Bioresour. Technol.*, *75*, 163–165.

Shahidi, F., & Ying Zhong., (2005). Citrus oils and essences. In: *Bailey's Industrial Oil and Fat Products*, Sixth Edition, John Wiley & Sons, Inc; pp. 49–66.

Shalini, R., & Gupta, D. K., (2010). Utilization of pomace from apple processing industries: a review. *J. Food Sci. Technol.*, *47*(4), 365–371.

Sherwood, S., (2007). Protein beverage and method for making the same. US Patent US 7906/160 B2.

Singh, B., Panesar, P. S., & Nanda, V., (2006). Utilization of Carrot Pomace for the Preparation of a Value Added Product. *World J. Dairy Food Sci.*, *1*(1), 22–27.

Singh, R., Kapoor, V., & Kumar, V., (2012). Utilization of agro-industrial wastes for the simultaneous production of amylase and xylanase by thermophilic actinomycetes. *Braz J. Microbiol.*, *43*(4),1545–1552.

Slavin, J., (2003). Why whole grains are protective: biological mechanisms. *Proc. Nutr. Soc.*, *62*, 129–34.

Soong, Y. Y., & Barlow, P. J., (2004). Antioxidant activity and phenolic content of selected fruit seeds. *Food Chem.*, *88*(3), 411–17.

Sruamsiri, S., & Silman, P., (2009). Nutritive value and nutrient digestibility of ensiled mango by-products. *Maejo Int. J. Sci. Technol.*, *3*(03), 371–378.

Sudha, M. L., Baskaran, V., & Leelavathi, K., (2007). Apple pomace as a source of dietary fiber and polyphenols and its effect on the rheological characteristics and cake making. *Food Chem.*, *104*(2), 686–692.

Teixeira, A., Baenas, N., Raul, D. P., Barros, A., Rosa, E., Diego, A. M., & Cristina, G. V., (2014). Natural Bioactive Compounds from Winery By-Products as Health Promoters: A Review. *Int J Mol Sci*, *15*(9),15638–15678.

Tran, A. V., (*2006*). Chemical analysis and pulping study of pineapple crown leaves. *Ind. Crops. Prod*, *24*, 66–74.

Upadhyay, A., Sharma, H. K., & Sarkar, B. C., (2010). Optimization of processing conditions for rice based extruded products from the incorporation of carrot pomace powder. *J. Food Quality*, *33*, 350–369.

Upadhyay, A., Sharma, H. K., & Sarkar, B. C., (2008). Characterization of dehydration kinetics of carrot pomace. *Agricultural Engineering International*, *10*, 1–9.

Veeriah, S., Kautenburger, T., Habermann, N., Sauer, J., Dietrich, H., Will, F., Pool-Zobel, B. L., (2006). Apple flavonoids inhibit growth of HT29 human colon cancer cells and modulate expression of genes involved in the biotransformation of xenobiotics. *Mol. Carcinog*, *45*, 164–174.

Waheed, A., Mahmud, S., Saleem, M., & Ahmad, T., (2009). Fatty acid composition of neutral lipid: Classes of Citrus seed oil. *J. Saudi Chem. Soc.*, *13*(3), 269–272.

Watson, H., (2006). Poultry meal vs poultry by-product meal. Dogs in Canada Magazine. Apex, Toronto, Canada.

Weinberg, J., & Kaltschmitt, M., (2013). Greenhouse gas emissions from first generation ethanol derived from wheat and sugar beet in Germany – Analysis and comparison of advanced by-product utilization pathways. *Appl. Energy*, *102*, 131–139.

World Bank, (2007). Environmental, health, & safety guidelines for poultry production., pp 1–18. International Finance Corporation, Washington DC.

Yamakoshi, J., Saito, M., Kataoka, S., & Tokutake, S., (2002). Procyanidin-rich extract from grape seeds prevents cataract formation in hereditary cataractous (ICR/f) rats. *J. Agric. Food Chem.*, *50*, 4983–4988.

CHAPTER 14

TRACEABILITY: SUSTAINABLE SOLUTION IN THE FOOD CHAIN

ROHANT KUMAR DHAKA, R. CHANDRAKALA, and
ASHUTOSH UPADHYAY

Department of Food Science and Technology, National Institute of Food Technology Entrepreneurship and Management (NIFTEM), Kundli, Haryana, India

CONTENTS

14.1   INTRODUCTION

In the era of food safety and quality, traceability has become a necessity for the food manufacturers to survive the fierce competition in food business. Olsen and Borit (2013) define traceability as "The ability to access any or all information relating to that which is under consideration, throughout its entire life cycle, by means of recorded identifications." Food traceability in particular may be defined as the ability to track any food or feed material throughout the food supply chain from farm to fork. Furthermore, Opara (2003) broadly classifies food traceability into:

a) **Product traceability:** It involves the tracking of physical location of a product at any stage in the supply chain, which in turn serves the purpose of logistics and inventory management, swift product recalls, and communication of relevant information to consumers.

b) **Process traceability:** This involves the agriculture practices and postharvest operations, which affect the product at different stages of its life cycle. Process variables significantly affect the product quality and safety; thus, the traceability system must be efficient enough to trace them.

c) **Genetic traceability:** Genetic traceability involves the tracking of genetic information. It includes information regarding type and source of genetically modified organisms and the materials used in the food.

d) **Inputs traceability:** Food receives various inputs during growth and postharvest operations, which include pesticides, fertilizers, feed, irrigation water, various food additives, etc. Proper documentation should be done regarding any input so that it can be traced back in times of emergency.

e) **Disease and pest traceability:** Foods of animal origin are prone to diseases like bird flu and mad cow disease. Consumption of foods affected by diseases or pests can result in huge disease outbreaks; thus, the food traceability system must be efficient enough to identify and isolate such cases at the first incidence.

f) **Measurement traceability:** If the measuring equipment used during processing is not calibrated according to national or international standards, it may significantly affect the product quality and safety. Thus, traceability of measuring instruments and standards is an integral part of food traceability.

Traceability systems ensure safety of consumers as the quality and safety issues can be traced during processing itself and fixed before it affects the general public. Producers gain from food traceability systems as it helps to maintain quality of raw materials, locate any problematic item and fix the root cause, ensure food safety, and prevent unfair competitive practices among producers (Espeneira and Santaclara, 2016). Food traceability also plays a major role in facilitating the international trade of food products. Disease outbreaks such as bovine spongiform encephalopathy (BSE) in early 2000s and consumer focus on food safety acted as a driving force for the implementation of food traceability systems. BSE outbreak forced EU to take more stringent measures (EU Regulation 178/2002) focusing on traceability. Livestock identification and meat traceability systems were implemented in order to prevent future outbreaks. Such traceability systems were designed in a way that the information regarding age, sex, breed, movements, and veterinarian treatment records could be easily recorded and maintained.

Traceability plays an important role in the investigation of disease outbreaks and food product recalls, enabling the investigative agencies to identify the root cause and take necessary measures to prevent such outbreaks in future. Modern traceability systems are efficient enough to track the cattle from birth to the finished product, which makes it easier to prevent diseases. Food shipments can be tracked to prevent tampering, and consumers can be well informed about product attributes like country of origin, animal welfare, and genetic composition.

Moreover, today's consumer is aware about food safety issues and considers many factors before choosing a product. These factors range from quality and safety to more specific ones like country of origin, farming practices, etc. The brand value of the manufacturer is affected by the food safety and traceability certifications. Generally, the news of product recalls attracts huge attentions by media, and thus, consumer is aware about the reputation of manufacturers regarding food safety and traceability. Consumer is more likely to choose the manufacturer who acts swiftly and efficiently in times of product recall (Figure 14.1).

Food traceability helps food manufacturers identify and isolate the problematic products at an early stage, thus reducing the production and distribution of unsafe and inferior quality products. This eventually saves manufacturers from product recalls, bad publicity, and legal actions. Food traceability also helps in market differentiation and segmentation. Manufacturers can trace back the origin of raw materials, which performed better for a particular segment of market and then target that market segment for profitability. Similarly, process attributes also affect the product quality and some process attributes may work wonders for a certain segment of market. The food traceability system enables manufacturers to analyze sales data and then track down the process attributes that generated more

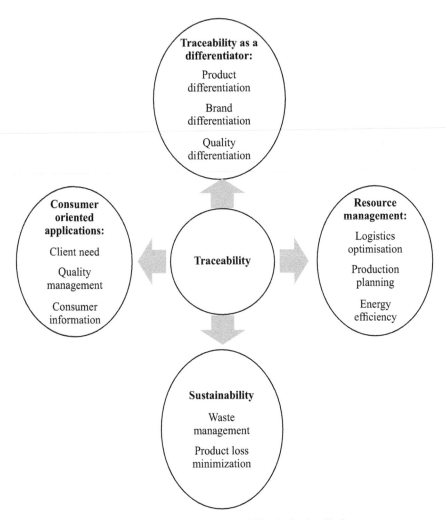

FIGURE 14.1   Relevance of traceability in the food industry.

revenues. Besides mutual interest of consumer and producer, legal compulsions also act as a driving force for implementation of food traceability systems.

Bosona and Gebresenbet (2013) reviewed the significance of traceability systems in logistics management and reported the following benefits of traceability in logistics management:

a) **Higher degree of consumer satisfaction:** Consumer satisfaction is related to the product information provided by the manufacture and food traceability system helps in organizing and communicating product information in a better way.

b) **Contribution to agricultural sustainability:** Traceability plays an important role in maintaining transparency in food production and sourcing practices which in turn helps in implementation of sustainability initiatives, especially at the farm level.

c) **Food supply chain management:** Food Traceability Systems help in reducing the supply activities-related costs, mainly the logistics costs.

d) **Better management during food crisis:** Traceability plays an important role in maintaining higher standards of food safety, thus minimizing the distribution of unsafe and poor quality products to public.

14.2 TRACEABILITY DRIVERS

Market for processed foods is growing at an encouraging rate due to several reasons like convenience, long shelf-life, variety, and taste. At the same time, consumers are also getting more aware about key food safety issues and are likely to scan the label before buying any food product (Figure 14.2).

According to Plessis and Rand (2011), consumers believe that the foods from the native country of origin are safer than imported foods. Consumers are more likely to opt for a product that highlights the relationship to production environment and practices to the quality and consequently identification of the product.

Other than consumer demands, legal requirements also act as a driver of traceability. Legal requirements are laid down by governing agencies in order to promote food safety and protect public health. Food recall procedures mentioned in section 28 of FSSA (2006) clearly state that if a food business operator considers or has reasons to believe that a food that has been processed, manufactured, or distributed is not in compliance with this Act, or the rules or regulations made there under, then it shall immediately initiate procedures to withdraw the food in question from the market and consumers, indicating reasons for its withdrawal and inform the competent authorities thereof. FDA Food Safety Modernization Act (FSMA) requires FDA to establish record-keeping requirements for high-risk foods to help in tracing products. General Food Law (GFL) of EU has also included important elements of traceability and withdrawal of risky products from market.

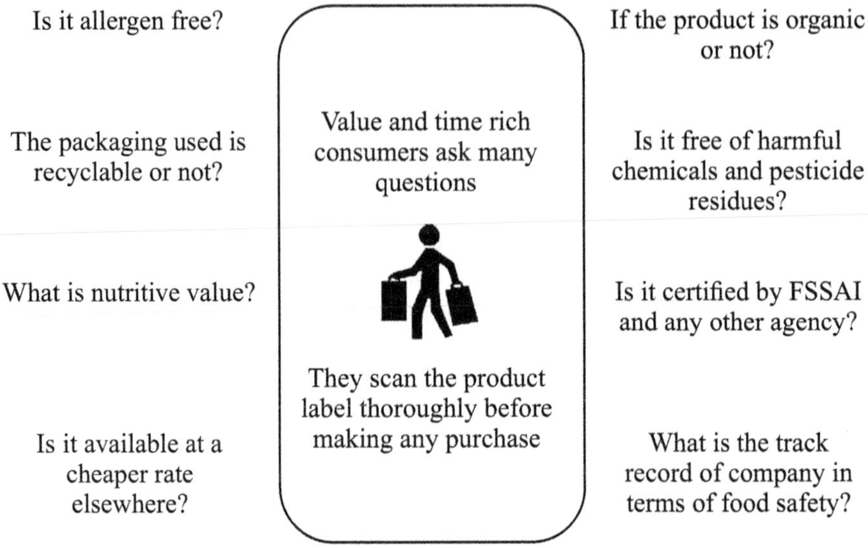

Is it allergen free?

If the product is organic or not?

The packaging used is recyclable or not?

Value and time rich consumers ask many questions

Is it free of harmful chemicals and pesticide residues?

What is nutritive value?

Is it certified by FSSAI and any other agency?

They scan the product label thoroughly before making any purchase

Is it available at a cheaper rate elsewhere?

What is the track record of company in terms of food safety?

FIGURE 14.2   Traceability drivers: consumer concerns and needs.

14.3   PRINCIPLES OF TRACEABILITY

14.3.1   TRACKING VERSUS TRACING

Tracking is the ability to follow the product from origin to its destination and is usually in forward direction, whereas tracing is the ability to:

1. recreate the history of a product along the original path.
2. identify the origin, movements, and desired information of a particular batch and/or unit of the product within the supply chain by reference to records held upstream.

14.3.2   TRACEABILITY CATEGORIES

Traceability can be broadly distinguished as the following types.

14.3.2.1   Internal Traceability

It is the ability of tracking product information internally, usually by a company through its own traceability system. Internal traceability possesses the following attributes:

- It belongs to a single company and one geographical location.
- It derives a lot of inputs from the company's production management systems and possesses some minor issues.

14.3.2.2   Chain Traceability

Chain traceability is related to the total supply chain that involves different stakeholders at different levels. It is the ability to trace the product through different levels of the supply chain. Stakeholders are supposed to share and receive product relevant data.

Chain traceability typically has the following characteristics:

- It is not company specific and involves different companies or even countries.
- It completely relies on the data generated through internal traceability.
- Here, the privacy issues are major.

Standards need to be followed for recording and exchange of data. But, the challenge lies in the dealing of the data because few companies in the chain may be using manual traceability systems in contrast to others using electronic traceability situations.

14.3.3   CRITICAL TRACKING EVENTS (CTE) AND KEY DATA ELEMENTS (KDE)

McEntire et al. (2010) describes critical tracking events (CTEs) and key data elements (KDEs) as follows:

Critical tracking events are the specific events in the supply chain, which significantly affect the original path of the product and define its ultimate path. Data collection at such points is necessary in order to implement traceability. Some of the CTEs are listed below:

- point of origin (attach a unique identifier as soon as the product enters the supply chain);
- aggregation (bring discrete items together);
- convert (repack or relabel);
- product receipt;
- product shipping;

- product sale (retail);
- product depletion (retail and food service).

Each CTE must be critically analyzed to ensure sufficient data to permit traceability. KDEs include data correlated to transactions and the actual product.

14.3.4   IMPLEMENTATION OF FOOD TRACEABILITY

Implementation is a complex process and involves many inter-related aspects that affect each other. Several hurdles can occur before, during, and after implementation of food traceability systems. Many authors have identified critical factors related to implementation of traceability system. Senneset et al. (2007) mentions the following eight factors related to the electronic traceability system for chain traceability in seafood supply chain:

1. procedures for chain traceability at the reception point,
2. procedures for chain traceability at the delivery point,
3. procedures for internal traceability,
4. software system to execute internal traceability,
5. automatic recording technology at reception,
6. automatic recording at delivery,
7. standardized identification of traceable units, and
8. software system.

Motivation plays the central role in effective implementation of a robust traceability system. The willingness and motivation of employees at all levels are required for successful implementation and functioning of the traceability system (Karlsen and Olsen, 2016).

14.4   TECHNIQUES USED FOR TRACEABILITY

Due to increasing awareness on health and constant outbreaks of foodborne illnesses, consumers are tempted to understand the history of the food they are consuming. Due to some public health issues resulting in fatality, legislative requirements are governing the animal identification and tracking of food for safety.

14.4.1   METHODS BASED ON ANALYSIS OF DNA

DNA is a molecule that holds genetic information of an organism. DNA of every human body is unique, and hence, analyzing this genetic material can lead to their identity. Currently, DNA analysis is one of the most reliable and accurate molecular method over protein-based methods. Boziaris (2014) **cites** the following reasons for the adoption of DNA-based methods:

i. DNA is more resistant to alteration during processing and is heat stable than protein.
ii. It is easy to amplify short segments of DNA and identify even when they are fragmented.
iii. It provides more information than proteins even after degeneration of genetic codes and presence of noncoding regions.
iv. Because DNA is present in all cells of an organism, it can be extracted from any organic substrate like muscle and blood.

The different molecular methods employed for food traceability based on the analysis of DNA include the following sections.

14.4.1.1   Polymerase Chain Reaction (PCR)

It is a novel technique for in vitro amplification and detection of DNA by an enzyme that uses defined segment in a strand of DNA as a template for assembling the complementary strand. It is known for its simplicity, sensitivity, and capability to detect even a single copy of DNA from a single cell sample (Chikuni et al., 1994). PCR amplification requires two oligonucleotide primers, four deoxy nucleotide triphosphates (dNTPS), magnesium ions, and thermostable DNA Tag polymerase. This technique includes repeated cycles made up of three defined steps.

1. Denaturation of double-stranded DNA: where template DNA is heated at excess temperature to separate the double-stranded DNA and produce two single strands
2. Annealing of primers: primers are attached to the dissociated DNA strands.
3. Primer extension: where the enzyme catalyzes and synthesizes new strands of DNA.

PCR applications are now increasingly explored in food traceability solutions for their simplicity in handling, being relatively quicker, and able to identify the species of origin in complex and processed foods too. In case of food tracing applications, specific detection of chicken meat in other meat products was studied using PCR. PCR-based assays are the precise, sensitive, and rapid method for routine assessment of authenticity and quality of meat and meat products (Mane et al., 2009). Species-specific DNA markers of mitochondrial origin with PCR-based protocols can be employed for the identification of chicken, duck, pigeon, and pig species (Haunshi et al., 2009). Species-specific PCR protocols have been developed for meat species identification and differentiation (Herman, 2001; Kesmen et al., 2007; Rodriguez-Lazaro and Hernandez, 2013). It can also be used for the identification of cow milk in sheep and goat milk by using primers targeting the mitochondrial 12S rRNA gene (Lopez-Calleja et al., 2004). Due to its sensitivity and rapidity, PCR application has also been widely employed for detecting the authenticity of other sea foods, fruits, and vegetables.

14.4.1.2 Polymerase Chain Reaction-linked Restriction Fragment Length Polymorphism (PCR-RFLP)

PCR-RFLP is also known as Cleaved Amplified Polymorphic Sequence (CAPS). This technique is based on endonuclease digestion of PCR-amplified DNA. Here, PCR is used initially to amplify the small amounts of DNA. Then, restriction endonuclease will specifically recognize and cleave the DNA in the region of mutation. Finally, electrophoresis is used to resolve these restriction fragments. It is the best technique for the identification of genetic constitution variation of specific species (Rasmussen, 2012).

Moreover, this technique is widely used in detecting meat adulteration where inferior or cheaper meat would be added with high quality meats. Hence, this technique can be commonly employed for species identification and differentiation (Girish et al., 2005). The advantages of this method are that it is simple, robust, inexpensive, and easy to design and there is no requirement of expensive instrument or expensive training of laboratory staff. The disadvantages of the technique include costly enzyme and longer time. There are several variants of PCR-RFLP like melting curve analysis of SNPs (McSNP) (Jahangir Tafrechi et al., 2007), amplified fragment length polymorphism (AFLP) (Meudt and Clarke, 2007), terminal restriction fragment length polymorphism (TRFLP) (Wang et al., 2010), and inverse

PCR-based amplified restriction fragment length polymorphism (iFLP) (Liu et al., 2004). These variants are used in DNA profiling like DNA fingerprinting of microbial population and authentication of foods.

Rapid, sensitive, and accurate detection of buffalo, cattle, and sheep's milk was carried out using species-specific PCR and PCR-RFLP techniques. DNA from a small amount of fresh milk was extracted to amplify the gene encoding species-specific repeat region (SSRR) and the mitochondrial DNA segment. PCR-RFLP is employed for differentiating buffalo's milk and cattle's milk (Abdel-Rahman and Ahmed, 2007). Authentication of meat species was studied using the PCR-RFLP assay targeting the mitochondrial cytochrome b gene. High precision and specificity of this method have resulted in a clear differentiation of beef, carabeef (meat of water buffalo), chevon (goat meat), mutton, and pork meat (Kumar et al., 2012). DNA barcoding equipped with PCR-RFLP is a cheap and rapid technique for identifying fish species in commercial raw and processed anchovy meats (Pappalardo and Ferrito, 2015). PCR-RFLP techniques are most commonly used in meat and seafood traceability.

14.4.1.3   Real-Time PCR (RT-PCR)

This method is also known as quantitative PCR or Real time quantitative PCR. This method was developed in order to overcome the post-PCR analysis problems. In the conventional method, gel electrophoresis is commonly employed for measuring the size of amplicons (piece of DNA or RNA) however; this method has limited specificity as different molecules of the same molecular weight cannot be distinguished. RT-PCR measures the progress of amplification by monitoring the changes in fluorescence (Figure 14.3). The fluorescent markers when bound to target DNA will emit fluorescence that could be amplified and detected. The amount of fluorescence produced is proportional to the amount of amplicon produced during PCR (Espiñeira and Santaclara, 2016). It can also be distinguished into nonspecific detection using DNA-binding dyes and target-specific probe-based detection. According to Shipley (2007), it is a continuous collection of fluorescent signals from one or more PCR over a range of cycles. Quantitative RT-PCR involves the conversion of these signals into a numerical value. The technique has several advantages such as high throughput, rapid cycling times, amplification and detection in an integrated system, constant reaction monitoring, and increased sensitivity.

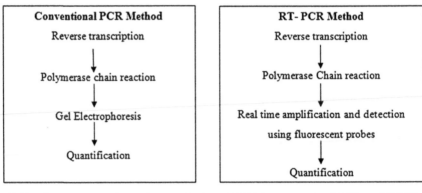

FIGURE 14.3   Detection and quantification of DNA by conventional PCR vs. RT-PCR.

The technique can be employed for the detection of species in highly processed food products, where the extreme temperature or pressure during processing greatly degraded DNA. Drummond et al. (2013) developed an RT-PCR assay for detecting bovine contamination in buffalo-derived dairy products. This study clearly indicated that RT-PCR can be helpful in identifying the adulteration in milk and dairy products. In the case of traceability in cereal products, Hernández et al. (2005) has studied RT-PCR systems for quantitative identification of barley, rice, sunflower, and wheat DNA in food products. Similarly, the common wheat adulteration in durum wheat pasta was detected by real-time PCR assay and quantified using the French official peroxidase marker method (Alary et al., 2002). The RT-PCR assay can be used as a tool for the authentication of olive oils (Gimenez et al., 2010) and DNA stability in olive oil using lambda markers (Spaniolas et al., 2008). Their application is extended to the field of soft fruit traceability, wherein the RT-PCR protocol was used to trace the different berries and other fruits in food matrices (Palmieri et al., 2009) (Table 14.1).

14.4.1.4   Forensically Informative Nucleotide Sequence (FINS)

FINS is a technique of genetic identification of species, which involves amplification of DNA, followed by sequencing and phylogenetic analysis. It is a robust and powerful technique to determine the inter-specific and intra-specific variability. Studies were conducted for identifying seven fish

TABLE 14.1 Molecular Methods and Their Applications in Traceability

Purposes	Molecular Method	References
Meat industry		
Meat species identification	PCR, PCR-RLFP, FINS, RT-PCR	Lago et al. (2011a); Han et al. (2013); Amaral et al. (2014); Okuma and Hellberg (2015)
Breed differentiation/ geographical origin of meat	SNP and Microsatellites	Dalvit et al. (2008)
Sex differentiation	PCR, PCR-RLFP, RT-PCR	Gokulakrishnan et al. (2013)
Dairy industry		
Dairy Product traceability	PCR, PCR-RLFP, RT-PCR	Abdel-Rahman and Ahmed (2007)
Identification & quantification of species in dairy products	RT-PCR	Diaz et al. (2007)
Fruits and vegetables industry		
Detection and identification of species	PCR, RT- PCR	Ortola-Vidal et al. (2007)
Differentiation of varieties in vegetables	SNP and Microsatellites	Koppel et al. (2009)
Wine industry		
Differentiation of grape varieties in wine	Microsatellite DNA based methods	Sefc and Lefort (2001)
Fats/Oils industry		
Species identification in vegetable oils	PCR and RT-PCR	Zhang et al. (2009)
Discrimination of varietal olive oils	Microsatellite and SNPs	Vietina et al. (2011)
Cereals industry		
Gluten detection	RT- PCR	Allmann et al. (1993); Dahinden et al. (2001)
Identification and detection of adulterants	PCR	Alary et al. (2007)

*PCR: Polymerase chain reaction, PCR-RLFP: Polymerase chain reaction-linked restriction fragment length polymorphism, RT-PCR: Real time PCR, SNP: Single Nucleotide Polymorphism, DNA: Deoxy ribose nucleotide.

species of *Lophius* by the genetic identification method. Species belonging to the same genus *Lophius* were identified using a molecular marker
cytochrome oxidase subunit I gene. Moreover, PCR-RFLP and FINS were
extremely specific in detecting the deliberate or unintentional mislabeling of
seafood products (Espineria et al., 2008). FINS are used for species identification of cephalopods (Chapela et al., 2002), shark (Blanco et al., 2008),
sardines (Lago et al., 2011b), hake and codfish (Perez and Presa, 2008),
fresh water eels (Espiñeira and Vieites, 2012), etc. In case of meat traceability, illegal labeling or substitution of species and chemical and microbial
contamination have increased the quality concern about the integrity of the
meat. Barlett and Davidson (1992), proposed the sequence to be followed in
FINS techniques (Figure 14.4).

14.4.1.5   Digital PCR (dPCR)

Digital PCR is a more sensitive, specific and stable method for the detection of nucleic acids as compared to RT-PCR methods. It is based on the
detection of target locus of individual molecules wherein the sample is well
diluted and aliquoted into number of droplets. This decreases the probability

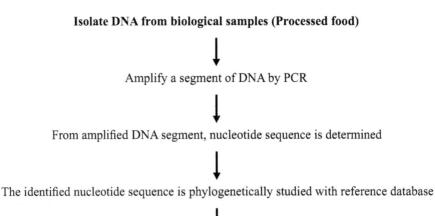

FIGURE 14.4   Forensically informative nucleotide sequence.

of occurrence of target molecule in every aliquot; thus, only few aliquots will be positive. Each of these positive aliquots is assumed to have only single target molecule in it (Xu, 2016). This takes place in a micro-droplet chip, where micro-fluidic devices and sensors are used for detecting the amplification product fluorometrically (Kuchta et al., 2014). It can be used in combination with Poisson distribution analysis to provide an accurate quantification. This technique is widely helpful in identifying rare variants like those that arise from DNA mutations. Specifically, it can be used for analyzing the food samples that contain mixture of species in a single assay, making multiple reactions in parallel by a nanofluidic chip (Espiñeira and Santaclara, 2016).

Similar to other PCR techniques, dPCR also finds application in meat species identification and serves as the best technique for quantifying absolute number of target molecules in meat products (Floren et al., 2015). The main application of dPCR lies in quantification of genetically modified (GM) foods due to high sensitivity and tolerability to PCR inhibitors. Several studies on this aspect also have proven that this technique can be employed for routine analysis of GM foods (Iwobi et al., 2016; Zhu et al., 2016). It can also be used for authenticating the purity of olive oils by defining its origin and varietal composition (Scollo et al., 2016).

14.4.1.6 Other DNA-Based Methods

There are several other molecular techniques such as microsatellite, which is a di-tri or tetra nucleotide tandem repeats in nuclear DNA with large mutation capacity (Rezaei et al., 2011). These are employed for determining the individual genetic profile of single species. Single nucleotide polymorphism (SNP), which involves the analysis of single nucleotide variation in a DNA sequence, helps in comparing the genomic variation between members of the same individuals (Espiñeira and Santaclara, 2016). Apart from these, the most trending and new generation techniques are next generation sequencing (NGS). It is based on fragmentation of DNA template followed by immobilization on a solid support system. Instead of sequencing the single fragment of DNA, NGS will sequence millions of fragments in a massive parallel fashion (Espiñeira and Santaclara, 2016).

14.4.2   RADIO FREQUENCY IDENTIFICATION (RFID)

Universal product codes (UPC) barcodes are traditionally used to standardize and automate the identification process, which require a line of sight between a tag and the reader. Radio frequency identification (RFID) is an electronic method of product automatic identification that uses radio waves to automatically identify objects. There are several methods of identification, but the most common is to store a serial number that identifies an object and also other information on a microchip that is attached to an antenna (the chip and the antenna together are called an RFID transponder or an RFID tag). It is a technique for the identification of an object from a distance even without requiring a line of sight.

RFID is fabricated with four basic components, namely transponder, reader, data accumulator, and processing software (Grooms, 2007). Transponders commonly employed for RFID animal identification systems are called as inductively coupled transponders. The transponder (Tag) contains a microchip with a unique identification number stored in it. Dabbene et al. (2016) has categorized these tags into passive, semi-passive, and active tags. Passive tags do not contain battery, and the power is supplied by electrical currents produced by the electromagnetic field from the transponder. Semi-passive tags contain a battery that is used to power the tag and is not for communication. Active tag also has a battery that supplies power for all functions. Transponders can be embedded in ear tags, leg tag, ruminal boluses, and injectables.

RFID has numerous applications in food traceability; specifically, it helps in process control in the food industry. The complete information about the status of the product, right from production to processing to delivery can be managed and put in place (Dabbene et al., 2016). It can also be used in supply chain management by tracking products during distribution and storage. The RFID system accurately maintains the database of inventory and alerts the warehouse management when inventories become low. Walmart, the world largest retailer in 2003, has instructed its suppliers to place RFID tags in pallets. RFID has greater speed and efficiency in stock operations, with greater benefits in food retailing. They control inventory management with attendant cost-saving, reduced labor cost, improvement in customer service, clear tracking of customers, tracking of their purchasing behavior, etc. (Peter et al., 2005).

The shelf-life of foods can be monitored, where the RFID tag senses the temperature of the product and correlate it with time to determine the shelf-life of the product. Food safety applications of RFID ensures whether the

products like meat, fruits, and vegetables are maintained at safe temperature during transportation and storage. The RFID reader with an onboard
micro-machined metal oxide gas sensor can monitor the fruit quality and
help in conservation of apples (Vergara et al., 2007). RFID temperature
logging provides complete temperature history of any product to which it is
tagged. It is being employed for cold chain monitoring to ensure the safety
of perishable commodities, thereby reducing their wastage (Grunow and
Piramuthu, 2013).

Recently, RFID **is** used in combination with smart packaging by providing complete information about light exposure, relative humidity, temperature, pressure, pH, etc. RFID tags inserted in polystyrene boxes can be used
to monitor the temperature and RH in packed fish supply chain (Trebar et al.,
2013). It is an anti-counterfeit device to prevent frauds in supply chain (Li,
2013). However, the main disadvantage of this technique is the high cost of
tag when compared with the barcode label.

14.4.3   NUCLEAR MAGNETIC RESONANCE (NMR) SPECTROSCOPY

Application of nuclear magnetic resonance (NMR) spectroscopy in food
quality analysis and traceability has been drastically increasing. NMR spectra can act as a fingerprint to compare, discriminate, and classify samples.
It depends on the ability of matter to interact with electromagnetic radiation. NMR is the absorption of radiofrequencies by atomic nuclei within a
sample that is placed in a magnetic field. Due to its high accuracy and good
repeatability, NMR is being used in microbial analysis and nondestructive
testing of fruits. In food analysis, NMR can provide information about the
internal structure of the foods; physical status of water, fat, starch, and protein in emulsions; and its molecular composition. Sacchi and Paolillo (2007)
discussed different types of NMR techniques with specific applications in
food science like solution state quantitative high-resolution NMR, solid state
high-resolution NMR, magnetic resonance imaging (MRI), site-specific
natural isotope fractionation (SNIF)-NMR, diffusion NMR, and relaxation
time NMR. High frequency NMR has interesting and specific applications
in detecting the adulterants in foods. It aids in differentiating foods obtained
using different food processing techniques. For geographic traceability of
fruit juice, oils and wines, SNIF-NMR is being employed. MRI with suitable
NMR fingerprints was found to be best suited to identify different origins of

tea (Zhang et al., 2011). Similarly, proton NMR spectra fingerprinting iden-tified the origin of beer. Proton and ^{13}C- NMR has been commonly used for oils and lipids for estimating their fatty acid composition and oil quality and authenticity (Sacchi et al., 1996). However, the main disadvantage of NMR is the high cost of instruments and trained personnel requirement along with specialized software and algorithms.

14.4.4   CHROMATOGRAPHIC ANALYSIS

Gas chromatography involves separation and analyses of compounds by vaporizing without decomposition. The compounds in the mixture are sepa-rated between a liquid stationary phase and a gaseous mobile phase. The analyte compound is vaporized and introduced into a stream of carrier gas (inert gas). The gaseous compound being analyzed will interact with the walls of the column coated with different stationary phase. Thus, each com-pound will elute at different times known as retention time. It has the advan-tage of high resolution, high sensitivity, and short time.

GC-MS and LC-MS are commonly employed for detecting veterinary and pesticide residues, food additives, and adulterants. Acrylamide forma-tion occurs when the food is processed at high temperature, which is highly toxic and carcinogenic. Acrylamide in different types of snacks can be detected using GC-MS and LC-MS (Ono et al., 2010). Jun and Bin (2009) developed a protocol for simultaneous detection of pesticide residues from tea by GC-MS. Specifically, in the case of detection of organochlorine pesti-cides, GC-MS with electron impact ionization has better performance, How-ever, LC-MS has better scope for the analysis of all pesticides (Alder et al., 2006). GC-MS is more commonly used in detecting the additives added in food with the higher detection rates.

14.4.5   OTHER TECHNIQUES

Other techniques such as inductively coupled plasma mass spectrometry (ICP-MS), near infrared spectroscopy (NIS), UV-Visible light spectroscopy, infrared spectroscopy, fluorescence emission spectroscopy, electronic noses, differential scanning calorimetry, isotopic analysis, electrophoresis, and immunological tests are being used, which can assist in the implementation of traceability in food systems.

14.5 TRACEABILITY FOR SUSTAINABILITY AND FOOD SAFETY

Sustainability may be defined as the ability to satisfy the needs of people in such a way that the ability of forthcoming generations to meet their own needs is not compromised. Most of the market and regulatory sustainability drivers demand for the sustainability of food supply chain, which can be implemented through the effective use of food traceability systems. Other than cost-based performance indicators, sustainability widely includes major concerns like environmental issues and social dimensions (Kleindorfer et al., 2005). Social dimension may be described as employees' health and safety, animal welfare, and ethical trading in procurement of raw materials. According to Xiaoshuan (2013), some of the current challenges faced by the food industry are as follows:

a) Rising concerns over food safety like genetically modified organisms (GMOs), mislabeling, product recalls etc.

b) Changing supplier-retailer relationship:
   - Power shift from manufacturer to retailer
   - Inventory control, order cycle time reduction (the order cycle is the time required by industry to complete an order, i.e., the time period between an order arrival and its delivery to the supplier. The traceability system enables effective tracing of product throughout the order cycle, thus reducing the time to complete it.)

c) Regulatory pressure and social responsibility;

d) Changing demographics and lifestyles;

e) Globalization that demands tailoring taste to local needs.

In terms of processing, preference shall be given to locally grown produce as it will reduce energy consumption required for transport and storage. Similarly, seasonal and non-seasonal aspect of raw ingredients shall be considered to reduce energy requirement for storage. Food traceability systems can effectively assist in making decisions related to sustainability factor of raw ingredients.

Moreover, wastage is also an important factor related to sustainability. Most of the countries waste around 30% of their produce due to lack of proper food supply chain (Chapman, 2010). Supply chain covers the product from farm to fork, and most companies offer services to customers in order to assist them regarding quality and safety issues. This also helps in receiving genuine feedbacks from consumers and scout for ideas for the development of new products. Food traceability is the backbone of a strong supply chain. Thus, food traceability is directly linked to the sustainability of the food processing sector. Traceability plays the following functional roles in food supply chain:

a) Robust traceability system is required for swift product recalls in times of disease outbreaks.

b) Residues such as pesticides and veterinary drugs need to be monitored through traceability in order to keep them under legal limits.

c) Traceability system gathers the data required for risk assessment of food additives, contaminants, etc.

d) Traceability system plays an important role in preventing false labelling, thus protecting consumers' rights.

e) Effective traceability can reduce fraud practices like black marketing and theft of food items.

f) Food traceability leads to an effective supply chain, thus reducing wastage by significant amounts.

Food safety can be defined as the science of keeping the food safe from potential hazards in order to provide wholesome food to the consumers. Worldwide, food safety has become a major issue for governing agencies due to public health risks associated with it. Industrialization and pollution in recent times have exposed our food to newer kinds of food safety hazards. In order to tackle this situation, food safety systems have undergone major developments in the last decade. Food safety systems have evolved from reactive to preventive mode, and food traceability system (FTS) is the fundamental preventive method of mitigating food safety risks.

Export of Indian food items like alphonso mangoes, taro plant, eggplant, snake gourd, and bitter gourd were temporarily banned by EU in 2014 due to various food safety issues like antibiotic residues, fruit flies, presence of cadmium, and *Vibrio* contamination (Sonwalkar, 2014; Dandage et al., 2016).

Recent controversies like lead contamination in noodles gained worldwide media attention, and the company had to suffer huge losses due to product recall. Peanut Corporation of America (PCA) had to recall 2100 products across 200 companies in 2009, which allegedly contained *Salmonella*-infested peanut products. Such incidences not only harm the public health but also the business and repute of food processing companies. Thus, implementation of a proper food traceability system can help manufacturers identify the root cause of quality and safety problems at the first instance and fix them early so as to prevent any further contamination and quality loss. This will not only ensure safe food for consumers but also the sustainable development of the food industry.

14.6 TRACEABILITY IN THE FOOD INDUSTRY

Apart from addressing the major concerns like food safety and sustainability, food traceability plays an important role in the optimization of business performance. Some of the case studies highlighting the importance of traceability are given as follows:

14.6.1 TRACEABILITY IN FISH SUPPLY CHAIN

Mai et al. (2007) investigated the results of traceability implementation in the fish supply chain. The surveyed companies admitted improved supply chains as one of the most significant contributions of traceability. Benefits of traceability as interpreted by survey results can be broadly classified into qualitative and quantitative.

*   **Qualitative benefits:** Traceability improved the efficiency of supply chain. Other major qualitative benefits perceived by companies were ability to retain customers, lesser consumer complaints, product differentiation, and product quality improvement.
*   **Quantitative benefits:** Quantitative benefits such as market growth, reduction in product recall cases, savings on labor costs and inventory management costs, spoilage reduction, process improvement, and product quality improvement were reported.

14.6.2 GENERAL SPECIFICATION STANDARD (GS1) INITIATIVE FOR TRACEABILITY IN THE HORTICULTURAL SECTOR

GS1 is a neutral, non-profit organization that facilitates collaboration amongst trading partners and technology providers in order to resolve challenges; it control standards and ensures safety, visibility, and efficiency in the entire value chain. It was known as European Article Numbering-Uniform Code Council (EAN-UCC). EAN-UCC was a supply chain standard family name that included barcodes for products ranging from FMCG to pharmaceuticals and e-commerce standards. In 2005, EAN-UCC changed its name to GS1 and is now headquartered at Brussels, Germany, with a global network of over 110 member organizations across the world. GS1 plays a significant role in design and implementation of global standards and solutions for improving the efficiency of supply chains. GS1 standards include GS1 identifiers, GS1 barcodes, GS1 EPC global standards (for RFID technology) etc. These standards provide a global language of business that facilitates information sharing throughout the chain in a seamless, structured, uniform, and consistent manner.

14.6.3   CHAINFOOD

Chainfood, from the Netherlands, is a small innovative company that provides Internet-based traceability solutions. The company has developed software, which improves the collaboration and transparency between all partners in the chain, thereby delivering real quality and sustainability benefits whilst reducing risk and cost. The company relies on the principle of chain quality management, i.e., recording information from individual chain partners on a single platform and then utilizing it to optimize the performance of total chain and the individual partners. Benefits of the Chainfood model includes the following:

a)   Data providers are in full control and authority over their data. They have choice regarding sharing of data with other partners.

b)   Company provides additional functions like tracking and tracing (logistics and product/process information), supply chain management, and collaborative business intelligence, which provide an opportunity to unravel previously undiscovered useful data patterns.

c)   Other than the above listed benefits, new traceability benefits can easily be added due to generic data model.

d)   Convenience, speed, and accuracy.

Currently, the system has been customized for pork, dairy, and fruit and vegetable applications.

14.6.4   GRAPENET AND ANARNET

Agricultural and Processed Foods Export Development Authority is a body under Indian government, which works for the welfare of farmers and exporters of agricultural produce by assisting them in residue monitoring and traceability of their produce. In the past, grape farmers and exporters were suffering huge losses due to rejection of consignments on grounds of pesticide residues, plant growth regulator residues, etc. India's image of a quality fresh product supplier went through a setback every time such incidence occurred. Taking lessons from such incidences, APEDA set up Internet-based traceability systems like GRAPENET and ANARNET that cover all major stakeholders in the export supply chain of grapes and pomegranates, like farmers, exporters, accredited laboratories, phytosanitary certification departments, and APEDA. These initiatives have proved to be effective in reducing export consignment rejections as the traceability system enables APEDA to track the consignment from farm to port and block it in the case of any irregularity with norms. GS1 India played a significant role in this project.

14.7 CONCLUDING REMARKS

In recent times, traceability has become a necessity for food supply chain. Traceability benefits all the stakeholders of food chain by offering advantages like logistics and inventory management, contribution to agricultural sustainability, higher degree of consumer satisfaction, and mitigation of food safety-related risks. Moreover, legal compulsions and consumer expectations act as a driving force behind implementation of the traceability system. Advent of modern techniques like RFID and DNA profiling have further strengthened the traceability systems. The importance of traceability for safe foods is a proven fact, but the processes involved in the implementation need to be explored so that those could be made cost effective and within the reach of all the organizations.

KEYWORDS

- food safety
- sustainability
- traceability
- traceability case studies
- traceability drivers
- traceability principles
- traceability techniques

REFERENCES

Abdel-Rahman, S. M., & Ahmed, M. M. M., (2007). Rapid and sensitive identification of buffalo's, cattle's and sheep's milk using species-specific PCR and PCR-RFLP techniques. *Food Control, 18*, 1246–1249.

Alary, R., Buissonade, C., Joudrier, P., & Gautier, M. F., (2007). Detection and discrimination of cereal and leguminous species in chestnut flour by duplex PCR. *Eur. Food Res. Technol., 225*, 427–434.

Alary, R., Serin, A., & Duviau, M. P., et al., (2002). Quantification of Common Wheat Adulteration of Durum Wheat Pasta Using Real-Time Quantitative Polymerase Chain Reaction (PCR). *Cereal Chem., 79*, 553–558.

Alder, L., Greulich, K., Kempe, G., & Vieth, B., (2006). Residue analysis of 500 high priority pesticides: Better by DC-MS or LC-MS/MS. *Mass Sepctrometry*, Rev *25*, 838–865.

Allmann, M., Candrian, U., Höfelein, C., & Lüthy, J., (1993). Polymerase chain reaction (PCR): A possible alternative to immunochemical methods assuring safety and quality of food Detection of wheat contamination in non-wheat food products. *Z Lebensm Unters Forsch., 196*, 248–251.

Amaral, J. S., Santos, C. G., & Melo, V. S., et al., (2014). Authentication of a traditional game meat sausage (Alheira) by species-specific PCR assays to detect hare, rabbit, red deer, pork and cow meats. *Food Res. Int., 60*, 140–145.

Barlett, S., & Davidson, W., (1992). FINS (forensically informative nucleotide sequencing): a procedure for identifying the animal origin of biological specimens. *Biotechniques, 12*, 408–411.

Blanco, M., Pérez-Martín, R. I., & Sotelo, C. G., (2008). Identification of shark species in seafood products by forensically informative nucleotide sequencing (FINS). *J. Agric. Food Chem., 56*, 9868–9874.

Boziaris, I. S. (2014). *Seafood Processing, Technology, Quality and Safety*, Wiley Blackwell.

Chapela, M. J., Sotelo, C. G., Calo-Mata, P., et al., (2002). Identification of cephalopod species (Ommastrephidae and Loliginidae) in seafood products by forensically informative nucleotide sequencing (FINS) *Introduction. Food Chem. Toxicol., 67*, 1672–1676.

Chapman, P. A., (2010). Reducing product losses in the food supply chain. In: *Delivering Performance in Food Supply Chains*, Mena, C., & Stevens, G., Woodhead, Cambridge, pp. 225–242.

Chikuni, K., Tabata, T., Kosugiyama, M., et al., (1994). Polymerase chain reaction assay for detection of sheep and goat meats. *Meat Sci., 37*, 337–345.

Dabbene, F., Gay, P., & Tortia, C., (2016). Radio-frequency identification usage in food traceability. In: *Advances in Food Traceability Techniques and Technologies*, 1st edn., Woodhead, Cambridge, pp. 67–89

Dahinden, I., Von Büren, M., & Lüthy, J., (2001). A quantitative competitive PCR system to detect contamination of wheat, barley or rye in gluten-free food for coeliac patients. *Eur Food Res Technol., 212*, 228–233.

Dalvit, C., De, Marchi, M., & Targhetta, C., et al., (2008). Genetic traceability of meat using microsatellite markers. *Food Res. Int., 41*, 301–307.

Dandage, K., (2016). Indian perspective in food traceability: A review. *Food Control*, doi:10.1016/j.foodcont.2016.07.005

Díaz, I. L. C., Alonso, I. G., Fajardo, V., et al., (2007). Application of a polymerase chain reaction to detect adulteration of ovine cheeses with caprine milk. *Eur. Food Res. Technol., 225*, 345–349.

Drummond, M. G., Brasil. B. S. A. F., Dalsecco, L. S., et al., (2013). A versatile real-time PCR method to quantify bovine contamination in buffalo products. *Food Control, 29,* 131–137.

Espiñeira, M., & Santaclara, F. J., (2016). What Is Food Traceability? In. *Advances in Food Traceability Techniques and Technologies,* 1ˢᵗ edn., Woodhead, Cambridge, pp. 3–8.

Espiñeira, M., & Santaclara, F. J. (2016). The Use of Molecular Biology Techniques in Food. In: *Advances in Food Traceability Techniques and Technologies,* 1ˢᵗ edn., Woodhead, Cambridge, pp. 91–118.

Espineria, M., Lavin, Nerea, G., Vieites, J. M., & Santaclara, F. J., (2016). Authentication of Anglerfish Species (*Lophius* spp.) by Means of Polymerase Chain Reaction – Restriction Fragment Length Polymorphism (PCR-RFLP) and Forensically Informative Nucleotide Sequencing (FINS) Methodologies. *J. Agric. Food Chem., 56,* 10594–10599.

Floren, C., Wiedemann, I., Brenig, B., et al., (2015). Species identification and quantification in meat and meat products using droplet digital PCR (ddPCR). *Food Chem., 173,*1054–1058.

Giménez, M. J., Pistón, F., Martín, A., & Atienza, S. G., (2010). Application of real-time PCR on the development of molecular markers and to evaluate critical aspects for olive oil authentication. *Food Chem, 118,* 482–487.

Girish, P. S., Anjaneyulu, A. S. R., Viswas, K. N., et al., (2005). Meat species identification by polymerase chain reaction-restriction fragment length polymorphism (PCR-RFLP) of mitochondrial 12S rRNA gene. *Meat. Sci., 70,* 107–112.

Gokulakrishnan, P., Kumar, R. R., Sharma, B. D., et al., (2013). Determination of sex origin of meat from cattle, sheep and goat using PCR based assay. *Small Rumin. Res., 113,* 30–33.

Grooms, D., (2007). Radio Frequency Identification (RFID) Technology for Cattle. *Ext. Bull.,* E-2970, 7–8.

Grunow, M., & Piramuthu, S., (2013). RFID in highly perishable food supply chains – Remaining shelf life to supplant expiry date? *Int J Prod Econ,* 146, 717–727.

Han, S., Park, S., Oh, H., et al., (2013). PCR-RFLP for the Identification of Mammalian Livestock. *Animal Species., 28,* 355–360.

Haunshi, S., Basumatary, R., Girish, P. S., et al., (2009). Identification of chicken, duck, pigeon and pig meat by species-specific markers of mitochondrial origin. *Meat Sci, 83,* 454–459.

Herman, B. L., (2001). Determination of the animal origin of raw food by species-specific PCR. *J Dairy Res., 68,* 429–436.

Hernández, M., Esteve, T., & Pla, M., (2005). Real-time polymerase chain reaction based assays for quantitative detection of barley, rice, sunflower, and wheat. *J Agric Food Chem., 53,* 7003–7009.

Iwobi, A., Gerdes, L., Busch, U., & Pecoraro, S., (2016). Droplet digital PCR for routine analysis of genetically modified foods (GMO) – A comparison with real-time quantitative PCR. *Food Control., 69,* 205–213.

Jahangir, Tafrechi, R. S., Van de Rijke, F. M., Allallou, A., et al., (2007). Single-cell A3243G mitochondrial DNA mutation load assays for segregation analysis. *J. Histochem. Cytochem., 55,* 1159–66.

Jun, shu, J., & Bin, Z., (2009). Simultaneous determination of 36 pesticide residues in tea by GC-MS. *Foo Sci., 30,* 276–330.

Kesmen, Z., Sahin, F., & Yetim, H., (2007). PCR assay for the identification of animal species in cooked sausages. *Meat Sci., 77,* 649–653.

Kleindorfer, P.R., Singhal, K., & Van Wassenhove, L. N., (2005). Sustainable operations management. *Product Oper Manage.*, *14*(4), 482–492.

Köppel, R., Dvorak, V., Zimmerli, F., et al., (2009). Two tetraplex real-time PCR for the detection and quantification of DNA from eight allergens in food. *Eur. Food Res. Technol.*, *230*, 367–374.

Kuchta, T., Knutsson, R., Fiore, A., et al., (2014). A decade with nucleic acid-based microbiological methods in safety control of foods. *Lett, Appl, Microbiol.*, *59*, 263–271.

Kumar, D., Singh, S. P., Karabasanavar, N. S., et al., (2012). Authentication of beef, carabeef, chevon, mutton and pork by a PCR-RFLP assay of mitochondrial cytb gene. *J. Food Sci. Technol.*, *51*, 3458–3463.

Lago, F.C., Herrero, B., & Madriñán, M., et al., (2011). Authentication of species in meat products by genetic techniques. *Eur. Food Res. Technol.*, *232*, 509–515.

Lago, F. C., Herrero, B., Vieites, J. M., & Espiñeira, M., (2011). FINS methodology to identification of sardines and related species in canned products and detection of mixture by means of SNP analysis systems. *Eur. Food Res. Technol.*, *232*, 1077–1086.

Li, L., (2013). Technology designed to combat fakes in the global supply chain. *Bus Horiz, 56*,167–177.

Liu, W. H., Kaur, M., Wang, G., et al., (2004). Inverse PCR-Based RFLP Scanning Identifies Low-Level Mutation Signatures in Colon Cells and Tumors. *Cancer Res.*, *64*, 2544–2551.

Lopez-Calleja, I., Gonzalez, I., Fajardo, V., et al., (2004) Rapid detection of cow's milk in sheep's and goat's milk by a species-specific polymerase chain reaction technique. *J. Dairy Sci. 87*, 2839–2845.

Mai, N., (2010). Benefits of traceability in fish supply chains – case studies. *British Food Journal.*, 976–1002.

Mane, B. G., Mendiratta, S. K., & Tiwari, A. K., (2009). Polymerase chain reaction assay for identification of chicken in meat and meat products. *Food Chem.*, *116*, 806–810.

McEntire, Jennifer, C., et al., (2010). Traceability (product tracing) in food systems: an IFT report submitted to the FDA, volume 1: technical aspects and recommendations. *Comprehensive Reviews in Food Science and Food Safety.*, *9*(1), 92–158.

Meudt, H. M., & Clarke, A. C., (2007). Almost Forgotten or Latest Practice? AFLP applications, analyses and advances. *Trends Plant Sci, 12*, 106–117.

Okuma, T. A., & Hellberg, R. S., (2015). Identification of meat species in pet foods using a real-time polymerase chain reaction (PCR) assay. *Food Control.*, *50*, 9–17.

Olsen, P., & Borit, M., (2013). How to define traceability. *Trends Food Science and Technology*, *29*(2), 142–150.

Ono, H., Chuda, Y., Ohnishi-Kameyama, M., et al., (2010). Analsyis of acrylamide by LC-MS/MS and Gc-MS in processed Japenese foods. *Jouranal food Addit Contam.*, *20*, 215–220.

Opara, L. U., (2003). Traceability in agriculture and food supply chain: A review of basic concepts, technological implications, and future prospects Traceability in agriculture and food supply chain: a review of basic concepts, technological implications, and future prospects. *Food Agriculture and Environment, 1*(1), 101–106.

Ortola-Vidal, A., Schnerr, H., Rojmyr, M., et al., (2007). Quantitative identification of plant genera in food products using PCR and pyrosequencing technology. *Food Control, 18*, 921–927.

Palmieri, L., Bozza, E., & Giongo, L., (2009). Soft Fruit traceability in food matrices using Real-Time PCR. *Nutrients.*, *1*, 316–328.

Pappalardo, A. M., & Ferrito, V., (2015). A COIBar-RFLP strategy for the rapid detection of *Engraulis encrasicolus* in processed anchovy products. *Food Control, 57*, 385–392.

Pérez, M., & Presa, P., (2008). Validation of a tRNA-Glu-cytochrome b key for the molecular identification of 12 hake species (*Merluccius* spp.) and atlantic cod (*Gadus morhua*) using PCR-RFLPs, FINS, and BLAST. *J. Agric. Food Chem., 56*, 10865–10871.

Peter, J., Clarke-Hill, C., Comfort, D., et al., (2005). Radio frequency identification and food retailing in the UK. *British Food Journal., 107*(6), 356–360.

Plessis, H. J. & Rand, G. E., (2012). The significance of traceability in consumer decision making towards Karoo lamb. *Food Research International., 47*, 210–217.

Rezaei, M., Shabani, A., Shabanpour, B., & Kashiri, H., (2011). Microsatellites reveal weak genetic differentiation between Rutilus frisii kutum (Kamenskii, 1901) populations south of the Caspian Sea. *Animal Biology, 61*, 469–483.

Rodriguez-Lazaro, D., & Hernandez, M., (2013). Real-time PCR in Food Science: Introduction. *Curr Issues Mol Biol., 15*, 25–38.

Sacchi, R., & Paolillo, L., (2007). NMR for Food Quality and Traceability. In: *Advances in Food Diagnostics*; Hui, Y. H., Ed., Blackwell Publishing, pp. 101–117.

Sacchi, R., Patumi, M., Fontanazza, G., et al., (1996). A high-field 1H nuclear magnetic resonance study of the minor components in virgin olive oils. *J. Am. Oil Chem. Soc., 73*(6), 747–758.

Scollo, F., Egea, L. A., Gentile, A., et al., (2016). Absolute quantification of olive oil DNA by droplet digital-PCR (ddPCR): Comparison of isolation and amplification methodologies. *Food Chem., 213*, 388–394.

Sefc, K. M, Lefort, F., et al., (2001). Microsatellite markers for grapevine: a state of the art. In: *Molecular Biology & Biotechnology of the Grapevine*, Roubelakis-Angelakis, K. A., Springer, pp. 433–463.

Senneset, G., Foras, E., & Fremme, K. M., (2007). Challenges regarding implementation of electronic chain traceability. *British Food Journal., 109*(10), 805–818.

Shipley, G. L., (2007). An introduction to real-time PCR. In: *Real -time PCR*. Tevfik Dorak, M., Ed., Taylor and Francis Group, pp. 1–37.

Sonwalkar, P., (2014). European Union bans Indian mangoes, vegetables due to concerns over pests. Hindustan Times, Retrieved from http://www.hindustantimes.com/world/european-union-bans-indian-mangoes-vegetables-due-to-concerns-over-pests/story-2edh0nXSLhUPSEqiYk5TjN.html (Accessed Oct 23, 2016.)

Spaniolas, S., Bazakos, C., Ntourou, T., et al., (2008). Use of lambda DNA as a marker to assess DNA stability in olive oil during storage. *Eur. Food Res. Technol., 227*, 175–179.

Espiñeira, M., Vieites, J. M., & Lago, F. C., (2012). Authentication of the most important species of rockfish by means of fins. *Eur. Food Res. Technol., 235*, 929–937.

Trebar, M., Lotric, M., Fonda, I., et al., (2013). RFID data loggers in fish supply chain traceability. *Int. J. Antennas. Propag.*

Vergara, A., Llobet, E., Ramírez, J. L., et al., (2007). An RFID reader with onboard sensing capability for monitoring fruit quality. *Sensors Actuators. B. Chem., 127*, 143–149.

Vietina, M., Agrimonti, C., Marmiroli, M., et al., (2011). Applicability of SSR markers to the traceability of monovarietal olive oils. *J. Sci. Food Agric., 91*, 1381–1391.

Wang,Q., Zhang, X., Zhang, H. Y., et al., (2010). Identification of 12 animal species meat by T-RFLP on the 12S rRNA gene. *Meat Sci., 85*, 265–269.

Xiaoshuan, Z., (2013). Food supply chain traceability and sustainable development. EU document.https://eeas.europa.eu/delegations/china/documents/eu_china/science_tech_environment/20131108_3.3_5.pdf. (Accessed Oct 25, 2016).

Xu, W., (2016). Functional Nucleic Acids Detection. In: *Food Safety: Theories and Applications*, 1st edn., Springer, pp 37–61.

Zhang, J., Zhang, X., Dediu, L., & Victor, C., (2011). Review of the current application of fingerprinting allowing detection of food adulteration and fraud in China. *Food Control.*, *22*, 1126–1135.

Zhang, L. I., Wu, G., Wu, Y., et al., (2009). The gene MT3-B can differentiate palm oil from other oil samples. *J. Agric. Food Chem.*, *57*, 7227–7232.

Zhu, P., Fu, W., Wang, C., et al., (2016). Development and application of absolute quantitative detection by duplex chamber-based digital PCR of genetically modified maize events without pretreatment steps. *Analytica. Chimica. Acta.*, *916*, 60–66.

INDEX

Printed and bound by CPI Group (UK) Ltd, Croydon, CR0 4YY

23/10/2024

01777705-0006